Nondigestible Carbohydrates and Digestive Health

IFT PRESS

The *IFT Press* series reflects the mission of the Institute of Food Technologists – to advance the science of food contributing to healthier people everywhere. Developed in partnership with Wiley-Blackwell, *IFT Press* books serve as leading-edge handbooks for industrial application and reference and as essential texts for academic programs. Crafted through rigorous peer review and meticulous research, *IFT Press* publications represent the latest, most significant resources available to food scientists and related agriculture professionals worldwide.

Founded in 1939, the Institute of Food Technologists is a nonprofit scientific society with 22,000 individual members working in food science, food technology, and related professions in industry, academia, and government. IFT serves as a conduit for multidisciplinary science thought leadership, championing the use of sound science across the food value chain through knowledge sharing, education, and advocacy.

WILEY-BLACKWELL

A John Wiley & Sons, Ltd., Publication

Nondigestible Carbohydrates and Digestive Health

Editors

Teri M. Paeschke

William R. Aimutis

WILEY-BLACKWELL
A John Wiley & Sons, Ltd., Publication

IFT PRESS

Edition first published 2011

Blackwell Publishing was acquired by John Wiley & Sons in February 2007. Blackwell's publishing program has been merged with Wiley's global Scientific, Technical, and Medical business to form Wiley-Blackwell.

Editorial Office
2121 State Avenue, Ames, Iowa 50014-8300, USA

For details of our global editorial offices, for customer services, and for information about how to apply for permission to reuse the copyright material in this book, please see our Website at www.wiley.com/wiley-blackwell.

Library of Congress Cataloging-in-Publication Data

Digestive health and nondigestible carbohydrates / edited by Teri M. Paeschke & William R. Aimutis.
p. cm. – (IFT Press series)
Includes bibliographical references and index.
ISBN 978-0-8138-1762-0 (hardback : alk. paper)
1. Fiber in human nutrition. 2. Digestion. I. Paeschke, Teri M. II. Aimutis, William R.
QP144.F52D547 2011
612.3–dc22

2010018386

A catalog record for this book is available from the U.S. Library of Congress.

Set in 11.5/13.5 Times NR PS by Aptara® Inc., New Delhi, India
Printed and bound in Malaysia by Vivar Printing Sdn Bhd

1 2011

Titles in the *IFT Press* series

- *Accelerating New Food Product Design and Development* (Jacqueline H. Beckley, Elizabeth J. Topp, M. Michele Foley, J.C. Huang and Witoon Prinyawiwatkul)
- *Advances in Dairy Ingredients* (Geoffrey W. Smithers and Mary Ann Augustin)
- *Bioactive Proteins and Peptides as Functional Foods and Nutraceuticals* (Yoshinori Mine, Eunice Li-Chan and Bo Jiang)
- *Biofilms in the Food Environment* (Hans P. Blaschek, Hua H. Wang, and Meredith E. Agle)
- *Calorimetry in Food Processing: Analysis and Design of Food Systems* (Gönül Kaletunç)
- *Coffee: Emerging Health Effects and Disease Prevention* (YiFang Chu)
- *Food Carbohydrate Chemistry* (Ronald E. Wrolstad)
- *Food Ingredients for the Global Market* (Yao-Wen Huang and Claire L. Kruger)
- *Food Irradiation Research and Technology* (Christopher H. Sommers and Xuetong Fan)
- *Foodborne Pathogens in the Food Processing Environment: Sources, Detection and Control* (Sadhana Ravishankar, Vijay K. Juneja and Divya Jaroni)
- *High Pressure Processing of Foods* (Christopher J. Doona and Florence E. Feeherry)
- *Hydrocolloids in Food Processing* (Thomas R. Laaman)
- *Improving Import Food Safety* (Wayne C. Ellefson, Lorna Zach and Darryl Sullivan)
- *Microbial Safety of Fresh Produce* (Xuetong Fan, Brendan A. Niemira, Christopher J. Doona, Florence E. Feeherry and Robert B. Gravani)
- *Microbiology and Technology of Fermented Foods* (Robert W. Hutkins)
- *Multiphysics Simulation of Emerging Food Processing Technologies* (Kai Knoerzer, Pablo Juliano, Peter Roupas and Cornelis Versteeg)
- *Multivariate and Probabilistic Analyses of Sensory Science Problems* (Jean-François Meullenet, Rui Xiong, and Christopher J. Findlay
- *Nanoscience and Nanotechnology in Food Systems* (Hongda Chen)
- *Natural Food Flavors and Colorants* (Mathew Attokaran)
- *Nondestructive Testing of Food Quality* (Joseph Irudayaraj and Christoph Reh)
- *Nondigestible Carbohydrates and Digestive Health* (Teresa M. Paeschke and William R. Aimutis)
- *Nonthermal Processing Technologies for Food* (Howard Q. Zhang, Gustavo V. Barbosa-Cànovas, V.M. Balasubramaniam, C. Patrick Dunne, Daniel F. Farkas, James T.C. Yuan)
- *Nutraceuticals, Glycemic Health and Type 2 Diabetes* (Vijai K. Pasupuleti and James W. Anderson)
- *Organic Meat Production and Processing* (Steven C. Ricke, Michael G. Johnson and Corliss A. O'Bryan)
- *Packaging for Nonthermal Processing of Food* (Jung H. Han)
- *Preharvest and Postharvest Food Safety: Contemporary Issues and Future Directions* (Ross C. Beier, Suresh D. Pillai, and Timothy D. Phillips, Editors; Richard L. Ziprin, Associate Editor)
- *Processing and Nutrition of Fats and Oils* (Ernesto M. Hernandez, and Afaf Kamal-Eldin)
- *Processing Organic Foods for the Global Market* (Gwendolyn V. Wyard, Anne Plotto, Jessica Walden and Kathryn Schuett)
- *Regulation of Functional Foods and Nutraceuticals: A Global Perspective* (Clare M. Hasler)
- *Resistant Starch: Sources, Applications and Health Benefits* (Yong-Cheng Shi and Clodualdo Maningat)
- *Sensory and Consumer Research in Food Product Design and Development* (Howard R. Moskowitz, Jacqueline H. Beckley, and Anna V.A. Resurreccion)
- *Sustainability in the Food Industry* (Cheryl J. Baldwin)
- *Thermal Processing of Foods: Control and Automation* (K. P. Sandeep)
- *Trait-Modified Oils in Foods* (Frank T. Orthoefer and Gary R. List)
- *Water Activity in Foods: Fundamentals and Applications* (Gustavo V. Barbosa-Cànovas, Anthony J. Fontana Jr., Shelly J. Schmidt, and Theodore P. Labuza)
- *Whey Processing, Functionality and Health Benefits* (Charles I. Onwulata and Peter J. Huth)

Table of Contents

Preface

Carbohydrates are a diverse set of compounds that are well known as nutritive ingredients to provide energy such as sugars and starches. Lesser known, but emerging, are carbohydrates that provide digestive health benefits such as enhanced immune system, mineral adsorption, and colonic health in general. These types of carbohydrates are not digested by either the stomach or small intestine and reach the colon intact or only partially digested. The chemical structure and bonding arrangement of these carbohydrates play a key role in their efficacy as prebiotic or fermentable substrates. Evolving science regarding prebiotics such as fructo-, galacto-, and xylooligosaccharides, as well as fermentable carbohydrates such as pectin, arabinoxylans, and resistant starch demonstrates the diversity of carbohydrates and their function in digestive health. This book features authors from academia as well as industry to provide a broad view of carbohydrates for digestive health.

This book evolved from several Symposia we organized at the 2008 Institute of Food Technologists (IFT) Annual Meeting in Las Vegas. The Symposia covered the importance of carbohydrates and digestive health in one manner or another, and provided a good start to this book. All of the contributors are internationally recognized experts in their fields that have spoken and published on these topics numerous times in the past. We intentionally challenged the authors to give us different perspectives in their areas, and even to add a "little edge" to some theories if they so chose. We think some did accomplish this in the end.

This text is a useful reference for all who are interested in carbohydrates and digestive health including food scientists (product developers) whom are often called upon to formulate products with "fiber" and need a background to understand the nutritional aspects and implications. Additionally, nutrition scientists will find this book useful as it provides a good background for understanding carbohydrates and their function in digestive health for nutrition scientists that have little background in this area.

Our deepest appreciation and respect goes to the chapter authors who contributed significant amounts of time, knowledge, talents, and expertise to bring this book from a concept to hardcover in a very rapidly changing field of study. We also thank Mr. Mark Barrett and Ms. Susan Engelken at Wiley-Blackwell for their guidance, advice, and assistance in the preparation and completion of this book.

We also must give our most heartfelt thanks and appreciation to our spouses for their understanding while we were trying to finish this book. They sacrificed many hours of family time in the process.

Finally, we dedicate the book to all those who have researched and devoted their talents and instincts to understanding the interactions of carbohydrates and the intestinal tract in healthy and diseased individuals. A special dedication to those patients that have suffered from some of the debilitating diseases discussed in this book. You have sacrificed your comfort at times in the name of science to better understand if carbohydrates were friend or foe to your digestive problems. We thank you for this graciousness.

Teri M. Paeschke
William R. Aimutis

Contributors

Aimutis, William R.
Cargill, Inc., Wayzata, MN, USA

Arrieta, Marie-Claire
Department of Medicine, University of Alberta, Edmonton, Canada

Bondia-Pons, Isabel
Department of Clinical Nutrition, University of Eastern Finland, Kuopio, Finland

Faber, T.A.
Department of Animal Sciences, University of Illinois, Urbana, IL, USA

Fahey, G. Jr.
Department of Animal Sciences, University of Illinois, Urbana, IL, USA

Field, Catherine
Department of Agricultural, Food and Nutritional Science, University of Alberta, East Edmonton, Canada

Hamaker, Bruce R.
Department of Food Science, Purdue University, West Lafayette, IN, USA

Janes, Marlene
Department of Food Science, Louisiana State University AgCenter, Baton Rouge, LA, USA

June Zhou
Laboratory of Geriatric Endocrinology and Metabolism, Veterans Affairs Medical Center, Washington, DC, USA

Keenan, Michael
Department of Human Nutrition and Food, Louisiana State University AgCenter, Baton Rouge, LA, USA

Kolehmainen, Marjukka
Department of Clinical Nutrition, University of Eastern Finland, Kuopio, Finland

Lappi, Jenni
Department of Clinical Nutrition, University of Eastern Finland, Kuopio, Finland

Martin, Roy J.
Pennington Biomedical Research Center, Lousiana State University System, Baton Rouge, LA, USA

Meddings, Jon
Faculty of Medicine, University of Calgary, Calgary, Canada

Mykkänen, Hannu
Department of Clinical Nutrition, University of Eastern Finland, Kuopio, Finland

Paeschke, Teri
Des Plaines, IL, USA

Polzin, Kayla
Cargill, Inc., Waukesha, WI, USA

Poutanen, Kaisa
VTT Technical Research Centre of Finland, Espoo, Finland

Rastall, R.A.
Department of Food Biosciences, University of Reading, Whiteknights, Reading, UK

Rose, Devin J.
Department of Food Science and Technology, University of Nebraska, NE, USA

Selinheimo, Emilia
VTT Technical Research Centre of Finland, Espoo, Finland

Senevirathne, Reshani
Department of Food Science, Louisiana State University AgCenter, Baton Rouge, LA, USA

Slavin, Joanne
Department of Food Science and Nutrition, University of Minnesota, St. Paul, MN, USA

Stewart, Maria
Department of Human Nutrition, Food and Animal Science, University of Hawaii at Manoa, Honolulu, HI, USA

Tzortzis, George
Clasado Ltd, Milton Keynes, England, UK

Venema, Koen
Department of BioSciences, TNO Quality of Life, AJ Zeist, The Netherlands

Chapter 1

Introduction to Fiber and Nondigestible Carbohydrates: Definition, Health Aspects, and Perspectives

Teri M. Paeschke and William R. Aimutis

Introduction

Carbohydrates for digestive health and/or fiber have come a long way since the first definition of dietary fiber was created by Hipsley in 1953 and expanded by Trowell and others in the early 1970s (Devries et al. 1999). These first definitions described the nondigestible components of plant cell wall material and associated compounds. In the following years, much information has been discovered with research regarding nondigestible carbohydrates and/or fiber (as well as other nondigestible substances) and how they nourish the body through fermentation in the large intestine. This is a very exciting time as more relationships between nondigestible carbohydrates, the human or the animal host, and gut microbiota are explored and brought to light. There is much more research necessary for further discoveries and aid in both human and animal digestive health.

We are using the term *nondigestible carbohydrate* in the contents of this book instead of or in addition to the term *fiber*, because some definitions of fiber preclude smaller nondigestible carbohydrates such as lactulose or galactobiose. These are classified as disaccharides and under most current regulatory rules, they qualify as sugar and not fiber, regardless of their digestibility or fermentability in the colon or glycemic impact. Furthermore, some nondigestible carbohydrates do not have

Nondigestible Carbohydrates and Digestive Health. Edited by T. M. Paeschke and W. R. Aimutis

analytical methods to quantify these substances as fiber using official American Organization of Analytical Chemists (AOAC) methods, although some of these may have the same benefits for digestive health as fibers that are quantified by AOAC methods. As an example, pectic and xylooligosaccharides have been suggested as possible prebiotics (see Chapter 9; Roberfroid 2009), but current AOAC fiber method analyses will not quantify these substances.

Fiber and nondigestible carbohydrates are tremendously difficult analytes because they are a group of compounds composed of a large number of carbohydrates with very different physical properties. Solubility in water and solvents of varying polarity can be very different, fiber structure can be highly branched or linear and vary by orders of magnitude in molecular weight and other factors cause difficulty in finding one analytical method that can analyze all types of fiber.

Fiber and nondigestible carbohydrates can be defined as those which are not digested (or minimally digested) in the stomach or small intestine and are carried on into the large intestine where they might be utilized by gut microbiota as substrates resulting in a number of metabolites. Some of these metabolites are utilized by the human host as energy, hormonal response stimulators, facilitators of mineral uptake, immune function enhancers, and other roles (Roberfroid 2009). Other metabolites are further utilized by gut microbiota themselves in cross-functional relationships where one species creates a metabolite and another uses that metabolite to create a different compound. These relationships are very complex and we are just beginning to understand the true role of the gut in human and animal health. Regardless of the definition or the ability to analyze nondigestible carbohydrates, the primary goal in studying them is to further enhance and develop human and animal health.

Physiological versus Analytical Definitions of Fiber

A good point to start exploring this topic is to consider the evolution of the definition of fiber, how it relates to health benefits, and the development of analytical methods. The definitions of fiber (as well as prebiotic) over time always stem from investigations into the potential health benefits. As more health benefits come to light and other nondigestible carbohydrates are identified as fermentable and/or prebiotic, better analytical methods will be developed. The need for better methods is driven by the definition and the definition is ultimately driven by health benefits. Excellent references in the history and development of the definition of fiber are available from many sources and a few key ones are cited here (Devries et al. 1999; Dietary Fiber Definition Committee 2001; Slavin 2003; Cui and Roberts 2009; Mann and Cummings 2009).

Table 1.1 contains some of the milestones in development of a fiber definition. Please note that not all researchers agreed on the definitions over time and it is still a matter of debate.

Hipsley is credited for creating the term *dietary fiber* in 1953, defining it as lignin, cellulose, and hemicellulose. The paper presented data on diets from several countries and time periods attempting to correlate fiber ingestion with the incidence of pregnancy toxemia, showing that diets high in fiber reduced incidence of the condition. The definition of fiber was broadened by Trowell, Burkitt, Painter, and others with research on diverticulitis (Painter and Burkitt 1971), transit time (Burkitt et al. 1972), cardiovascular disease (Trowell 1972) in subsequent years when a more encompassing definition was published in 1974 and expanded in 1976 as detailed in Table 1.1. The definitions were driven by Trowell's hypothesis and research on the health benefits of fiber on cardiovascular disease proposing that "Experiments in animals and man may be interpreted to support a suggestion that dietary fiber decreases the reabsorption of bile salts, increases fecal excretion, and reduces hyperlipidemia" (Trowell 1972). A few years later, Jenkins et al. (1980a, 1980b, 1980c) were experimenting with viscous soluble fiber (guar gum) and modification of glycemic response finding that guar gum at high levels greater than 20 g/day moderated glucose and insulin levels in diabetic patients and normal subjects. Clearly, the science around the health benefits of soluble fiber was advancing, showing health benefits of fiber beyond cellulose, hemicellulose, and lignin defined as dietary fiber by Hipsley in 1953.

Along with advances in scientific evidence of the health benefits of fiber, it became clear that better analytical methods needed to be developed as the current methods of the time were inadequate. Until 1985, there was no AOAC method specifically designed to analyze dietary fiber. Crude, acid, and neutral detergent methods for fiber were used to digest (hydrolyze) proteins, sugars, starch, and pectins from plant cell wall material. These are analytical wet chemistry methods for

Table 1.1. Selected history of proposed definitions of fiber.

1953 Hipsley	Dietary fiber is lignin, cellulose, and hemicellulose.
1974/1976 Trowell et al.	1974 Structural polysaccharides: celluloses, also homopolysaccharides and heteropolysaccharides, formerly inaccurately classified as hemicelluloses. These unavailable carbohydrates are bound as a fine lattice work with gums and lignins: polymers based on phenylpropane units. Includes "Unavailable lipids: waxes and cutins, associated with fibre." 1976 . . . dietary fiber should be redefined equally simply as the plant polysaccharides and lignin which are resistant to hydrolysis by the digestive enzymes of man. Dietary fiber includes storage polysaccharides, mucilages, and algal polysaccharides.
1985 Working definition of fiber (Devries et al. 1999)	AOAC Official Method of Analysis 985.29, Total Dietary Fiber in Foods—Enzymatic–Garavametric Method Adopted. Method and the equivalent AACC Approved method 32-05 become the *defacto working definition of fiber*.
2000 AACC definition issued	"Dietary fiber is the edible parts of plants or analogous carbohydrates that are resistant to digestion and absorption in the human small intestine with complete or partial fermentation in the large intestine. Dietary fiber includes polysaccharides, oligosaccharides, lignin, and associated plant substances. Dietary fibers promote beneficial physiological effects including laxation, and/or blood cholesterol attenuation, and/or blood glucose attenuation."
2002 Institute of Medicine	*Dietary Fiber* consists of nondigestible carbohydrates and lignin that are intrinsic and intact in plants. Functional Fiber consists of isolated, nondigestible carbohydrates that have beneficial physiological effects in humans. *Total Fiber* is the sum of *Dietary Fiber* and *Functional Fiber*.

Table 1.1. (*Continued*)

2009 Codex definition	Dietary fiber means carbohydrate polymers[1] with ten or more monomeric units, [2] which are not hydrolyzed by the endogenous enzymes in the small intestine of humans and belong to the following categories: Edible carbohydrate polymers naturally occurring in the food as consumed; Carbohydrate polymers, which have been obtained from food raw material by physical, enzymatic, or chemical means and which have been shown to have a physiological effect of benefit to health as demonstrated by generally accepted scientific evidence to competent authorities; Synthetic carbohydrate polymers which have been shown to have a physiological effect of benefit to health as demonstrated by generally accepted scientific evidence to competent authorities.

[1]When derived from a plant origin, dietary fiber may include fractions of lignin and/or other compounds when associated with polysaccharides in the plant cell walls and if these compounds are quantified by the AOAC gravimetric analytical method for dietary fiber analysis: Fractions of lignin and the other compounds (proteic fractions, phenolic compounds, waxes, saponins, phytates, cutin, phytosterols, etc.) intimately "associated" with plant polysaccharides are often extracted with the polysaccharides in the AOAC 991.43 method. These substances are included in the definition of fiber insofar as they are actually associated with the poly- or oligosaccharidic fraction of fiber. However, when extracted or even reintroduced into a food containing nondigestible polysaccharides, they cannot be defined as dietary fiber. When combined with polysaccahrides, these associated substances may provide additional beneficial effects (pending adoption of Section on Methods of Analysis and Sampling).
[2]Decision on whether to include carbohydrates with monomeric units *lower than* ten should be left to national authorities.

estimating fiber, not what is actually digested by mammalian enzymes in the upper gastrointestinal tract or the fraction that progresses undigested to the large intestine. These methods are still in use today for analyzing livestock feed.

The Trowell definition (1974, 1976) gained wide acceptance through international collaboration and became the basis definition for the first official AOAC International method (Devries et al. 1999) issued in 1985 as AOAC 985.29 and became the "defacto working definition

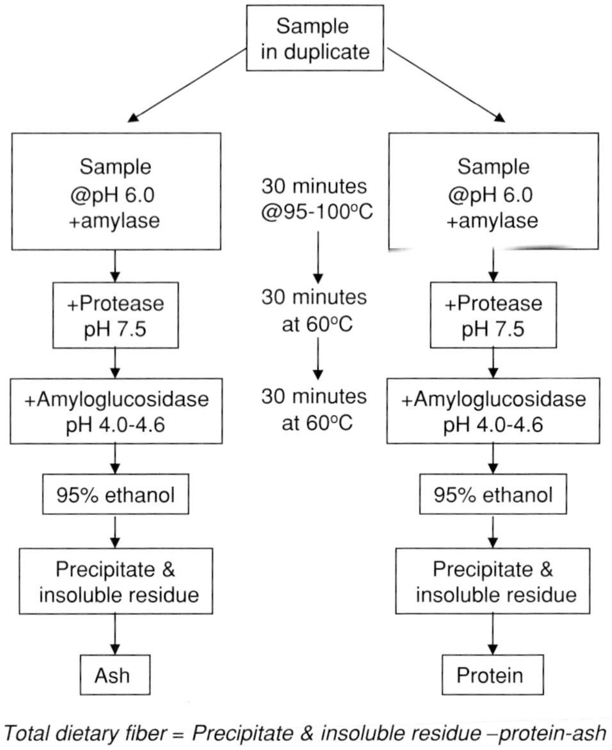

Figure 1.1. AOAC method 985.29: enzymatic/gravimetric method for fiber analysis.

of fiber" (Devries et al. 1999). As is shown in Figure 1.1 and Table 1.2, AOAC 985.29 is an enzymatic/gravimetric method using amylase, amyloglucosidase, and protease to digest what is not fiber.

It does not represent mammalian digestion, but it does constitute a standard method to analytically determine total dietary fiber (TDF) for labeling purposes and is still in use today. AOAC 985.29 does not distinguish between soluble and insoluble fibers, but simply measures TDF. As the evidence for benefit of soluble fibers increased over the years with research on oat beta glucan (Wood 1991; Brennan and Cleary 2005), and other soluble fibers such as pectin, guar, and psyllium (Theuwissen and Mensink 2008), a method was developed that quantified insoluble and soluble fiber (AOAC 991.43; see Figure 1.2).

Again, it is an enzymatic/gravimetric method, using similar enzymes to digest carbohydrates and proteins. The viscous soluble fibers (higher

Table 1.2. AOAC methods for fiber analysis.

AOAC method	Analyzes	
985.29	Total dietary fiber	Enzymatic/gravimetric
991.43	Insoluble and soluble fiber	Enzymatic/gravimetric
997.08	Fructans	Water extraction/enzyme/LC analysis/high performance anion-exchange chromatography with pulsed amperometric detection (HPEAD-PAD)
2001.03	Dietary fiber with resistant maltodextrin	Enzymatic/gravimetric/LC specific for resistant maltodextrin
2000.11	Polydextrose	Water extraction/enzyme /LC analysis HPAED-PAD
2001.02	Transgalactooligosaccharides	Water extraction/enzyme/LC analysis HPAED-PAD
2002.02	Resistant starch	Enzyme/glucose oxidase-peroxidase/ spectrophotometric
2009.01	Total dietary fiber (high molecular weight + low molecular weight)	Enzymatic/gravimetric /LC

molecular weight) are precipitated with ethanol and acetone, but unfortunately lower molecular weight moieties such as inulin and resistant maltodextrin are not analyzed with this method because they are soluble in ethanol. There has been some debate over exactly what molecular weight of fiber will precipitate with ethanol or acetone fraction and what will stay in solution. Some researchers feel that a degree of polymerization (DP) of 10 is the cutoff point, meaning that all fibers with a DP of over 10 will precipitate and those under 10 will stay in solution. The authors of this chapter are of the opinion that the molecular structures of particular nondigestible carbohydrate dictate what precipitates and what does not and that the DP of what does or does not precipitate in this method can be variable, depending on molecular structure. Branching and bonding as well as monosaccharide composition of nondigestible low molecular weight carbohydrates influences solubility in both water

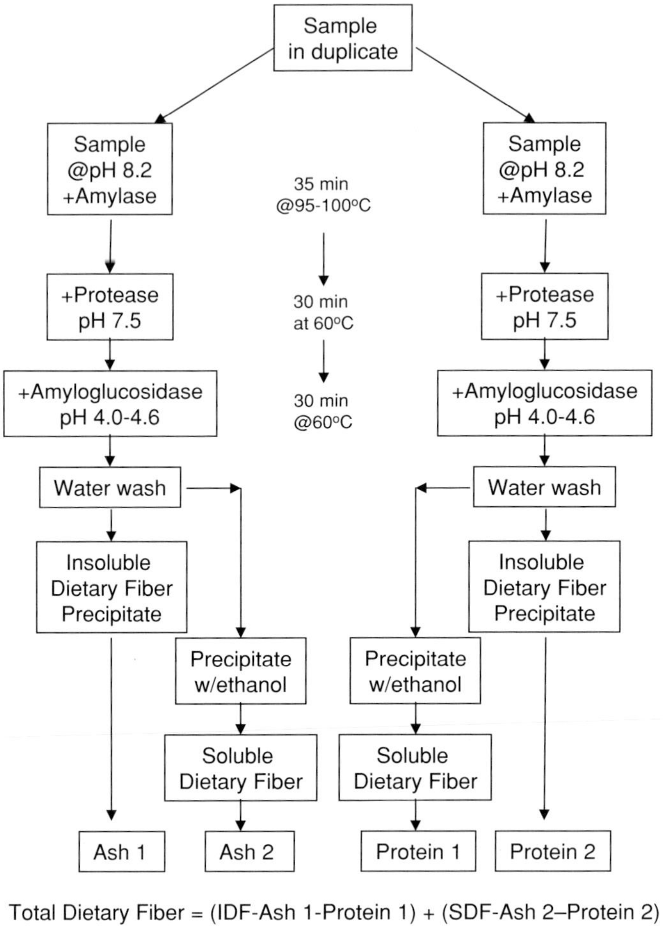

Figure 1.2. AOAC method 991.43: enzymatic/gravimetric method for fiber analysis.

and alcohol. It is not the purpose of this chapter to debate on this issue, simply to bring it to light and let the reader form his or her own opinion in this matter.

Evidence for the health benefits of lower molecular weight soluble fibers and other nondigestible carbohydrates as fermentable substrates started to develop in the late 1980s and early 1990s. An often-quoted review paper by Gibson and Roberfroid (1995) proposed a definition of prebiotic stating, "A prebiotic is a nondigestible food ingredient that beneficially affects the host by selectively stimulating the growth and/or activity of one or a limited number of bacteria in the colon, and thus improves host health." Later criteria were added for resistance to digestion, expanding the 1995 definition (Roberfroid 2009).

Roberfroid (2009) recently summarized the state of the science detailing functional effects and disease risk reduction, with functional effects such as intestinal colonic function, bioavailability of minerals, and satiety and disease risk reduction such as metabolic syndrome, obesity, and colon cancer in the book *Handbook of Prebiotics* (2009). Other authors in this book go into much greater detail regarding health benefits and mechanisms of lower molecular weight nondigestible carbohydrates. As evidence mounted for fermentability and/or prebiotic nature of lower molecular weight nondigestible carbohydrates and was amassed, more AOAC methods were created to analyze the components separately, using liquid chromatography and some spectrophotometric methods. The evolution of fiber analysis in Table 1.2 shows the complexity of analyzing for constituents such as inulin or polydextrose and the variety of methods used to do so.

Of particular note here is resistant starch. Other authors in this book detail the types and physiological functions in much greater detail but from an analytical viewpoint, it is not quite insoluble or soluble fiber and the quantity can be misrepresented by traditional fiber analysis methods. Both the 985.29 and 991.43 AOAC methods digest substrates at a temperature of 100°C for 30 minutes using nonmammalian enzymes, conditions that do not reflect human digestion. Type 2 and 3 resistant starches are either naturally inherent crystalline material (type 2) or created by retrogradation (type 3), may be subject to partial gelatinization and solubilization in solution at the high temperature, and can be lost in the analysis. AOAC 985.29 and 991.43 fiber analytical methods are standardized and do provide a good estimation of fiber for a wide range of food products, but it was clear that better methods needed to be developed and was the subject of great debate at both AACC and International Food Technologists (IFT) meetings for a number of years.

Recently, a new definition of fiber was put forth by the Food and Agriculture Organization (FAO) and the World Health Organization at the Codex Alimentarius Commission during the 32nd session at FAO Headquarters, Rome, June 29–July 4, 2009 (ALINORM 09/32/A) as detailed in Table 1.1. Accompanying the new definition, an analysis was developed by Barry McCleary (in collaboration with others) recently as AOAC 2009.01 method issued late in 2009. A summary of this method is shown in Figure 1.3.

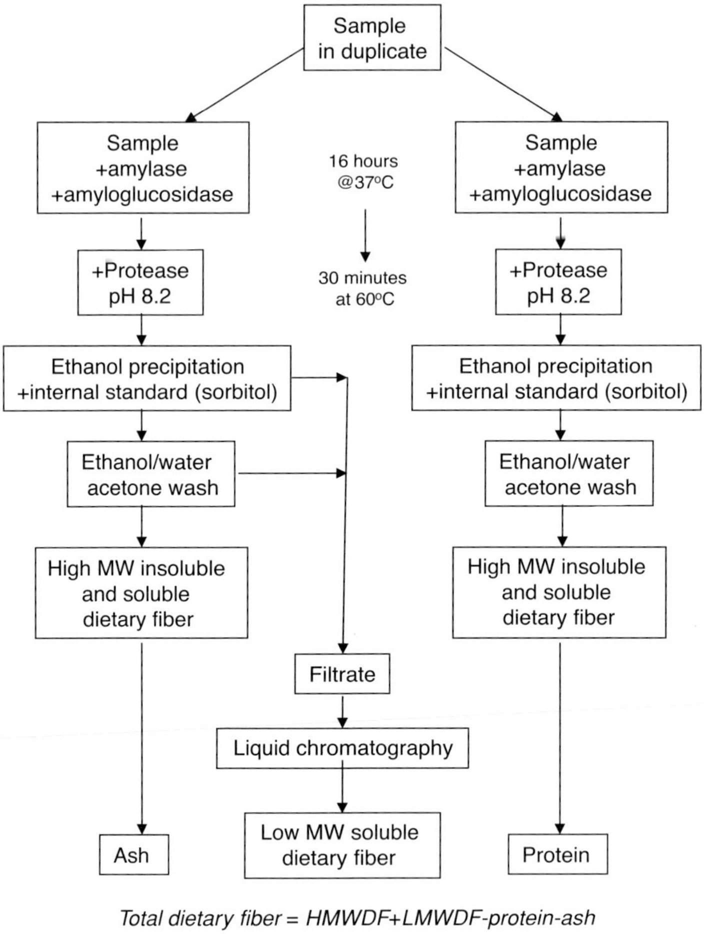

Figure 1.3. AOAC method 2009.01: enzymatic/gravimetric method for fiber analysis.

This method more closely reflects human carbohydrate digestion by providing a 16-hour digestion for the starch portion of digestible carbohydrate at 37°C, rather than 30 minutes at 100°C as per the 985.29 and 991.43 earlier methods. AOAC 2009.01 also uses a porcine pancreatic amylase, rather than the fungal-derived heat stable amylase in 985.29 and 991.43, again closer to mammalian digestion. Undoubtedly, this is an improvement over previous methods, although this method will only report high molecular weight fiber and low molecular weight fiber and does not separate out the soluble viscous portion as does 991.43. As this method gains acceptance and adoption by the food industry, we can hope to see more consistent and less confusing results when it comes to testing for fiber. Although AOAC 2009.01 more closely reflects mammalian digestion, it is still wise to note that it is an analytic method (as are AOAC 985.29 and 991.43) meant to mimic (and thus is not an exact replicate of) digestion and fermentation. Care must be taken not to interpret results as completely accurate representation of mammalian digestion. Official AOAC methods are meant to standardize an analysis, making it easy to replicate as cheaply as possible. To understand true digestion and fermentation, researchers must use animal or human data.

A problematic issue with fiber fermentation research and in vitro methods is how to prepare samples. Substrates such as resistant starch, oligosaccharides (with residual sugars from manufacturing), brans, and legumes have components that are completely digestible in the upper gastrointestinal tract.

Digestible starch and sugars can give misleading results as they are readily available for microbial fermentation. Some researchers attempt to predigest substrates with various procedures to simulate digestions, some of which are based on the AOAC methods because they are standardized and kits are easily obtained. Some researchers (especially in earlier work) do not predigest samples. This variable alone can cause significantly different results between researchers when studying the same substrates. When the variability of individual animal or human fecal flora is compounded by inadequate or erroneously digested samples using in vitro methods, comparing data between research studies is impractical. Again, use of fiber analysis methods or variations for predigestion is best used as an approximation of digestion, with full knowledge of the flaws of these methods.

As is shown in Tables 1.1 and 1.2, the analytical methods for fiber reflect developments in the discovery of health benefits associated fiber and nondigestible carbohydrates and our ability to identify the components responsible for the effect.

These will change in the future as more research is published regarding the health benefits of nondigestible carbohydrates and fiber. In coming years, the editors of this book hope that these issues are resolved to the extent they can be and it is important to remember the ultimate goal for all concerned is to improve public health. In the mean time, it is hoped that the reader realizes the definitions and health benefits of nondigestible carbohydrates and/or fiber are an emerging science that will continue to evolve over time and that a certain amount of debate is to be both welcomed and expected.

It is very well-established fact that people do not eat enough fiber and nondigestible carbohydrates globally. Increasing the body of research regarding the science surrounding the health benefits of these substances enables the development of more consumer-friendly forms that will encourage fiber consumption.

References

Brennan, C.S., Cleary, L.J. (2005) The potential use of cereal (1-3, 1-4)-beta-D-glucans as functional food ingredients. *J. Cereal Sci.* 42, 1–13.

Burkitt, D.P., Walker, A.R.P., Painter, N.S. (1972) Effect of dietary fiber on stools and transit-times, and its role in the causation of disease. *Lancet* Dec 30, 1408–1410.

Codex Alimentarius Commission (2009) *32nd Session at FAO Headquarters*, Rome, June 29–July 4, 2009 (ALINORM 09/32/A).

Cui, S.W., Roberts, K.T. (2009) Dietary fiber: Fulfilling the promise of added-value formulations. In: Kapsis, S., Norton, I., Ubbink, J. (editors), *Modern Biopolymer Science*. Elsevier, Amsterdam.

Dietary Fiber Definition Committee, American Association of Cereal Chemists (2001) The definition of dietary fiber. *Cereal Food World* 48(3), 112–126.

Devries, J.W., Prosky, L., Li, B., Cho, S. (1999) A historical perspective on defining dietary fiber. *Cereal Food World* 44(5), 367–368.

Gibson, G.R., Roberfroid, M.B. (1995) Dietary modulation of the human colonic microbiota: Introducing the concept of prebiotics. *J. Nutr.* 125, 1401–1412.

Hipsley, E.H. (1953) Dietary "fibre" and pregnancy toxaemia. *BMJ* 2, 420.

Jenkins, D.J.A., Wolever, T.M.S., Nineham, R. et al. (1980a) Improved glucose tolerance four hours after taking guar with glucose. *Diabetologia* 19, 21–24.

Jenkins D.J.A., Wolever T.M.S, Bacon S., et al. (1980b) Diabetic diets: high carbohydrate combined with high fiber. *Am. J. Clin. Nutr.* 33, 1729–1733.

Jenkins, D.J.A., Wolever, T.M.S., Taylor, R.H., Reynolds, D., Nineham, R., Hockaday, T.D.R. (1980c) Diabetic glucose control, lipids, and trace elements on long-term guar. *Br. Med. J.* 280, 1353–1354.

Mann, J.I., Cummings, J.H. (2009) Possible implications for health of the different definitions of dietary fibre. *Nutr. Metab. Cardiovas.* 19, 226–229.

Painter, N.S., Burkitt, D.P. (1971) Diverticular disease of the colon: A deficiency disease. *BMJ* 2, 450–454.

Roberfroid, M.B. (2009) General introduction: Prebiotics in nutrition. In: Gibson, G.R., Roberfroid, M.B. (editors), *Handbook of Prebiotics*. CRC Press, Taylor and Francis Group, Boca Raton, FL.

Slavin, J. (2003) Impact of the proposed definition of dietary fiber on nutrient databases. *J. Food Compos. Anal.* 16, 287–291.

Theuwissen, E., Mensink, R.P. (2008) Review: Water soluble fibers and cardiovascular disease. *Physiol. Behav.* 94, 285–292.
Trowell, H. (1972) Ischemic heart disease and dietary fibre. *Am. J. Clin. Nutr.* 25, 926.
Trowell, H.C. (1974) Definition of dietary fibre. *Lancet*, 503.
Trowell, H.C., Southgate, D.A.T., Wolever, T.M.S., Leeds, A.R., Gassull, M.A., Jenkins, D.J.A. (1976) Dietary fibre redefined. *Lancet a*, 967.
Wood, P.J. (1991) Oat beta glucan-physiochemical properties and physiological effects. *Trends Food Sci. Tech.* 2, 311–314.

Chapter 2

The Gastrointestinal Tract and Its Microflora

William R. Aimutis and Kayla Polzin

Introduction

Individuals studying the impact of nondigestible carbohydrates and fiber on human physiology must possess a solid understanding of the anatomy and physiology of the gastrointestinal tract (GIT). Long considered by researchers as a dark, moist, and somewhat odiferous organ within the human body, the GIT in recent years has become a site of intensive study especially as to how its actions contribute to comfort or distress in its human host. In reality, the GIT, or gut, is a series of organs that begin the digestive process from the moment a particle of food enters the mouth until the final waste products exit the body as feces. In adult humans, the GIT is approximately 6.5–8 meters long and consists of upper [mouth, stomach, and duodenum (proximal to the ligament of Treitz)] and lower sections (duodenum, jejunum, ileum, cecum, and colon) each with its own specific role in the digestion of food ingested by the human host. Several other organs act as supporting members in the digestive process by secreting organic and inorganic compounds to regulate the digestive process. Although not considered a specific organ, the microbiota (see Table 2.1 for a listing of key terms used in this chapter) of the human intestinal tract in recent literature is being viewed as having an important role in maintaining comfort for the human host. Methods to study the microbiome have rapidly evolved in

Nondigestible Carbohydrates and Digestive Health. Edited by T. M. Paeschke and W. R. Aimutis

Table 2.1. List of terms and definitions used in this chapter.

Term	Definition
Microbiome	Collective genomes within a microbiota
Microbiota	Collective community of microorganisms within a habitat
Commensal	Relationship in which one partner benefits without detriment to the other
Facultative anaerobe	An organism that grows equally well under in the presence or absence of oxygen
Opportunistic pathogen	An organism that exists harmlessly as part of the normal human body environment and does not become a health threat until the body's immune system fails
Contig	Contiguous sequence of DNA created by assembling overlapping sequenced fragments of a chromosome
Germ-free	Animals raised in the absence of all microorganisms
Conventionalized	Animals derived germ-free and later colonized with a microbiota harvested from conventionally raised donors
Metabolomics	Identification and quantification of host and microbial metabolites in a particular host compartment (e.g., the intestinal tract)

recent years to allow researchers to better understand the complexity and function of this large microbial population. This chapter offers a brief overview into the GIT microflora, methods to study the microbial population, and a brief introduction to intestinal diseases.

The Intestinal Microbiota

The fetal GIT is sterile until the infant begins leaving the womb during natural childbirth. As a baby journeys through the vaginal canal, the infant begins inhaling and ingesting microorganisms that constitute the normal or indigenous microflora of the mother's vagina and feces (Mandar and Mikelsaar, 1996). In contrast, infants delivered by cesarean section have an initial intestinal microflora much different than a vaginally born infant (Gronlund et al., 1999). The colonizing flora in these infants comes from the environment where they were born, and is typically characterized by the following predominant species: *Enterococcus faecalis*, *Escherichia coli*, *Enterobacter cloacae*, *Klebsiella pneumoniae*, *Staphylococcus epidermidis*, and *Staphylococcus haemolyticus* (Fanaro et al., 2003). But as the infant is united with the mother, the initial colonizing flora quickly evolves to be more similar to a vaginally delivered infant.

The infant's intestinal flora continually evolves during the first 2–4 years of life (Mackie et al., 1999), but most radically changes in the first few weeks after birth. If the infant is breast fed, a microbial community dominated by *Lactobacillus* and *Bifidobacterium* is established. Breast-fed infants have high levels of acetate and lactate in their feces, and lower levels of other short-chain fatty acids. With the introduction of solid food, the microflora of breast-fed infants rapidly changes to a very mixed flora and is characterized by significant decreases in acetate and lactate, while propionate significantly increases (Scholtens et al., 2004). Comparatively, formula-fed infants establish a very mixed population characteristic of the population the child will have after weaning and similar in composition to the birth mother's intestinal microflora (Ley et al., 2005). Their feces contain higher concentrations of propionate and butyrate than breast-fed infant's feces (Knol et al., 2005). Inclusion of galactooligosaccharides and/or fructooligosaccharides in infant formula establishes a different intestinal microflora exemplified by the increased concentration of acetate and lower levels of propionate in the feces.

The adult intestine contains over a 100 trillion microbes which exceeds the total number of cells in the body by greater than ten-fold (Figure 2.1). The intestinal microbiota is dominated by two major divisions of bacteria—Bacteroidetes and Firmicutes. In addition, one member of Archaea, *Methanobrevibacter smithii* is frequently identified (Backhed et al., 2005). There have been over 1,000 species identified in the GIT (Egert et al., 2006), but this seems to only be a fraction of the total population as much of this microbial community is difficult to culture and characterize by traditional phenotypic methodology (Eckburg et al., 2005). Approximately 80% of all microbial members of the gut microbiota cannot be cultured outside the GIT. The factors for selecting a specific microbial community within a human host are not known today, but animal models are being used to elucidate the

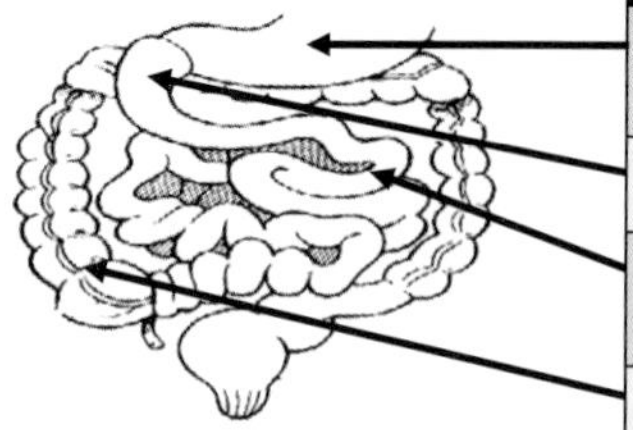

Intestinal anatomy	Microbial population (CFU/mL)	Role in digestion
Stomach	10^1–10^3	Particle size reduction, lipid emulsification, and initial protiein hydrolysis
Duodenum	10^1–10^3	Amino acid homeostasis and osmotic regulation
Jejunum/ileum	10^4–10^7	Amino acid homeostasis and osmotic regulation; lipid metabolism
Colon	10^{11}–10^{12}	Protein and carbohydrate metabolism; muscular activity and osmotic protection

Figure 2.1. This figure illustrates anatomy and digestive characteristics of the human intestinal tract and microbial population density in each section.

contribution of host genotype and diet to gut microbial diversity (Camp et al., 2009).

Elderly persons tend to have more impaired gastrointestinal function. Changes in the intestinal microflora and reduced intestinal immunity in this age group could have clinical consequences that favor gastrointestinal infections. Geriatric patients have a decreased population of anaerobes and bifidobacteria, and an increase of enterobacteria (Hébuterne, 2003).

The use of animal models and antibiotics targeted at specific microbial populations are providing insights into functionality of the microflora. The use of amoxicillin in rat pups showed depletion in *Lactobacillus* species and microarray analysis showed significant changes in many genes associated with development (Schumann et al., 2005). In another study, Gram positive bacteria were depleted in mice by vancomycin, and metabolomic studies showed that carbohydrate metabolism was disrupted and resulted in increased concentrations of unfermented oligosaccharides and reduced concentrations of short-chain fatty acids in the feces (Yap et al., 2008). These studies indicate that significant changes in the gastrointestinal microbiota population can significantly impact intestinal functionality.

Colonization

The intestinal environment is harsh to invading microorganisms presumably even those that eventually do establish colonization. Those microbes that survive the initial exposure to gastric acids and bile in the stomach will be forced to compete for nutrients with other microorganisms for survival. Additionally, there are many other factors that do not favor unwanted microbial visitors to the intestinal tract. The host intestinal cells will be protecting themselves by secretion of mucin to line the epithelial cells and guard against microbes that do not have the necessary adherence factors and metabolic machinery to sustain life (Figure 2.2). Additionally, the intestine is continually in peristaltic motion and able to washout unwanted nonindigenous microbes. The host can deliver a final line of defense in the complex immune response via the gastrointestinal-associated lymphoid tissue (GALT) to "seek and destroy" unwanted microbes. Nonetheless, the microbiome that does establish itself is genetically equipped for survival in the intestinal environment and remains relatively stable throughout the host from 4 years of age until late in life.

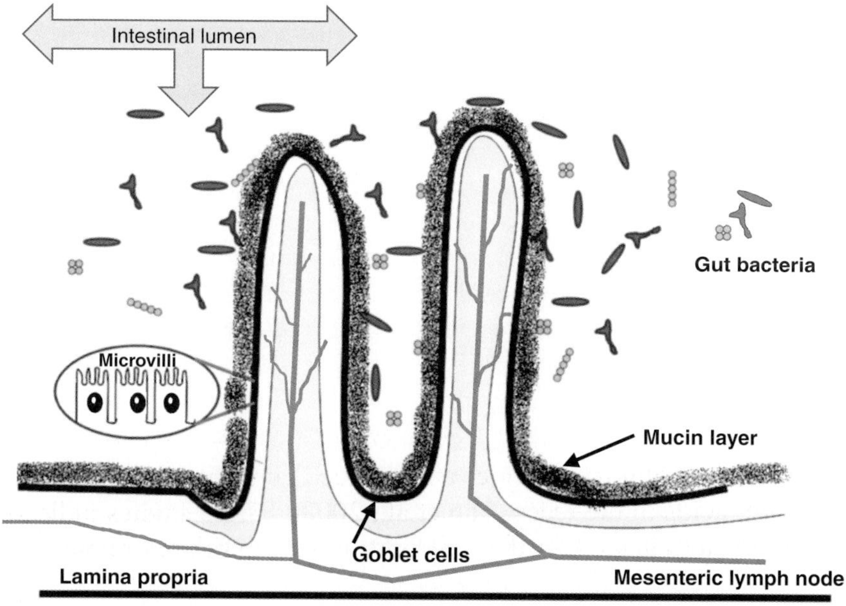

Figure 2.2. This cartoon shows a cross section of human small intestine demonstrating villi and microvilli structure in the epithelial layer. Mucin is secreted in the crypts of the villi and bathes the epithelial cells lining the villi. Mucin acts as a lubricant to move food particles and other digestive contents through the tract. As it becomes sloughed off the villi, indigenous bacteria also use it as a nutrition source. The mesenteric lymph node is responsible for delivering neutrophils to the epithelial surface to fight off bacterial foes.

The gut microbiome is beneficial to the host in several respects, and accomplishes this by microbe–microbe and microbe–epithelial cell cross talk (Walker, 2009). Cross talk is a means of in vivo communication to establish a symbiotic relationship between the microbiota and the host. Cross talk begins at parturition as the newborn develops epithelial and immune intestinal defenses and the intestinal microflora is rapidly establishing. Infants born prematurely, delivered by cesarean section, or receiving perinatal antibiotics do not benefit from receiving intact microbial inocula from their mother at parturition (Gronlund et al., 1999). As a result, the microbial–epithelial cross talk necessary for intestinal defense maturation is compromised. Intestinal defense maturation being delayed, or in some instances permanently dysfunctional, leads to an increased incidence of immunomediated diseases including allergy and autoimmune diseases (Walker, 2009). It is through this intimate microbial–epithelial cell communication that the host recognizes "friend from foe," and the infant is protected presumably for their entire life from a variety of intestinal maladies including inflammatory bowel disease (IBD).

Actual colonization and permanent residency in the GIT is genetically controlled by microbial-assisted molecular patterns (MAMP) on the surface of colonizing bacteria interacting with pattern recognition receptors (PRRs) on eukaryotic cells. The resultant signal transduction and activation of transcription factors activates numerous genes that affect immune and epithelial barrier functions (Kohler et al., 2003; Sansonetti, 2004). Toll-like receptors (TLRs) on the epithelial cell surface are the most characterized of the PRRs, and are recognized as receptors for ligands on colonizing bacteria. The resulting interaction between the colonizing bacteria and TLRs allows the bacteria to establish themselves in the gut and avoid potential washout. Similarly, bacterial ligands can attach to sloughed epithelial cells or secreted mucin and metabolize them as nutrients. In addition, colonizing bacteria secrete molecules that mediate anti-inflammatory factors and prevent chronic gut inflammation (Neish et al., 2000).

The Indigenous Microflora of the Human Intestinal Tract

The human GIT can be viewed as three distinct compartments in terms of microbial population—stomach, small intestine, and large intestine

(Figure 2.1). Although the human intestinal tract begins at the mouth, microorganisms indigenous to the oral cavity probably play little role in digestion. The low pH and rapid gastric motility of a healthy stomach limits the bacterial colonization in this compartment. Nonetheless, acid-tolerant species of *Lactobacillus* and *Streptococcus* can be found in the stomach approaching a population of 100–10,000 organisms/mL of contents (Guarner and Malagelada, 2003). In addition, *Helicobacter pylori* is isolated from stomach contents though this organism is not viewed as commensal to the host (Marshall, 1994).

As one transcends the intestinal tract to the small intestine, the environment becomes more anaerobic and bacterial populations approach 10^6–10^8 organisms/mL of contents (Simon and Gorbach, 1984). The microflora of the small intestine is much more diverse than the stomach and numerous species of obligate anaerobes, Gram-negative facultative anaerobes, lactobacilli, and streptococci are isolated. As mentioned earlier, many of these species cannot be cultured and characterized by traditional microbiological methods.

The large intestine is the most densely populated compartment of the intestinal tract (Simon and Gorbach, 1984). The higher environmental pH, strict anaerobic conditions, slow motility, and nutrient rich environment leads to microbial numbers that are 10^{11} to 10^{12} organisms/mL of contents. The microbial biomass comprises over 50% of colonic contents, and has plenty of substrates available for growth. In fact, the large quantity of nutrients probably accounts for the large diversity of species observed in the colon. Substrates not only come from the diet as undigested carbohydrates and proteins, but also from the host themselves in the form of mucin. Carbohydrates and undigested protein escape digestion in the small intestine to provide growth nutrients for fermentation by the microflora in the large intestine. Undigested carbohydrates are composed mainly of resistant starches, plant-origin nonstarch polysaccharides and dietary fiber, and nondigestible oligosaccharides (MacFarlane et al., 1992).

A healthy, adult large intestine contains a highly diverse microflora dominated by species of strict anaerobes from the genera *Atopobium*, *Bacteroides*, *Bifidobacterium*, *Clostridium*, *Butyrovibrio*, *Eubacterium*, *Fusobacterium*, *Peptostreptococcus*, and *Ruminococcus* (Moore and Holdeman, 1975). Facultative anaerobe species from *Lactobacillus* and *Streptococcus* inhabit this region. Interestingly, several species present in the colon can become opportunistic pathogens to the host if an imbalance occurs. Such disturbances can occur from antibiotics, various disease states, stress, substance abuse, and other factors known and unknown.

Carbohydrate Metabolism by the Gut Microbiota

Humans do not ingest enough carbohydrate daily to sustain the energy they require without a symbiotic relationship with the microbiome. Indigenous gut flora ferment dietary nondigestible carbohydrates, exfoliated epithelial cells, and endogenous epithelial-secreted mucous as a source of carbon and nitrogen elements (Figure 2.3). In a true symbiotic relationship, bacteria utilize the carbon and nitrogen not only to sustain themselves, but also to supply a source of energy back to the human host in the form of short-chain fatty acids (SCFAs) and lactate. Humans rely on bacterial enzymes to complement their own biochemical catalog of enzymes necessary to hydrolyze complex carbohydrates. Intestinal microorganisms possess glycoside hydrolases and polysaccharide lyases which humans do not have for digesting complex carbohydrates. This process is especially apparent in the large intestine where polysaccharides and fibers, not digested prior to entering this region of the intestinal

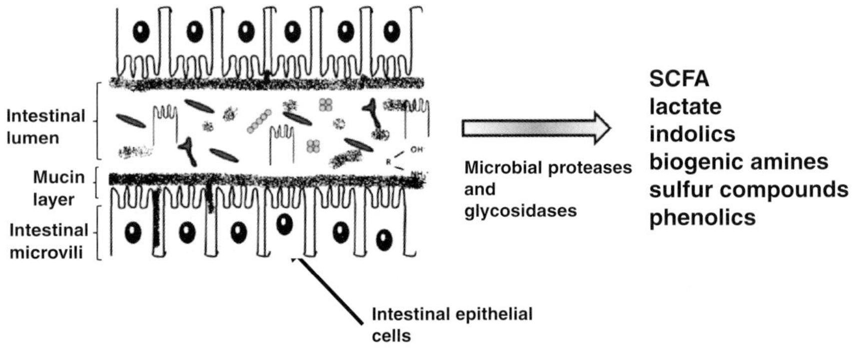

Figure 2.3. This diagram illustrates products of microbial metabolism in the gastrointestinal tract. The lower small intestine and upper small intestine sees nutrients from a variety of sources including sloughed epithelial cells, mucin, amino acids and peptides, hydrolyzed oligosaccharides, polysaccharides, and some simple sugars and lipids. Microbial protease and glycosidases are important to the microflora for hydrolyzing the complex molecules to simpler forms for absorption and metabolism by the microbial flora.

tract, are utilized to sustain a large microbial population. The microbiome preferentially ferments carbohydrates before proteins, and will switch their metabolism to protein fermentation once the carbohydrates have been exhausted (Ouwehand et al., 2005). Unfortunately, protein metabolism sometimes leads to production of toxic metabolites such as ammonia, phenolics, indolics, and sulfur-containing compounds. Some investigators have hypothesized that toxic metabolites from protein metabolism may be responsible for IBD and colon cancer (Van Nuenen et al., 2004). Carbohydrate fermentation tends to occur in the proximal part of the large intestine and protein fermentation in the distal colon (Guarner and Malagelada, 2003).

Bacterial fermentation of dietary fiber and other organic materials in the large bowel, especially by species of *Bacteroides, Eubacterium,* and *Bifidobacterium*, leads to increased concentrations of SCFAs (Englyst and MacFarlane, 1986). SCFAs (especially acetic, propionic, and butyric) are rapidly absorbed by the colonic mucosa to contribute to host energy requirements (Cummings and MacFarlane, 1991). SCFA metabolism by the host occurs in different regions of the body. Acetate is metabolized by muscle, heart, kidney, and brain tissues; propionate is removed by the liver. Butyrate is metabolized by the colonic epithelium and regulates cell growth and differentiation. Proteins, especially those high in branched-chain amino acids, are metabolized by colonic microflora to branched-chain fatty acids (isobutyrate and isovalerate) which are important energy sources for the host.

SCFA have other functions which benefit the human host. Their presence improves gastric motility rate by improving colonic water and sodium absorption by an unknown mechanism (Elsen and Bristrian, 1991). They also regulate gut function by reducing intestinal permeability (Rosenfeldt et al., 2004). Other researchers have shown that increased concentrations of acetic, propionic, and butyric acids improve mineral absorption (Younes et al., 2001).

Obesity and the Microbiota

Can the microbiota be linked to obesity? Recent studies have investigated the role of the intestinal microflora at being too efficient in removing energy from the diet, and consequently having a link to human obesity. Substantial changes were observed in the composition and metabolic function of gut microbiota in obese mice (Backhed et al., 2004). Germ-free mice conventionalized with a cecum-derived microbial community from obese conventional mice produced a dramatic increase in total body fat composition despite no increase in caloric load or decrease in energy expenditure. In addition, it appeared that the gut microbiota was responsible for increased monosaccharide uptake from the gut with subsequent increases in hepatic production of triglycerides. Ultimately, the mice developed insulin resistance.

Mice with a dominant gene (*ob/ob*) for obesity have a gut microbiota different than their lean littermates that were gene recessive or more wild type for obesity (*ob/+*) (Ley et al., 2005, 2006). All mice were fed a polysaccharide-rich diet, and their microbiota was characterized by 16S rRNA sequencing. Obese mice showed a 50% reduction in the Bacteroidetes and an equivalent increase in the Firmicutes. In addition, these mice had higher populations of methanogenic bacteria. It is believed this bacterial population difference accounts for increased efficiency of bacterial fermentation in obese mice. A supporting study was done with *Methanobacter smithii* and *Bacteroides thetaiotaomicron* (Samuuel and Gordon, 2006) in co-colonization of mice. Results indicated enhanced efficiency and specificity of bacterial polysaccharide fermentation to salvage more energy, and an increase in adiposity compared with mice colonized with either organism alone.

The feeding of high-fat diets also rearranges the gut microflora with a decrease in the *Bifidobacterium* population and elevated levels of pro-inflammatory cytokines (Cani et al., 2007a, 2007b). Obesity and metabolic disorders (Type 2 diabetes and insulin resistance) are strongly linked to inflammation (Kahn and Flier, 2000; Wellen and Hotamisligil, 2005). Dietary intervention by feeding fructooligosaccharides to increase the bifidobacteria population was shown to improve glucose tolerance in mice fed a high-fat diet, and also lowered the inflammatory response. This is further evidenced by the fact that the gut microbiome influences the pathophysiological regulation of inflammation, glucose tolerance, and insulin secretion.

With confirmation of the relationship of gut microbiota to energy capture and obesity in mice, researchers have now studied humans to see if similar relationships exist. Adult monozygotic and dizygotic twin pairs concordant for leanness or obesity, and their mothers were studied to determine if host genotype, environmental exposure, and host adiposity may exhibit differences in the gut microbiome (Turnbaugh et al.,

2008). Obese individuals had lower Bacteroidetes and higher Acintobacteria populations than lean individuals. The study also showed that the gut microbiome is shared among family members with a comparable degree of covariation between adult monozygotic and dizygotic twin pairs. This supports the belief that the gut microbiota is inherited from maternal microflora.

The gut microbiota has an important role in harvesting energy for the host from carbohydrates in the diet. However, more research needs to be conducted to better understand individual variation in the efficiency of energy harvest and how it relates to obesity.

Molecular Methods to Study the Gut Microbiota

While traditional microbiological plating techniques can be used to detect and enumerate the populations of some gut microorganisms, complexity of the gut microbiota and the fact that as much as 80% of gut microbes are not culturable (Eckburg et al., 2005) has caused scientists studying the gut microbiome to turn to molecular-based techniques developed for the study of other microbial ecologies. These methods not only enable detection of nonculturable organisms, but also can be highly selective and so enable the researcher to monitor changes in even a single species of gut microorganism (Hoffman et al., 2009). All of these methods rely on species-specific differences in the DNA sequence of the 16s rRNA gene.

One of the earliest methods used to monitor changes in the gut microbiome is denaturing gradient gel electrophoresis (DGGE) (Muyzer et al., 1993). A variant of this method is called temperature gradient gel electrophoresis (TGGE). In DGGE, DNA is extracted from a mixed population and a variable region of the 16s rRNA gene is amplified by polymerase chain reaction (PCR) to create a mixture of equal-sized fragments differing in sequence. Because all of the fragments are of equal size, they cannot be separated on a normal agarose gel. However, DGGE gels incorporate a gradient of a denaturing agent (such as urea) (or in the case of TGGE, a temperature gradient), which causes DNA fragments to melt as they are moving through the gel. At a specific point in the gel, a particular fragment will melt depending on its DNA sequence. The fragment is prevented from melting completely due to addition of a "GC clamp" during PCR amplification; thus when the

16s rRNA portion of the fragment melts, it produces a partial single-stranded structure that migrates much more slowly through the gel than the double-stranded molecules. That this happens at different times for the different sequenced fragments results in a sequence-specific separation of DNA fragments in the gel. These bands can be correlated to specific species by running known control species, although in some cases two different species can migrate similarly in the gel due to having the same melting point. Therefore, the best method of correlating the bands to species is by cutting the bands out of the gel and sequencing them. DGGE can be used to detect bacterial species that make up as little as 1% of the bacterial population (Muyzer et al., 1993). However, differences in rRNA copy number between species, PCR bias (i.e., some sequences amplifying more efficiently than others), and unequal DNA extraction (i.e., DNA is not equally recoverable from all species) means DGGE can only give a general idea of the proportion of a particular species in a mixed population (Polz and Cavanaugh, 1998; Farrelly et al., 1995; Kunin et al. 2008).

The development of high throughput sequencing techniques such as 454 pyrosequencing and powerful bioinformatics tools has enabled a new method of monitoring the gut microbial population called metagenomics. In metagenomics, a large amount of DNA sequence, covering either the entire extracted DNA or just the 16s rRNA gene (following PCR amplification), is sequenced to give an idea of either the gene content or the species content, respectively, of a particular gut microbiome (Kurokawa et al., 2007; Hoffman et al., 2009; Turnbaugh et al., 2009). In cases where the total extracted DNA is sequenced, bioinformatics tools can be used to construct gene contigs and partial genomic sequences. This method has been used in a number of recent studies to detect differences in the microbiomes of obese versus lean individuals (Turnbaugh et al., 2008b) or in individuals with Crohn's disease (Manichanh et al., 2006), to name only a few applications. However, while metagenomics can give an idea of the relative abundance of different microbial species, it suffers from some of the same limitations as DGGE in this respect.

While DGGE and metagenomics can give an idea of what different types of microbes are present in the gut and also an idea of their relative abundance, fluorescent in situ hybridization (FISH) can be used to determine the location of and associations among gut microorganisms (Moter and Göbel 2000; Duncan et al., 2008). In FISH, species-specific fluorescently labeled DNA oligonucleotide probes based on the

16s rRNA gene are hybridized to gut microbes on sections of gut tissue. Fluorescent microscopy is then used to visualize location of the target bacteria in the gut. The use of multiple probes labeled with different fluorophores allows detection of multiple species simultaneously and can reveal colocalization of different species that may indicate a dependent relationship (Kunin et al., 2008). Because FISH is dependent on the design of probes to species known to be present in the gut, it can be used to learn more about species that were previously identified using DGGE or metagenomics. In addition, FISH allows visualization of intact, individual bacteria allowing enumeration of the number of bacteria in a sample. This could potentially give a general idea of species density in a particular gut region. However, this would require counting numerous samples to confirm that the specific microbe is uniformly distributed.

Better enumeration can be obtained when fluorescent probe hybridization or fluorescently labeled antibodies are combined with flow cytometry (Alvarez-barrientos et al., 2000; Collado et al., 2008). In flow cytometry, single microbial cells pass through an orifice where they are subjected to laser excitation of the fluorophore. Cells tagged with fluorophore emit fluorescent light in response and are counted by the instrument. Some instruments also allow fluorescently tagged cells to be sorted (i.e., separated) from untagged or differently tagged cells allowing them to be purified for further study (fluorescence-activated cell sorting or FACS). Tagging different probes with different fluorophores enables the counting and separation of multiple species simultaneously. The specific tags can also be coupled with fluorescent dyes such as *live/dead* stains or stains that fluoresce in response to different metabolic activities, thus revealing the physiological state of targeted cells (Amor et al., 2002).

Common Intestinal Diseases Impacted by Fiber, Prebiotics, and Probiotics

Acute and chronic digestive diseases cost Americans over $100 billion/year in direct medical expenses. Roughly 60–70 million Americans are affected by digestive symptoms and diseases that are associated with significant morbidity, mortality, and a loss of quality of life. Individuals of all ages, race, ethnicity, sex, and socioeconomic status are affected by digestive diseases. These symptoms and diseases are disorders manifested by erroneous gastric functionality and motility, infections in the GIT, and inflammatory bowel disorders. These disorders are impacted by diet, and can somewhat be mitigated by fiber, prebiotics, or probiotics singly or in combination. In this section, we provide a brief overview into these disorders and their symptoms for the reader to gain an understanding of the etiology. For information on the impact of fiber, prebiotics, and/or probiotics, the reader is referred to other chapters in this book.

Irritable Bowel Syndrome

Irritable bowel syndrome (IBS) has a myriad of symptoms and no known cause or effective remedy. This debilitating condition is probably the number one disease for which individuals seek treatment from gastroenterologists. The disease affects approximately 10–22% of the population. Often, these patients have no other apparent medical issues. IBS has been called several other names—spastic colon, colitis, mucous colitis, spastic bowel, and functional bowel disease. Most just call it miserable.

Fortunately, the majority of patients (approximately 70%) have mild symptoms. The remaining unfortunate population has symptoms ranging from moderate (approximately 25%) to severe (approximately 5%). The disease begins in young adults most often in their 20s to 30s, but some children have also been reported to have the disease. Symptoms include abdominal cramps, loose and frequent stools, a bloated feeling, and excessive gas build-up. These symptoms can dissipate and the patient becomes constipated only to have the symptoms return. This cycle has no set pattern. In some patients it takes only days, but in others it can take weeks. Some patients feel they have recovered only to be struck by the unfortunate symptoms again.

Gastroenterologists and other medical professionals have not identified a known cause for this disease. In most patients there is no known organic problem or physical abnormality. IBS is a functional disorder of the intestinal tract, and many health professionals speculate it involves disturbances in the muscles or nerves of the gut. Symptoms often appear during times of stress, emotional upset, or other psychological factors. Despite the link to the brain, there is no doubt the symptoms are real and not psychosomatic.

Bacterial fermentation in the gut is increased in IBS patients (King et al., 1998). A significant increase in colonic anaerobes, *E. coli*, and *Bacteroides* species were reported in IBS patients (Swidinski et al., 1999). Concurrently, coliforms, lactobacilli, and bifidobacteria decrease in the microflora. There is a possibility that radical changes in the gut microflora could lead to mucosal inflammation and be a cause of IBS.

Most physicians treat the symptoms in this disease since there is no recommended therapy. Diet is usually the first item most clinicians review. The increase of fiber intake is often recommended to increase stool's bulk and slow gastric motility. However, this diet recommendation is not successful at eliminating the symptoms totally. Other dietary recommendations are more from a restrictive perspective and include limiting the diet from caffeine, chocolate, non-nutritive sweeteners (especially certain polyols like sorbitol), some dairy products, and gas-producing vegetables (like cabbage, broccoli, and beans).

Constipation

Constipation is another digestive symptom that commonly afflicts most people at some time during their life, and can be a chronic problem to some. The discomfort is another outcome of functionality and motility in the gut characterized by hard, dry stools and reduced defecation frequency. Constipation tends to affect more women than men, and is common in older people also.

Gastric motility reduction leads to the colon absorbing too much water from the large intestinal lumen, and results in dry, hard stools. Many individuals suffer cramps during constipation because colon motility will return to normal and the colon tries to move the stool on its own. Fecal flora does not seem to be impacted during constipation episodes. Total bacterial counts, SCFA concentrations, and pH remains constant (Kleessen et al., 1997).

Regularity in individuals is difficult to define as what is normal for one person may be different for another. Many individuals do not have bowel movements on a daily basis and are not considered constipated. Similarly, an individual that has three movements a day does not necessarily have diarrhea. We all experience variations in our regularity at given times. Diet, stress, physical exercise, travel, and changes in personal routine can all have effects on our regularity. We can mitigate the chances of being constipated by increasing the fiber in our diet, drinking plenty of fluids, regular exercise, and maintaining our daily routines.

Gastrointestinal Infections

Gastrointestinal infections are most commonly accompanied by diarrhea. Diarrhea is characterized by loose, watery stools occurring very frequently in short to extended time periods. The severity of symptoms can be quite wide-ranging from stools that have little change in appearance, other than being more watery than usual, to severe with mucous and blood presence. Most individuals that get diarrhea do not have a serious infection. The gut uses diarrhea often times to eliminate unwanted antagonizers or as a cleansing system. When individuals have diarrhea that lasts more than three-quarters of the time and lasts for longer than 3 months, then there is a functional problem in the GIT.

Diarrhea can be caused by numerous objects including food, medication (especially antibiotics), certain medical conditions, bacteria, viruses, parasites, and stress. The etiological agent will usually dictate the severity of symptoms. Generally, food, medication, stress, and certain other physiological conditions will be less severe than a bacterial, viral, or parasitical infection. Individuals with acute mild, diarrhea should be treated at onset for dehydration as the body is rapidly losing fluids which would normally be absorbed through the gut. Chronic or severe diarrheal patients should seek medical attention to determine the cause of this condition. Traveler's diarrhea is not uncommon in many individuals. This is usually caused by a combination of stress, uncommon foods being ingested, and a change in one's routine. On rare occasions, a traveler may contract diarrhea from contaminated water or foods. Numerous studies have looked at using probiotics to mitigate traveler's diarrhea, but the results have been very inconsistent. Diets high in fiber may contribute to slowing gastric motility, and allow the body to defend itself from rapid fluid loss.

Inflammatory Bowel Disease

IBD is two related but diverse diseases—Crohn's disease and ulcerative colitis. Both diseases cause chronic inflammation of the intestinal tract, and can also cause problems for other organs (Rutgeerts et al., 2009). IBD usually inflicts young people between the ages of 10 and 30 years. However, individuals aged between 50 and 60 years can also develop IBD but at a lower incidence rate. IBD does show increased incidence rates in certain races and ethnicities with Caucasians and Jewish people having higher risk for the diseases. The incidence rate in the United States is 10–20 cases per 100,000 people.

An overactive immune system is thought to be responsible for the uncontrolled inflammation seen in this disease. Crohn's disease is driven by production of interleukin-12 (IL-12) and interferon-γ (IFN-γ), whereas ulcerative colitis is driven by IL-13 (Bouma and Strober, 2003). Most likely there is a triggering event in one's lifetime such as a bacterial or viral infection that sensitizes an immune response in the gut. Once activated, the immune cells continue to mount an immunological attack on normal intestinal cells even after the triggering substance has been eliminated resulting in prolonged inflammation of the bowel mucosal lining and underlying tissues. The exact mechanism for why this happens is unknown at this time, but there does appear to be a genetic predisposition to the disease.

Crohn's disease and ulcerative colitis are two different diseases both characterized by inflammation in the lower GIT (Rutgeerts et al., 2009). Crohn's is differentiated from ulcerative colitis in that Crohn's causes inflammation anywhere in the intestinal tract usually in patches. Commonly, Crohn's affects the terminal ileum, cecum, perianal, and colon. Crohn's can spread very deep into the affected tissues. Ulcerative colitis affects the innermost linings of the large intestine and rectum. Both diseases can cause irreversible impairment of gastrointestinal structure and function.

Symptoms for both diseases are similar with patients experiencing extreme discomfort from cramping, diarrhea (more severe in ulcerative colitis), blood in the stools, and frequent fever. Crohn's disease can also cause ulcers that penetrate deep into intestinal tissue. Both diseases can go in and out of remission. Severe cases of both diseases cause inflammation in joints, eyes, skin, mouth, and liver. Patients with severe ulcerative colitis, and to a less extent Crohn's, can develop colon carcinoma.

The intestinal bacterial microflora is implicated in the etiology of IBD. Genomic technology was used to demonstrate that microflora diversity in Crohn's disease patients was reduced by 50% compared with controls and 30% in ulcerative colitis patients (Ott et al., 2004). The reduction in diversity was due to loss of normal anaerobic bacteria including *Bacteroides* species, *Eubacterium* species, and *Lactobacillus* species. This reduction in microbial diversity contributes to mucosal inflammation.

Symptoms of both diseases can be eased by use of anti-inflammatory and immunosuppression agents. Severe cases may require surgery to save a patient due to other complications because IBD in itself is not associated with high fatality. Diets high in fiber and probiotics have been reportedly successful in reducing severity of both diseases.

Conclusion

Our understanding of the GIT is continually being demystified as investigators better understand the microbiota and its interaction with the human host. The intestinal microbiota is a complex community with responsibility for maintaining comfort in the human host, but if an imbalance occurs it can mischievously antagonize the host causing mucosal inflammation that leads to a myriad of disease symptoms. Diet, environmental, and genetic factors all have a role in maintaining equilibrium in the microbiota. Interestingly, the intestinal tract has mechanisms to discriminate between commensal, symbiotic microflora that should be present, and exogenous pathogens (Blum et al., 1999). Although the mechanisms are still being identified and understood, there is still a great deal we do not understand. The intestinal epithelial cells are guardians that rapidly discriminate friend (commensal) from foe (pathogens) and maintain homeostasis.

Molecular analyses of the intestinal microbiome will allow investigators to assess the impact of different diet constituents, including prebiotics and probiotics, on the normal flora. Once baseline knowledge has been established about the intestinal flora of a healthy or diseased individual, the impact of attempting to influence the gut microbiome can be followed during and after treatments including dietary changes. This will allow development of better therapeutic agents to target diseases such as IBS, IBD, and other gastrointestinal infections.

References

Alvarez-Barrientos, A., Arroyo, J., Cantón, R., Nombela, C., Sánchez-Pérez, M. (2000) Applications of flow cytometry to clinical microbiology. *Clin. Microbiol. Rev.* 13, 167–195.

Amor, A., Santos, V.E., Verbaarschot, P., et al. (2002) Multiparametric flow cytometry and cell sorting for the assessment of viable, injured, and dead *Bifidobacterium* cells during bile salt stress. *Appl. Environ. Microbiol.* 68, 5209–5216.

Backhed, F., Ding, H., Wang, T., et al. (2004) The gut microbiota as an environmental factor that regulates fat storage. *Proc. Indian. Natl. Sci. Acad. Phys. Sci.* 101, 15718–15723.

Backhed, F., Ley, R.E., Sonnenburg, J.L., Peterson, D.A., Gordon, J.I. (2005) Host-bacterial mutualism in the human intestine. *Science* 307, 1915–1920.

Blum, S., Alvarez, S., Haller, D., Perez, P., Schiffrin, E.J. (1999) Intestinal microflora and the interaction with immunocompetent cells. *Antonie. Leeuwenhoek* 76, 199–205.

Bouma, G., Strober, W. (2003) The immunological and genetic basis of inflammatory bowel disease. *Nat. Rev. Immunol.* 3, 521–533.

Camp, J.G., Kanther, M., Semova, I., Rawls, J.F. (2009) Patterns and scales in gastrointestinal microbial ecology. *Gastroenterology*. 136, 1989–2002.

Cani, P.D., Amar, J., Iglesias, M.A., et al. (2007a) Metabolic endotoxemia initiates obesity and insulin resistance. *Diabetes*. 56, 1761–1772.

Cani, P.D., Neyrinck, A.M., Fava, F., et al. (2007b) Selective increases of bifidobacteria in gut microflora improve high-fat-diet-induced diabetes in mice through a mechanism associated with endotoxaemia. *Diabetologia*. 50, 2374–2383.

Collado, M.C., Isolauri, E., Laitinen, K., Salminen, S. (2008) Distinct composition of gut microbiota during pregnancy in overweight and normal weight women. *Am J Clin Nutr*. 88, 894–899.

Cummings, J.H., MacFarlane, G.T. (1991) The control and consequences of bacterial fermentation in the human colon: a review. *J Appl Bacteriol*. 70, 443–459.

Duncan, S.H., Lobley, G.E., Holtrop, G., et al. (2008) Human colonic microbiota associated with diet, obesity and weight loss. *Int. J. Obes*. 32, 1720–1724.

Eckburg, P.B., Bik, E.M. (2005) Diversity of the human intestinal microbial flora. *Science*. 308, 1635–1638.

Egert, M., de Graaf, A.A., Smit, H., de Vos, W.M., Venema, K. (2006) Beyond diversity: functional microbiomics of the human colon. *Trends Microbiol*. 14, 86–91.

Elsen, R.J., Bristrian, B.R. (1991) Recent developments in short chain fatty acid metabolism. *Nutrition*. 7, 7–10.

Englyst, H.N., MacFarlane, G.T. (1986) Breakdown of resistant and readily digestible starch by human gut bacteria. *J Sci Food Agr*. 37, 699–706.

Fanaro, S., Chierici, R., Guerrini, P., Vigi, V. (2003) Intestinal microflora in early infancy: composition and development. *Acta Paediatr*. 92, 48–55.

Farrelly, V., Rainey, F.A., Stackebrandt, E. (1995) Effect of genome size and *rrn* gene copy number on PCR amplification of 16s rRNA genes from a mixture of bacterial species. *Appl and Environm Microbiol.* 61, 2798–2801.

Gronlund, M.M., Lehtonen, O.P., Eerola, E., Kero, P. (1999) Fecal microflora in healthy infants born by different methods of delivery: permanent changes in intestinal flora after cesarean delivery. *J of Pediatr Gastroenterol Nutr.* 28, 19–25.

Guarner, F., Malagelada, J.R. (2003) Gut flora in health and disease. *Lancet.* 361, 512–519.

Hebuterne, X. (2003) Gut changes attributed to ageing: effects on intestinal microflora. *Curr Opin Clin Nutr Metab Care.* 6, 49–54.

Hoffman, C., Hill, D.A., et al. (2009) Community-wide response of the gut microbiota to enteropathogenic *Citrobacter rodentium* infection revealed by deep sequencing. *Infect Immun.* 77, 4468–4678.

Kahn, B.B., Flier, J.S. (2000) Obesity and insulin resistance. *J Clin Invest.* 106, 473–481.

King, T.S., Elia, M., Hunter, J.O. (1998) Abnormal colonic fermentation in irritable bowel syndrome. *Lancet.* 352, 1187–1189.

Kleessen, B., Sykura, B., Zunft, H.J., Blaut, M. (1997) Effects of inulin and lactose on fecal microflora, microbial activity, and bowel habit in elderly constipated persons. *Am J Clin Nutr.* 65, 1397–1402.

Knol, J., Scholtens, P.A.M.J., Kafka, C., et al. (2005) Colon microflora in infants fed formula with galacto- and fructo-oligosaccharides: more like breast-fed infants. *J Pediatr Gastroenterol Nutr.* 40, 36–42.

Kohler, H., McCormick, B.A., Walker, W. (2003) Bacterial-enterocyte crosstalk: cellular mechanisms in health and disease. *J Pediatr Gastroenterol Nutr.* 36, 175–185.

Kunin, V., Copeland, A., Lapidus, A., Mavromatis, K., Hugenholtz, P. (2008) A bioinformatician's guide to metagenomics. *Microbiol Mol Biol Rev.* 72, 557–578.

Kurokawa, K., Itoh, T., Kuwahara, T., et al. (2007) Comparative metagenomics revealed commonly enriched gene sets enriched in the human gut microbiomes. *DNA Res.* 14, 169–181.

Ley, R.E., Backhed, F., Turnbaugh, P., Lazupone, C.A., Knight, R.D., Gordon, J.I. (2005) Obesity alters gut microbial ecology. *Proc Natl Acad Sci USA.* 102, 11070–11075.

Ley, R.E., Turnbaugh, P.J., Klein, S., Gordon, J.I. (2006) Microbial ecology: human gut microbes associated with obesity. *Nature.* 444, 1022–1023.

MacFarlane, G.T., Gibson, G.R., Cummings, J.H. (1992) Comparison of fermentation reactions in different regions of the human colon. *J Appl Bacteriol.* 72, 57–64.

Mackie, R.I., Sghir, A., Gaskins, H.R. (1999) Developmental microbial ecology of the neonatal gastrointestinal tract. *Am J Clin Nutr.* 69, 1035S–1045S.

Mandar, R., Mikelsaar, M. (1996) Transmission of mother's microflora to the newborn at birth. *Biol Neonate.* 69, 30–35.

Manichanh, C., Riggotier-Gois, L., Bonnaud, E., et al. (2006) Reduced diversity of faecal microbiota in Crohn's disease revealed by metagenomic approach. *Gut.* 55, 205–211.

Marshall, B.J. (1994) Helicobacter pylori. *Am J Gastroenterol.* 89(Suppl.), S116–S128.

Moore, W.E.C., Holdeman, L.V. (1975) Discussion of current bacteriological investigations of the relationships between intestinal flora, diet, and colon cancer. *Cancer Res*. 35, 3418–3420.

Moter, A., Göbel, U.B. (2000) Fluorescence in situ hybridization (FISH) for direct visualization of microorganisms. *J. Microbiol. Methods* 41, 85–112.

Muyzer, G., De Waal, E.C., Uitterlinden, A.G. (1993) Profiling of complex microbial populations by denaturing gradient gel electrophoresis analysis of polymerase chain reaction-amplified genes coding for 16s rRNA. *Appl Environm Microbiol*. 59, 695–700.

Neish, A.S., Gerwitz, A.T., Zeng, H., et al. (2000) Prokaryotic regulation of epithelial responses by inhibition of Iκβα ubiquination. *Science*. 289, 1560–1563.

Ott, S.J., Musfeldt, M., Wenderoth, D.F., et al. (2004) Reduction in diversity of the colonic mucosa associated bacterial microflora in patients with active inflammatory bowel disease. *Gut*. 53, 685–693.

Ouwehand, A.C., Derrien, M., deVos, W., Tiihonen, K., Rautonen, N. (2005) Prebiotics and other microbial substrates for gut functionality. *Curr Opin Biotechnol*. 16, 212–217.

Polz, M.F., Cavanaugh, C.M. (1998) Bias in template-to-product rations in multitemplate PCR. *Appl Environm Microbiol*. 64, 3724–3730.

Rosenfeldt, V., Benfeldt, E., Valerius, N.H., Paerregaard, A., Michaelsen, K.F. (2004) Effect of probiotics on gastrointestinal symptoms and small intestinal permeability in children with atopic dermatitis. *J Pediatr*. 145, 612–616.

Rutgeerts, P., Vermeire, S., Van Assche, G. (2009) Biological therapies for inflammatory bowel diseases. *Gastroenterology*. 136, 1182–1197.

Samuuel, B.S., Gordon, J.I. (2006) A humanized gnotobiotic mouse model of host archaeal-bacterial mutualism. *Proc Natl Acad Sci USA*. 103, 10011–10016.

Sansonetti, P.J. (2004) War and peace at mucosal surfaces. *Nat Rev Immunol*. 4, 953–964.

Scholtens, P.A.M.J., Alles, M.S., Van der Linde, E.G.M., Knol, J. (2004) Disturbance of the intestinal microflora of fully breast fed infants that are introduced to solid weanling foods. *J Pediatr Gastroenterol Nutr*. 39, S488–S489.

Schumann, A., Nutten, S., Donnicola, D., et al. (2005) Neonatal antibiotic treatment alters gastrointestinal tract developmental gene expression and intestinal barrier transcriptome. *Physiol Genomics*. 23, 235–245.

Simon, G.L., Gorbach, S.L. (1984) Intestinal flora in health and disease. *Gastroenterology*. 86, 174–193.

Swidinski, A., Klilkin, M., Ortner, M., Swidinski, G., Friedrichshain, K.H., Jutta, W., (1999) Alteration of bacterial concentration in colonic biopsies from patients with irritable bowel syndrome (IBS). *Gastroenterology*. 116, A1.

Turnbaugh, P.J., Backhed, F., Fulton, L., Gordon, G.I. (2008) Diet-induced obesity is linked to marked but reversible alterations in the mouse distal gut microbiome. *Cell Host Microbe*. 3, 213–223.

Turnbaugh, P.J., Hamady, M., Yatsunenko, T., et al. (2008b) A core gut microbiome in obese and lean twins. *Nature*. 457, 480–484.

Turnbaugh, P.J., Ridaura, V.K., Faith, J.J., Rey, F.E., Knight, R., Gordon, J.I. (2009) The effect of diet on the human gut microbiome: a metagenomic analysis in humanized gnotobiotic mice. *Sci Transl Med.* 1, 1–10.

Van Nuenen, M., Venema, K., Van Der Woude, J.C.J., Kuipers, E.J. (2004) The metabolic activity of fecal microbiota from healthy individuals and patients with inflammatory bowel disease. *Dig Dis Sci.* 49, 485–491.

Walker, W.A. (2009) Bacterial colonization, probiotics, and the development of intestinal host defense. *Funct Food Rev.* 1, 13–19.

Wellen, K.E., Hotamisligil, G.S. (2005) Inflammation, stress, and diabetes. *J Clin Invest.* 115, 1111–1119.

Yap, I.K., Li, J.V., Saric, J., et al. (2008) Metabonomic and microbiological analysis of the dynamic effect of vancomycin-induced gut microbiota modification in the mouse. *J Prot Res.* 7, 3718–3728.

Younes, H., Coudray, C., Bellanger, J., Demigne, C., Rayssiguier, Y., Remesy, C. (2001) Effect of two fermentable carbohydrates (inulin and resistant starch) and their combination on calcium and magnesium balance in rats. *Br J Nutr.* 86, 479–485.

Chapter 3

The Immunomodulatory Effects of Dietary Fiber and Prebiotics in the Gastrointestinal Tract

Marie-Claire Arrieta, Jon Meddings, and Catherine J. Field

Introduction

One of the main functions of the gastrointestinal tract is to selectively absorb nutrients. These processes are selective and allow the passage of only certain compounds from the intestinal lumen across the intestinal epithelium. The lumen of the intestine consists of a milieu of food components, endogenous compounds, and microorganisms, all of which coexist in a complex but coordinated environment. Nutrient absorption represents the careful, and selective, movement of some compounds across the epithelium and the selective exclusion of others. This process requires constant vigilance from the gut-associated immune system (gut-associated lymphoid tissues, GALT). This particular "section" of our immune system is only partially developed at birth and it matures once the epithelium is in contact with food and bacteria. The GALT is able to discriminate between essential and nonessential dietary components as well as bacterial signals in the gut lumen and can be modulated by changes in these signals.

Dietary fiber is a nonessential component in mammalian diets. Dietary fiber are carbohydrate polymers with ten or more monomeric units that are not digestible by pancreatic or brush border enzymes in the intestine and arrive in the colon undigested, where they are subject to bacterial degradation (Cummings et al. 2009). Some of these fibers

Nondigestible Carbohydrates and Digestive Health. Edited by T. M. Paeschke and W. R. Aimutis

induce changes in the luminal microflora, which confer health benefits to the host. These fibers are referred as prebiotics and they include inulin (IN), oligofructose (OF), and galactooligosaccharides (GOS; de Vrese and Schrezenmeir, 2008).

The first section of this chapter provides an overview of the mammalian immune system and the GALT while the second provides a comprehensive review of the current experimental and clinical data regarding how dietary fiber and prebiotics influence immune components of the gut.

Overview of the Immune System

The immune system is defined as the host's defense against destructive forces from both outside (e.g., bacteria, viruses, parasites) and within (e.g., malignant and autoreactive cells) the body. Immune responses are generally classified as either innate (inborn components of the immune system) or acquired (adaptive). The components and cells that comprise these two arms of the immune system are presented in Table 3.1.

The innate immune system provides immunity to invading organisms without the need for prior exposure to these antigens and includes physical barriers such as the skin and mucous membranes, cell-mediated barriers including phagocytic cells, inflammatory cells, dendritic cells, and natural killer (NK) cells, and soluble mediators such as cytokines, complement and acute-phase proteins (Delves and Roitt, 2000a). This arm of the immune system provides the early phases of host defense that protect the organism during the 4–5 days it takes for lymphocytes to become activated. Macrophages and their precursor monocytes and the polymorphonuclear leukocytes (neutrophils) make up the major cellular component of the innate immune system (Table 3.1). Macrophages are essential not only in directly destroying organisms but also in processing and presenting antigens to helper T (Th) cells to initiate acquired immune defenses (Delves and Roitt, 2000a). NK cells are effective against self cells that have been transformed by viruses or DNA damage, and are also key players in the innate immune system (Delves and Roitt, 2000a).

The acquired, or adaptive, immune system develops over an individual's lifetime. Immune responses by this system generally occur after those of the innate immune system; they are antigen-specific and are

Table 3.1. The immune system.

Arm of Immune System	Defenses	Components	Functions
Innate immune system	Physical barriers	Skin	Prevent the entry of antigens into systemic circulation
		Mucous membranes	
	Cell-mediated barriers	Phagocytic cells, e.g., neutrophils, macrophages	Engulf foreign antigens
		Inflammatory cells, e.g., basophiles, mast cells	Release inflammatory mediators, e.g., histamine, prostaglandins
		Natural killer cells	Destroy infected or malignant cells
		Dendritic cells	Present antigens to lymphocytes
	Soluble factors	Cytokines	Activate/recruit other cells
		Complement	Enhance phagocytosis
		Acute-phase proteins	Promote repair of damaged tissue
Acquired immune system	B-lymphocytes	Plasma cells	Secrete antibody
	T-lymphocytes	CD4+ T-cells	Induce activation of lymphocytes
		Th1 cells	Promote cell-mediated responses
		Th2 cells	Promote humoral (antibody) responses
		Tregs	Peripheric tolerance
		CD8+ T-cells	Destroy infected or malignant cells
		Cytotoxic T-cells	Suppress activity of lymphocytes
		Suppressor T-cells	

more efficient upon secondary exposure to the pathogen (Goust and Bierer, 1993). Lymphocytes are an important cellular component of this arm of the immune system that modulate the function of other immune cells or directly destroy cells infected with intracellular pathogens (Table 3.1). Each developing T- or B-cell generates a unique receptor, or recognition molecule, by rearranging its receptor genes such that a set of cells expressing a vast array of diverse receptors is produced, allowing immune cells to selectively eliminate virtually any foreign antigen that enters the body (Delves and Roitt, 2000a). B-cells, abundant in lymph nodes, recognize foreign antigen through membrane-bound antibodies, or immunoglobulins (Ig), and upon activation become antibody-secreting plasma cells to effectively remove soluble bacteria/antigens (Delves and Roitt, 2000a). Antibodies are secreted in soluble form and bind foreign particles to facilitate clearance by phagocytes (Delves and Roitt, 2000b). B-cells can also serve as antigen-presenting cells (APCs) and in this respect influence T-cell function (Delves and Roitt, 2000a).

T-cells express a T-cell receptor that recognizes foreign antigen presented in complex with a major histocompatibility complex (MHC) molecule on the surface of an APC (Delves and Roitt, 2000a). Subpopulations of T-cells include the Th cells, which are identified by the presence of the membrane glycoprotein CD4, and cytotoxic/suppressor T-cells that express the CD8 glycoprotein (Delves and Roitt, 2000a). CD4+ cells recognize antigen in complex with MHC class II molecules (found primarily on APC such as macrophages and dendritic cells), whereas CD8+ cells recognize antigen in complex with MHC class I molecules (most nucleated cells in the body express MHC class I) (Delves and Roitt, 2000a). CD4+ cells secrete a number of cytokines that are important in the activation of B- and other T-cells, as well as cells of the innate immune system. On the basis of the types of cytokines these CD4+ cells produce, they are classified into a number of Th types (1, 2, or reg) (Chen et al., 1994; MacDonald, 1998; Delves and Roitt, 2000a). Th1 cells generally promote cell-mediated inflammatory responses via production of IFN-γ, TNF-α, IL-2, and IL-12, whereas Th2 cells support antibody (humoral) responses via production of IL-4, IL-5, and IL-13 (Delves and Roitt, 2000b). Treg cells secrete IL-10 and TGF-β, which actively suppress Th1 and Th2 responses. CD8+ cells include T-cells with cytotoxic or suppressor function. Cytotoxic T-cells destroy (by releasing granules or inducing apoptosis) intracellular bacteria/virus-infected cells and tumor cells (Delves and Roitt,

2000b). Less is known about CD8+ suppressor cells, but they are believed to suppress the activation or activities of other immune cells, and may play a role in immunological tolerance, such as the tolerance to foreign antigens encountered in the gut (Bloom et al. 1992; Green and Webb, 1993). Although there is considerably less known about these suppressor cells, they have been classified into different subtypes, based on the cytokines they produce (Fitch et al. 1995).

The Gut-Associated Lymphoid Tissues

A mucosal immune system is strategically placed in areas where external pathogens and antigens may gain access to the body. This includes the mucosal-associated lymphoid tissues, which protect sites such as the respiratory, urinary, and reproductive tracts, and GALT, which protects the intestine. As the intestine is the first line of defense from the environment, and must integrate complex interactions among diet, external pathogens, and local immunological and non-immunological processes, it is critical that protective immune responses are made to potential pathogens, yet it is equally important that hypersensitivity reactions to dietary antigens are minimized. Although there is little anatomical data available, it has been estimated that approximately 25% of the intestinal mucosa is made up of lymphoid tissue (Kagnoff, 1987). The GALT is composed of aggregated tissue in the form of Peyer's patches and solitary lymphoid follicles, and non-aggregated cells in the lamina propria and intraepithelial regions of the intestine, as well as mesenteric lymph nodes (Langkamp-Henken et al. 1992) (Figure 3.1).

Peyer's patches are aggregates of lymphoid follicles found throughout the mucosa and submucosa of the small intestine. These patches contain CD4+ and CD8+ T-cells, as well as naïve B-cells, plasma cells, macrophages, and dendritic cells (Langkamp-Henken et al. 1992). Overlying the Peyer's patches are specialized epithelial cells known as M-cells, which endocytose, transport, and release antigens from the gut into the Peyer's patches, where these antigens are presented on APC to T- and B-cells (Langkamp-Henken et al. 1992; Kagnoff, 1993). Upon activation, B-cells undergo class switching to produce immunoglobulin A (IgA) antibodies, a process that is facilitated by both activated CD4+ and CD8+ cells (Langkamp-Henken et al. 1992; Kagnoff, 1993). Activated immune cells exit the Peyer's patches via the mesenteric lymph

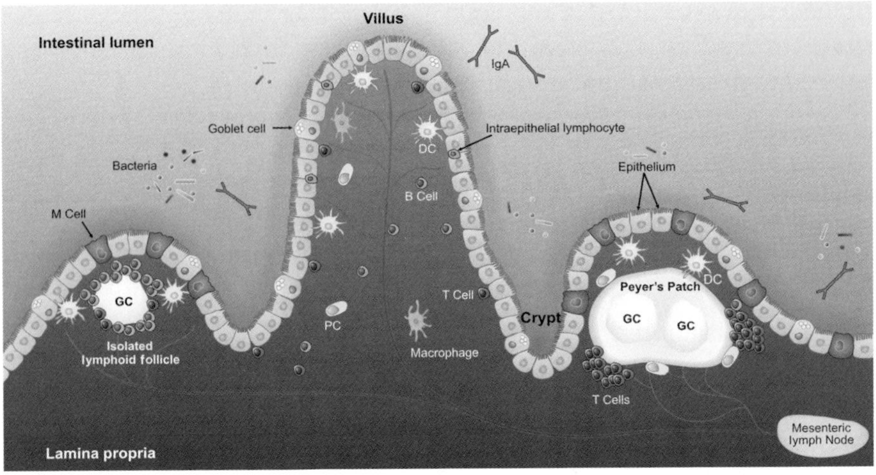

Figure 3.1. The gut-associated lymphoid tissues (GALT).

nodes, enter the systemic circulation by way of the thoracic duct, and then specifically home back to populate the lamina propria and intraepithelial regions of the intestine (Langkamp-Henken et al. 1992; Kagnoff, 1993; DeWitt and Kudsk, 1999). Thus, Peyer's patches represent a major "sampling" site for intestinal antigens. Solitary lymphoid follicles are present throughout the length of the intestinal tract, particularly in the colon and rectum, and have M-cells associated with the overlying epithelium (Laissue and Gebbers, 1992). At present, it is assumed that these follicles are functional equivalents of Peyer's patches (Laissue and Gebbers, 1992).

The lamina propria consists of a diffuse population of T- and B-cells, plasma cells, mast cells, and macrophages, all covered by a single layer of epithelial cells (Langkamp-Henken et al. 1992). The role of the intestinal epithelial cells (IECs) or enterocytes has evolved from cells involved mainly in digestion, absorption, and transport of nutrients, to also a main role in immune regulation of the underlying cells of the lamina propria.

IECs are known to produce different chemokines and cytokines depending on the type of microbial molecules that come in contact with the epithelium (Hurley and McCormick, 2004). IECs have been shown to condition the immune response of dendritic cells by inducing the secretion of tolerogenic signals (IL-10) over active immunity signals

(IL-12; Rimoldi et al. 2005). The enterocyte is also involved in antigen transport to underlying APCs, or they can present antigen themselves to lymphocytes. Moreover, the enterocyte is polarized in such a way that antigen is processed within the apical portion of the cytoplasm and can be presented in the context of MHC class II in the basolateral surface (Hershberg et al. 1998).

The majority of T-cells within the lamina propria are CD4+. Ninety percent of the plasma cells (mature B-cells) in the lamina propria secrete IgA (Laissue and Gebbers, 1992). Most of the IgA is secreted into the gut lumen where it takes the form of secretory IgA (sIgA), distinct from serum IgA. sIgA is a dimer of two monomeric IgA molecules with an attached secretory component (the cleaved extracellular domain of a transmembrane protein expressed on IECs). The secretory component facilitates the transport of IgA through the epithelium and into the gut lumen and protects IgA from degradation by intestinal enzymes and toxins (Laissue and Gebbers, 1992). The main function of IgA is to prevent the attachment of intestinal pathogens (Kagnoff, 1993).

Intraepithelial lymphocytes (IEL) are located in the interstitial spaces of the mucosal epithelium in a ratio of approximately 1 lymphocyte for every 6–10 epithelial cells (Langkamp-Henken et al. 1992), making IEL the largest immunocompetent cell pool in the body. Although IEL line both small and large intestines from the crypt base to the villus tip (Abreumartin and Targan, 1996), their exact biological function in the mucosal immune system is not clear. Unlike most lymphocytes that develop thymically, IELs are thought to develop within the lamina propria in areas known as cryptopatches (Suzuki et al. 2000). IELs include CD8$\alpha\alpha$+ and TCR$\gamma\delta$ lymphocytes and they are thought to be involved in induction of tolerance as they can prevent colitis in an adoptive transfer mouse model (Poussier et al. 2002) They are in continuous contact with luminal antigen through the epithelial layer, and it has been suggested that IEL may be the first compartment of the immune system that responds to gut-derived antigens (McKay and Perdue, 1993).

Although not situated within the intestinal mucosa, the mesenteric lymph nodes are considered part of the GALT. Mesenteric lymph nodes are composed of immune cells leaving and entering the gut and those that are part of the peripheral circulation. Immune cells drain to the intestinal lymphatics after differentiation in Peyer's patches, and pass through mesenteric lymph nodes en route to the thoracic duct and then again en route back to the lamina propria regions of the gut (Weiner, 1997).

Dietary Fiber, Prebiotics, and Immune Function

There is increasing evidence that the addition of fermentable fiber to the diet alters the function and structure of the gut, modifies the production of gut-derived hormones, and is associated with improved whole body glucose homeostasis even in the absence of disease (Mortensen and Clausen, 1996; Massimino et al. 1998). To date, relatively few studies have been conducted on the effects of dietary fiber on GALT. Because of the limited number of studies, it is not possible at this time to draw conclusions on the immune effects of different types of fermentable fibers. Therefore, we have combined all studies examining the immune effects of fermentable dietary fibers, defining fermentable fibers as those that can be metabolized by intestinal microorganisms. The specific fiber sources used in the studies can be found in Table 3.2.

Animal Studies

Most of the work on dietary fiber and prebiotics has been done in animal models. OF obtained from fermentation of sucrose was given to mice for a period of 4 weeks and did not induce significant effects on the systemic immune response. However, some local immune effects were observed, such as an increased size of the Peyer's patches obtained from mice fed the OF supplement. Ex vivo culture of cells from Peyer's patches of the OF treated mice with sonicates of *Bifidobacterium pseudocatenulatum* induced higher production of IgA and Th-2 type cytokines (IL-10, IL-5, and IL-6), suggesting that OF fiber stimulates immune cells from Peyer's patches to elicit a Th-2 type immune profile in the presence of bifidobacteria (Hosono et al., 2003). OF fiber has also been studied as a supplement in the diet of *Min* mice, which have a mutation in the *Apc* gene and are used as a model of intestinal cancer. Consistent with the previously mentioned study, the authors also observed an increase in size of the lymphoid nodules in the ileum and a reduction in the number of colon tumors (Pierre et al. 1997). Manhart et al. (2003) also reported that supplementation with OF for 16 days resulted in a higher number of cells in the Peyer's patches, that appeared to be due to an increase in

Table 3.2. Modulation of the gastrointestinal immune system by dietary fibers and prebiotics—animal studies.

Reference	Animal/Cell Line/Tissue	Basal Diet	Experimental Fiber	Results
Buddington et al. (2002)	B6C3F1 mice	Cellulose, 100 g/kg	OF or IN (100 g/kg)	No effect in lung tumor incidence. Reduced colon crypt tumors in group with both IN and OF supplementation.
Daddaoua et al. (2006)	Wistar rats	Methylcellulose, 1% w/w	Oligosaccharide extracted from goat milk (500 mg/kg)	↑ body weight compared to TNBS-treated controls ↓ colonic IL-1β, inducible NO synthase, cycloxygenase 2, and mucin 3 ↑ trefoil factor 3
Femia et al. (2002)	F344 rats	High-fat diet (230 g maize oil/kg diet) with low cellulose content (20 g/kg diet)	Prebiotic group: IN enriched OF (100g/kg)	Both groups: reduced incidence in colon cancer.
			Synbiotic group: also received *Lactobacillus rhamnosus* GG and *Bifidobacterium lactis* Bb12	Prebiotic: Recovery of NK cell activity after treatment with AOM ↑ IL-10 secretion in PP ↑ cecal content of SCFA

(*Continued*)

Table 3.2. *(Continued)*

Reference	Animal/Cell Line/Tissue	Basal Diet	Experimental Fiber	Results
Field et al. (1999)	Adult mongrel dogs	Cellulose 8.3 g/kg	Fermentable fiber mixture (beet pulp, oligofructose*, gum Arabic) 8.7 g/kg	↑ CD8+ cells in IEL, PP, and LP ↑ CD4+ cells in MLN and peripheral blood Higher T cell mitogen responses in MLN and IEL
Fukuda et al. (2002)	Sprague-Dawley rats	Not specified	GBF	↓ histologic inflammation scores ↑ butyrate concentration in cecum
Gaskins et al. (1996)	C57BL/6NHsd mice	Ensure (low residue)	Oligofructose* 30 g/L drinking water	↑ cecal and colonic macrophages
Hoentjen et al. (2005)	HLA-B27 transgenic rats	Not specified	Long chain IN and OF (Synergy 1)	↓ gross cecal scores and histologic inflammation scores ↓ cecal IL-1β ↑ cecal TGF-β ↑ bifidobacteria and lactobacilli concentration

Hosono et al. (2003)	BALB/c mice	OF free diet	2.5% and 7.5% oligofructose and sonicated *Bifidobacterium pseudocatenulatum*	↑ fecal IgA in 2.5% group
	Ex vivo culture of PP cells			↓ serum IgG1, compared to control group
				Ex vivo culture:↑ IgA, IFN-γ, IL-5, IL-6, and IL-10 compared to group without OF supplementation.
Kanauchi et al. (2001)	Sprague-Dawley rats	Cellulose, 3% w/w	GBF or fiber fraction from GBF	GFB and GFB-fiber: ↓ colitis severity ↑ butyrate production ↑ epithelial repair, compared to DSS-treated controls.
Kaufhold et al. (2000)	Veal calves	Cellulose 5% w/w	Oligofructose* 10 g/d	↑ eosinophil granulocytes in blood
Kelly-Quagliana et al. (2003)	B6C3F1 rats	Cellulose, 10 g/100 g	OF or IN	Both fibers: ↑ splenic NK cell and peritoneal macrophage phagocytic activity. ↓ white blood cell counts, compared to controls.

(*Continued*)

Table 3.2. (*Continued*)

Reference	Animal/Cell Line/Tissue	Basal Diet	Experimental Fiber	Results
Kudoh et al. (1998)	Sprague-Dawley rats	Cellulose, 5% w/w	Arabic gum, Celfur (corn husk), lactulose†, or purified indigestible dextrin, 5% w/w	↑ κ-light chain- and IgA-positive cells in small intestine and cecal mucosa (all fibers) ↓ CD4+:CD8+ ratio in spleen (celfur, lactulose) ↓ CD3+ cells in spleen (arabic gum)
Kudoh et al. (1999)	Sprague-Dawley rats	Cellulose, 5% w/w	Celfur (corn husk), glucomannan, curdlan, or lactulose† 5% w/w	↑ IgA-positive cells in cecum (celfur, lactulose) ↑ IgA secretion into cecal contents (all fibers)
Lara-Villoslada et al. (2006)	Sprague-Dawley rats	Cellulose, 5% w/w	Mixture of goat milk oligosaccharides (20 g/kg)	Fiber supplementation prevented DSS-induced weight loss. ↓ clinical symptoms ↓ MPO and ↑ MUC-3 expression
Lim et al. (1997)	Sprague-Dawley rats	Cellulose, 5% w/w	Pectin, konjak mannan, or chitosan, 5% w/w	↑ MLN IgA and IgG (pectin, chitosan) ↓ serum and MLN IgE (all fibers) ↑ serum IgA and IgG (pectin) ↑ CD4+ T cells in MLN (pectin) ↑ IFN-γ in MLN (pectin)

Madar et al. (1998)	Sprague-Dawley rats	Cellulose, 3% w/w	Cellulose*, white grape, or tomato peel, 15% w/w	↑ area of lymphoid infiltrates in colonic mucosa close to tumor (white grape, tomato peel) ↑ Ki-67+ cells in colonic mucosa and tumors (all fibers)
Madsen et al. (1999)	IL-10 gene deficient mice	Not specified	Lactulose	↓ colonic TNFα, IFNγ, compared to controls ↓ levels of colonic mucosal adherent and translocated bacteria
Manhart et al. (2003)	BALB/c mice	Cellulose, 2% w/w	Oligofructose, 10% w/w	↑ B cells within Peyer's patches (PP).
	Ex vivo culture of PP cells			↑ CD4:CD8 ratio within PPs.
Nagai et al. (2000)	Wistar rats	Fiber-free	Sugar beet fiber, 10% w/w	↑ CD8+ IEL in colorectum
Nagendra and Venkat Rao (1994)	Wistar rats	Infant formula	Lactulose†, 0.5% of energy	↑ phagocytic function of intraperitoneal macrophages

(*Continued*)

Table 3.2. (*Continued*)

Reference	Animal/Cell Line/Tissue	Basal Diet	Experimental Fiber	Results
Pierre et al. (1997)	*Min* mice	Cellulose, 2% w/w	Oligofructose (from sucrose)*, wheat bran, or resistant starch, 5.8% w/w	↑ number of PP in small intestine (short-chain FOS)
Roller et al. (2004)	F344 rats	High-fat diet (230 g maize oil/kg diet) with low cellulose content (20 g/kg diet)	Prebiotic group: OF enriched IN (Synergy 1) 100 g/kg	Prebiotic supplementation ↑ IL-10 and IFN-γ in PP. ↑ IgA in cecum.
			A symbiotic group also received *L. rhamnosus* GG and *B. lactis* Bb12	Synbiotic supplementation ↑ IgA in ileum. Prevented loss of NK cytotoxicity in PP, compared to AOM treated controls. ↓ lymphocyte proliferation in PP.
Rumi et al. (2004)	Wistar rats	Not specified	Lactulose, 300, 600, and 1,000 mg/kg	Amelioration of DSS-induced colitis in a dose dependent manner. ↑ body weight, colon length. ↓ colonic MPO activity.

Videla et al. (2001)	Sprague-Dawley rats	Not specified	Inulin orally or by enema (400 mg)	Oral inulin ↓ inflammation scores and MPO, compared to controls. ↑ counts of lactobacilli. Enemas had no effects.
Yamada et al. (1999)	Sprague-Dawley rats	Cellulose, 5% w/w	PHGG, guar gum, HM pectin, or glucomannan 5% w/w	↑ IgA in spleen and MLN (all fibers) ↑ IgG in spleen (glucomannan, pectin) and MLN (all fibers) ↑ serum IgA (guar gum, glucomannan, pectin) and IgM (glucomannan)
Yun et al. (1997)	C57BL/6 mice (immuno-suppressed)	Not specified	Oat β-glucan, 3 mg every 48 hours	↑ non-specified and antigen specific IgG in serum ↑ IFN-γ and IL-4-secreting cells in spleen and MLN
Yun et al. (1998)	C57BL/6 mice	Not specified	Oat β-glucan, 3 mg every 48 hours	↑ intestinal antigen-specific IgA ↓ IL-4 secreting cells in MLN
Zusman et al. (1998)	Rats	Cellulose, 3% w/w	Cellulose, white grape or tomato peel, 15% w/w	↑ plasma cells in splenic red pulp (all fibers)

IEL, intraepithelial lymphocytes; PP, Peyer's patches; LP, lamina propria; MLN, mesenteric lymph nodes; Ig, immunoglobulin; IFN, interferon; IL, interleukin; PHGG, partially hydrolyzed guar gum; HM pectin, highly methoxylated pectin; SCFA, short-chain fatty acid; GBF, germinated barley foodstuff.

*Recognized prebiotic fiber.

†Potential prebiotic fiber.

the B cell population. These studies suggest that OF supplementation, although not apparently affecting systemic immune cell composition, modulate areas of the gut that are distant from the colon, where most of the fermentation of these fibers take place.

The immune effects of different types of prebiotic fibers have been compared within the same study. In a rat model of colorectal cancer, rats injected with B16F10 tumor-inducing cells were exposed to OF or IN supplementation. Tumor incidence was not different in the lungs among fiber treatments, but rats that received either of the prebiotic supplementation had significantly less aberrant foci in the colonic crypts. In the same study, rats were exposed to microbial pathogens and those that received prebiotic supplementation in their diets had a lower mortality rate than the control group, with the IN group providing the best resistance to infections (Buddington et al., 2002).

Further experiments by the same research group showed that both IN and OF treatment given for 6 weeks to mice increased NK cell activity and peritoneal macrophage activity, but did not change the relative lymphocyte population or fecal IgA concentration (Kelly-Quagliana et al. 2003). These studies provided evidence that diet supplementation with two types of prebiotic fiber may protect from colon cancer development and improve resistance to potentially lethal bacterial infections through modulating cells of the innate immune system.

Another study, performed in a rat model of colorectal cancer, looked at the effects the OF-enriched IN supplementation for 33 weeks (Femia et al. 2002). Ten days after the initiation of treatment, some rats were administered the carcinogen azoxymethane (AOM). The OF-enriched IN treatment lowered the incidence of colon cancer in these rats and recovered the NK cell activity that is partially lost with AOM treatment. Interestingly, the control rats that did not receive AOM also showed higher splenic NK cell activity in the prebiotic treated group, suggesting that long-term diet supplementation with these fibers has immunomodulatory effects on innate immune effector cells of the GALT and that this influences systemic immune function.

An interesting study by Roller et al. (2004) compared the immune effects of a diet supplemented with prebiotic fiber, probiotics, or synbiotics, the latter being a combination of prebiotics and probiotics. Mice were fed a supplemented diet with OF-enriched IN from chicory roots. The probiotic group was given a combination of *Lactobacillus rhamnosus* GG and *Bifidobacterium lactis* Bb12, and the synbiotic group a combination of both. Prebiotic supplementation increased the production of IL-10 and IFN-γ within the Peyer's patches, as well as the secretion of IgA in the cecum. Rats also treated with probiotics showed higher secretion of IgA in the ileum and a lower myeloperoxidase (MPO) activity, a marker of neutrophil activation. Probiotic supplementation alone had minor immune effects in this study compared to the control group. The results of this study suggest that the immunological effects of symbiotic supplementation are not simply the additive effects of prebiotics or probiotics supplements. Additionally, the administration of both prebiotics and probiotics appears to change the site of the gut where most of the immunomodulatory effects occur. This may be due to the fact that prebiotic fermentation will occur sooner and more proximally when given simultaneously with fiber-fermenting microorganisms.

Studies in our own laboratory conducted on adult dogs indicate that adding fermentable fiber to the diet can modulate the type and function of cells from different regions of the GALT (Field et al. 1999). In this study, 16 adult dogs (23 ± 2 kg) were fed (for 14 days) in a randomized crossover design, two isoenergetic, isonitrogenous diets containing 8.3 g/kg nonfermentable or 8.7 g/kg fermentable fibers for 2 weeks. The fermentable fiber diet was a mixture of plant fibers (beet pulp, OF powder, and gum arabic). The fiber content of the diet significantly altered the proportion of T-cells (CD4+ and CD8+) in GALT and their in vitro response to mitogens (Field et al., 1999). Specifically, we found that switching from a low to high fermentable fiber diet as compared to cells from dogs switched from a high to a low fermentable fiber diet resulted in higher ($P < 0.05$) mitogen responses in the predominantly T-cell tissues (mesenteric lymph nodes and IEL) and lower mitogen responses ($P < 0.05$) in areas involving B-cell function (lamina propria and Peyer's patches). After consuming the high fermentable fiber diet, there was a higher proportion of CD8+ T-cells among the IEL, lamina propria, and Peyer's patches, and a higher proportion of CD4+ T-cells in the mesenteric lymph nodes and peripheral blood (Field et al., 1999). Apart from a higher CD4:CD8 ratio (2.4 ± 0.2 vs. 1.7 ± 0.2, $P < 0.05$), switching to the high fermentable fiber diet did alter immune functions (^{3}H-thymidine uptake in response to mitogens or NK activity) in peripheral blood (Field et al., 1999). The observed effect of changes in the fermentable fiber content of the diet on the composition and function of GALT, but not peripheral immune cells, raises practical issues in assessing the role of diet on immune function since peripheral blood is the most frequently sampled immune site in human and large animal studies. Our results are consistent with studies demonstrating proportionately more CD4+ T-cells in mesenteric lymph nodes of rats fed a diet containing 5% w/w pectin as compared to cellulose (Lim et al. 1997), and with an increased proportion of CD8+ IEL in rats fed a diet supplemented with sugar beet fiber, compared to a fiber-free diet (Nagai et al. 2000).

Studies using prebiotics have been aimed at potentially treating gut inflammation and have been performed in animal models of inflammatory bowel disease (IBD). Lactulose has been shown to attenuate colitis in the IL-10 gene deficient mouse (Madsen et al. 1999), whereas IN and lactulose attenuated dextran sulfate sodium (DSS)-induced colitis in rats (Videla et al. 2001; Rumi et al. 2004). Additional studies have been made using the HLA-B27 rat model of IBD. A combination therapy of IN and OF was successful in preventing colitis in this animal model and this effect was parallel to an increase in the bifidobacteria and lactobacilli population in the gut (Hoentjen et al., 2005). Goat milk oligosaccharides were given to rats that received DSS treatment and it was shown to be beneficial in reducing neutrophil activity and increasing barrier function, as measured by mucin-3 (MUC-3) expression (Lara-Villoslada et al. 2006). The same oligosaccharide was effective in reducing colitis in a trinitrobenzene sulfonic acid (TNBS)-induced rat model (Daddaoua et al. 2006).

A few studies have tested the effect of nonprebiotic dietary fiber in animal models of gut inflammation. A mixture of glutamine-rich protein and hemicellulose-rich fiber known as germinated barley foodstuff (GBF) was fed to rats suffering from DSS-induced colitis. This treatment was found effective in reducing colonic inflammation and increasing fecal concentration of butyrate (Kanauchi et al. 2001; Fukuda et al. 2002). However, it is still unclear whether the effect observed with this diet were due to the fiber or protein content/type of the diet as the individual treatments were not included in the study.

Although exploration in this area is still in its infancy, animal studies have clearly demonstrated that dietary fiber content and type can modulate measures of immune function and have efficacy in the treatment of animal models of gut inflammation such as IBD and colon cancer.

Human Studies

Only a few clinical trials have evaluated the immune effects of prebiotic fiber (Table 3.3). A small randomized, double-blinded controlled trial using 18 patients with active ulcerative colitis were given a symbiotic treatment containing OF-enriched IN and *Bifidobacterium longum* for 1 month. Symbiotic treatment decreased colonic inflammation, as assessed by endoscopy and histology. Intestinal mRNA levels of proinflammatory cytokines TNF, IL-1, and human beta defensins 2, 3, and 4 were reduced in the treatment group (Furrie et al. 2005). Since there was not a group receiving prebiotics alone, it is difficult to speculate as to whether the effects observed by the authors were due to the administration of the probiotic, the prebiotic fiber, or both.

In a more recent study with ulcerative colitis patients, the authors compared the effects of providing prebiotic, probiotic, or symbiotic treatment on a marker of systemic inflammation and quality of life (Fujimori et al. 2009). Probiotic treatment consisted of a daily capsule with 2×10^9 colony-forming units (CFU) of *B. longum.* The prebiotic treated group received daily capsules of psyllium, and the symbiotic group received both. Serum C-reactive protein decreased significantly only in the group that received symbiotic treatment. Emotional function and bowel function were significantly improved in groups receiving probiotics and prebiotics, respectively. The authors concluded that the symbiotic treatment offered more quality-of-life benefits than the prebiotic or probiotic treatments.

The effect of prebiotic supplementation in the diet of Crohn's disease patients was studied in a small, nonplacebo controlled trial (Lindsay et al. 2006). In this study, in the patients who ingested 15 g of IN-enriched OF, there was a decrease in intestinal inflammation. This was measured as a significant increase in IL-10 positive dendritic dells and in dendritic cells expressing Toll-like receptor-2 and -4 (TLR). These changes in dendritic cell function were associated with a significant increase in fecal bifidobacteria concentration.

Another study with IBD patients tested the effects of the potential prebiotic carbohydrate lactulose (Hafer et al. 2007). Patients with ulcerative colitis or Crohn's disease received standard medication or the same medication combined with 10 g of lactulose daily for 4 months. There was no significant difference in the clinical activity index, endoscopic score, or immunohistochemical assessment between both groups. However,

Table 3.3. Clinical studies using prebiotics.

Author	Study Design	Prebiotic Therapy	Control	Results
Inflammatory bowel disease				
Fujimori et al. (2009)	Randomized, controlled	*Bifidobacterium longum* (2×10^9) daily capsule, OR Psyllium (daily capsule), OR Both	Placebo	↓ C-reactive protein and ↑ overall quality-of-life benefits in symbiotic group ↑ emotional function in probiotic group ↑ bowel function in prebiotic group
Furrie et al. (2005)	Double-blind, randomized, controlled	*B. longum* (2×10^9) and OF enriched IN (6 g)	Placebo	↓ mRNA levels of TNF-α, IL-1α, β-defensins-2, -3, and -4.
Hafer et al. (2007)	Open label	Conventional therapy + daily lactulose (10 g)	Conventional therapy	No significant difference in clinical activity index, endoscopic score or histology. Significant increase in quality-of-life index in the lactulose group.
Lindsay et al. (2006)	Open label	Fructooligosaccharides (15 g)	None	Improved endoscopic scores ↑ IL-10 positive intestinal dendritic cells ↑ intestinal dendritic cells expression TLR-2 and -4

Atopy				
Arslanoglu et al. (2008)	Double-blind, randomized, controlled	Formula supplemented with mixture of scGOS/lcFOS (8 g/L)	Placebo (8 g/L maltodextrin)	At 2 years of age: Reduced atopy, respiratory infections, and antibiotic prescriptions in prebiotic supplemented group.
Moro et al. (2006)	Double-blind, randomized, controlled	Formula supplemented with mixture of scGOS/lcFOS (8 g/L)	Placebo (8 g/L maltodextrin)	Reduced incidence in atopy and higher fecal bifidobacteria concentration at 6 months in prebiotic supplemented group.
VanHoffen et al. (2009)	Double-blind, randomized, controlled	Formula supplemented with mixture of scGOS/lcFOS (8 g/L)	Placebo (8 g/L maltodextrin)	↓ titers of total IgE, IgG1, IgG2, and IgG3. ↓ CMP-specific total Igs in prebiotic supplemented group. No change in Ig titers specific to the DPT vaccine between groups

there was a significant increase in the quality-of-life index between the ulcerative colitis group that received lactulose and the control group. As there were no apparent changes in gut histology, the mechanism for the effect of lactulose treatment on quality-of-life index is not clear.

The immune modulatory effects of prebiotics in cancer were assessed in a recent study. The authors investigated the immune effect of a symbiotic treatment (*L. rhamnosus* GG, *B. lactis* Bb12, and 10 g/day IN-enriched OF) in colon cancer patients who had undergone colonic resection (Seifert and Watzl, 2007). Peripheral blood mononuclear cells from patients who received symbiotic supplementation in their diets secreted more IFN-γ. Patients who had undergone a polypectomy and were receiving a placebo showed a decline in IL-2 production, whereas the symbiotic treatment prevented this decrease in IL-2 production.

The effect of prebiotics on childhood atopy has been assessed in several interesting trials. A double blind randomized controlled trial with 259 infants with parental history of atopic dermatitis (AD) looked at the incidence in atopy during the first 6 months of life (Moro et al. 2006). The children received a formula supplemented with a mixture of short-chain GOS and long-chain fructooligosaccharides (FOS) (8 g/L scGOS/lcFOS) or placebo (8 g/L maltodextrin) formula during the first 6 months of life. Children that received formula supplemented with prebiotics had a reduced incidence of atopy compared to the placebo group. Fecal samples from children that received prebiotic supplementation had higher concentration of bifidobacteria.

The authors later assessed 134 of these children at 2 years of age to establish if the protection from atopy lasted more than the 6 months of the prebiotic supplementation period (Arslanoglu et al. 2008). Similarly to the previous study, infants that drank prebiotic-supplemented formula during the first 6 months of life had a lower incidence of atopy, including AD, wheezing, and urticaria. Moreover, infants in the placebo group reported more respiratory infections, fever, and antibiotic prescriptions, suggesting that not only did prebiotic supplementation decrease the incidence of allergic episodes but also increased the resistance to subsequent infections.

In order to analyze the immune changes induced by the prebiotic-supplemented formula in these children, the authors later examined the levels of Ig and cow milk protein (CMP) in the serum of 6-month-old children that received the same treatment as the previous two studies (Van Hoffen et al. 2009). Total levels of IgE, IgG1, IgG2, and IgG3 were significantly reduced in the prebiotic supplemented group. CMP-specific Igs were also decreased in the treatment group. Interestingly, specific antibodies titers for the DPT vaccine, which was administered at 3 months of age, did not change between the treated and placebo groups, suggesting that prebiotic supplementation could be a safe option to prevent atopy in children and would not cause hyporesponsiveness to vaccines.

The authors of these trials did not examine the mechanisms by which atopy was prevented by the GOS/FOS supplementation. However, it may be related to the observed effect of diet (infant formula compared to breast milk) on modulating the intestinal bacterial flora very early in life. It is believed that this is a critical period in life when the intestinal flora is established (Palmer et al., 2007; Tannock, 2007). One might hypothesize that specific bacterial signals from probiotic-type species are necessary to "train" a healthy intestinal immune system to avoid overstimulation (atopy) or lack of stimulation (infection) and to function properly.

In summary, the human studies are promising but more are needed to specifically define how prebiotics modulate the intestinal immune components, and to establish whether prebiotics can be used to modulate these components to prevent or treat diseases. The evidence indicating that prebiotic supplementation early in life can induce significant and long-lasting effects in the infant immune system is compelling. However, more studies are needed to characterize the cellular mechanisms that lead to the prevention of atopy and infections early in life.

Mechanisms for the Effects of Dietary Fiber and Prebiotics on the Immune System

The mechanism for the effect of fermentable dietary fibers on immune function in the gut has not been established. A number of interesting hypotheses have been proposed until date and are listed in Table 3.4 and illustrated in Figure 3.2. It has been clearly established that prebiotic fibers are neither hydrolyzed nor absorbed in the upper part of the gastrointestinal tract, and they enter the colon where they become selective substrates for one or a limited number of beneficial colonic bacteria thereby altering the microflora of the gut (Gibson and Roberfroid, 1995). There is strong evidence indicating that consumption of

Table 3.4. Immunomodulatory effects of orally administered lactic acid bacteria.*

Immune Tissue	Effect	Documented in: Humans	Mice	References
Peyer's patches	↑ B cell number		•	De Simone et al., (1987b)
	↑ response to T- and B-cell mitogens		•	De Simone et al., (1987b); Shu et al. (2000)
	↑ proliferation of PP cells		•	Takahashi et al. (1993)
	↑ antibacterial activity of PP cells (via IgA)		•	De Simone et al. (1988)
	↑ IgA production		•	Takahashi et al. (1998)
	↑ total Ig production		•	Yasui et al. (1989)
	↑ IL-10		•	Calcinaro et al. (2005)
GALT[†]	↑ IgA in feces/intestinal contents	•	•	Perdigon et al. (1990); Fukushima et al. (1998), Fukushima et al. (1999); Tejada-Simon et al. (1999b)
	↑ IgA in small intestinal wall extracts		•	Takahashi et al. (1998)
	↑ total Ig in intestinal contents		•	Shu et al. (2000)
	↑ regulatory DCs, Tregs		•	Di Giancinto et al. (2005)
	↑ IL-10	•	•	Pathmakanthan et al. (2004); Lammers et al. (2005)
MLN	↑ response to mitogens		•	Shu et al. (2000)
Blood	↑ phagocytosis by blood leukocytes	•	•	Schiffrin et al. (1995); Pelto et al., (1998); Donnet-Hughes et al., (1999); Shu et al. (2000); Chiang et al. (2000); Gill et al., (2001); Gill, (2003)

↑ phagocytic function of RES		•	Perdigon et al. (1986a, 1986c, 1987, 1988, 1991)
↑ NK cell activity	•	•	Chiang et al. (2000); Gill et al. (2001b); Sheih et al. (2001); Cross, (2002); Olivares et al. (2006)
↑ B cell number	•		De Simone et al. (1992)
↑ IFN-γ production by lymphocytes	•		Halpern et al. (1991); Solis-Pereyra and Lemonnier, (1993)
↑ IgA/IgA-secreting cells	•	•	Yasui et al. (1989); Takahashi et al. (1993); Link-Amster et al. (1994); Malin et al. (1996); Tejada-Simon et al. (1999b); He et al. (2001); de Vrese et al. (2006)
↓ IgE production		•	Matsuzaki et al. (1998)
↑ total Ig production		•	Perdigon et al. (1988, 1991); Yasui et al. (1989); Kaila et al. (1992); Shu et al. (2000)
↑ IFN-γ, IL-2	•		De Simone et al. 1987b; Solis-Pereyra and Lemonnier, (1991); Wheeler et al. (1997); Halpern et al. (1991); Aattouri and Lemonnier, (1997); Arunachalam et al. (2000).

(*Continued*)

Table 3.4. *(Continued)*

Immune Tissue	Effect	Documented in: Humans	Mice	References
	↑ IFN-γ, TNF-α, IL-1b, IL-6, IL-10	•		Halpern et al. (1991); Aattouri and Lemonnier, (1997); Solis Pereyra and Lemonnier, (1993); Miettinen et al. (1996).
Spleen	↑ responses to T- and B- cell mitogens		•	De Simone et al. (1988); Shu et al. (2000)
	↑ IgM-secreting cells		•	Perdigon et al. (1986c, 1987)
	↓ IgE from stimulated spleen cells		•	Matsuzaki et al. (1998)
	↓ IL-4, IL-5, IL-6, IL-10		•	Matsuzaki et al. (1998)
	↑ IFN-γ, IL-2		•	Matsuzaki et al. (1998)
	↑ IL-10		•	Calcinaro et al. (2005)
Peritoneal cells	↑ macrophage phagocytic/enzymatic activity		•	Perdigon et al. (1986a, 1986b, 1986c, 1987, 1988); Shu et al. (2000)
	↑ IL-6, IL-12, IFN-γ, and nitric oxide		•	Tejada-Simon et al. (1999a)
Lung macrophages	↑ phagocytic function		•	Moineau and Goulet (1991)

PP, Peyer's patch; Ig, immunoglobulin; GALT, gut-associated lymphoid tissue; MLN, mesenteric lymph nodes; NK, natural killer; IFN, interferon; IL, interleukin; RES, reticuloendothelial system; DC, dendritic cell.

*Primarily species of lactobacilli and bifidobacteria.

†Data insufficient to classify region of GALT.

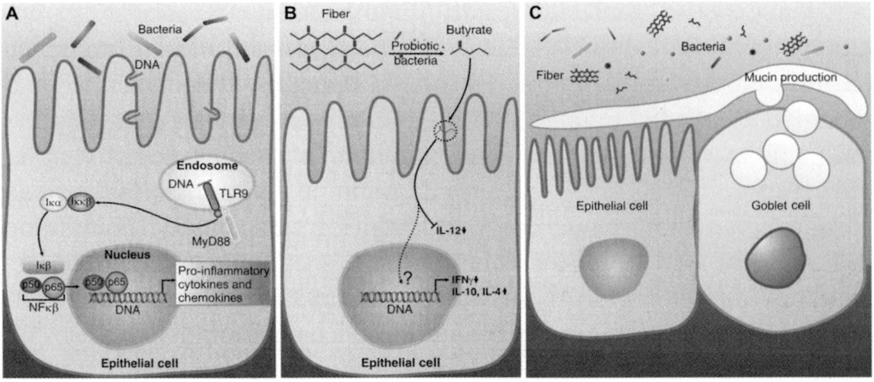

Figure 3.2. Proposed mechanisms for the effect of fermentable dietary fibers on immune function.

prebiotic fibers (IN and OF) increase the proportion of beneficial lactic acid bacteria in the human colon (Mitsuoka et al. 1987; Gibson et al., 1995; Bouhnik et al., 1996; Buddington et al., 1996; Kleesen et al., 1997; Menne et al., 2000). Other oligomers that may be prebiotics but for which more evidence is required are lactulose and oligosaccharides containing xylose, mannose, and galactose (Gibson, 1998). The studies which have examined the effects of prebiotic fibers on the immune system are reviewed in Tables 3.2 and 3.3. The studies conducted with recognized prebiotic fibers (OF) have shown increased lymphocyte and/or leukocyte numbers in GALT (Gaskins et al., 1996; Pierre et al., 1997; Field et al., 1999) and peripheral blood (Kaufhold et al., 2000). Additionally, studies have documented that feeding lactulose is associated with increases in IgA secretion or IgA+ cells in GALT (Kudoh et al., 1998, 1999), a decrease in the CD4+/CD8+ ratio in the spleen (Kudoh et al., 1998), and an increase in the phagocytic function of intraperitoneal macrophages (Nagendra and Venkat Rao, 1994).

Direct Contact of Probiotic Bacteria or Bacterial Products with Intestinal Immune Cells

It is often assumed that the consumption of prebiotics, through their effects on the colonic microflora, will have a similar effect as probiotics on immune function. In contrast to work on prebiotic fibers, there are more studies performed that have documented effects of feeding lactic acid bacteria (i.e., lactobacilli and bifidobacteria) on various parameters of immune function (Table 3.4). Oral administration of probiotic bacteria increased the production of Ig, especially IgA, in GALT and modulated both the number and the activity of Peyer's patch immune cells (Table 3.4). There are also a number of studies demonstrating effects of oral probiotics on systemic immune functions, and immune parameters in the lungs, peritoneum, and mesenteric lymph nodes (Table 3.4). The mechanism(s) by which probiotics consumed in the diet affect immune function have been largely speculative to date. One logical mechanism might be immune stimulation through direct contact of the colonic microflora with GALT. Small numbers of bacteria can cross the intestinal epithelial barrier into the Peyer's patches (Berg, 1985) inducing activation or leading to the activation of other immune cells (Berg, 1985; De Simone et al., 1987b; Link-Amster et al. 1994; Schiffrin et al. 1995). In vitro studies have supported this mechanism. In a study by Park et al. (1999), a macrophage cell line increased production of nitric oxide, H_2O_2, IL-6, and TNF-α after in vitro culture with bifidobacteria. Similarly, coculture with bifidobacteria significantly increased the production of TNF-α and IL-6 by macrophages and the production of IL-2 and IL-5 by stimulated CD4+ cells (Marin et al. 1997). Culturing murine Peyer's patch cells with bifidobacteria (*Bifidobacterium breve*) resulted in increased proliferation and antibody production by B-lymphocytes and activated macrophage-like cells (Yasui and Ohwaki, 1991).

Other authors have suggested that it is not the bacteria but microbial substances (e.g., cytoplasmic antigens, cell wall components) that penetrate the intestinal epithelia and this might activate GALT (De Simone et al., 1987a; Perdigon et al., 1988; Solis Pereyra and Lemonnier, 1993; Takahashi et al., 1993; Takahashi et al., 1998; Tejada-Simon et al., 1999a). In vitro, a macrophage cell line was stimulated, similar to that produced by whole bacteria, by incubation with cell-free extracts of both *B. longum* and *Lactobacillus acidophilus* (Hatcher and Lambrecht, 1993). Similarly, in vivo administered supernatants from cultures of *L. acidophilus* and/or *Lactobacillus casei* resulted in stimulated phagocytic activity of peritoneal and reticuloendothelial phagocytes and splenocyte activation similar to that produced by administration of the live bacteria (Perdigon et al., 1988). Cytoplasmic components of bacteria have also been demonstrated to produce some of the same immune effects (IgA production by Peyer's patch cells) as live bacteria (Takahashi et al., 1998). The mechanism by which cell wall components (such as peptidoglycans) or cytoplasmic antigens may activate immune cells is probably through pattern recognition receptors (PRRs) in epithelial and immune cells of the intestinal mucosa. These receptors can "read" differences in pathogen-associated molecular patterns (PAMPs) such as lipopolysaccharides (LPS), unmethylated CpG bacterial DNA, teichoic acids, etc (Akira et al., 2001). These receptors include the membrane bound TLRs and mannose receptor, and the cytoplasmic nucleotide-binding oligomerization domain receptors (NODs) (Girardin et al., 2001, Chamaillard et al., 2003). Bacterial and fungal components interact with these receptors and initiate intracellular signaling events that lead to the activation of the transcription factor NF-κB and the subsequent expression and secretion of inflammatory mediators like chemokines and cytokines (Abreu, 2002). It is still not known how cellular components of different bacterial species can elicit different immune responses that lead to tolerance of active immunity, but it has been shown to occur. Exposure of HT-29 colonic cells to DNA from pathogenic strains of *Salmonella* and *Escherichia coli* resulted in a significant increase in TLR9 mRNA expression. *Salmonella enterica* serovar Dublin DNA increased surface TLR9 protein and IL-8 secretion. There was no change in mRNA levels or localization of TLR9 in response to *B. breve* (Ewaschuk et al., 2007). These results demonstrate that IECs can recognize pathogenic bacterial DNA and respond by increasing surface localization and expression of TLR9, suggesting that the epithelial inflammatory response to pathogenic DNA is mediated at least in part by increased TLR9 expression.

Furthermore, peptidoglycans can bind to the CD14 cell surface antigen, and can stimulate mononuclear phagocytes and endothelial cells to release cytokines (Matsuzaki, 1998). Finally, there is some speculation that the immune effects observed with the administration of probiotic bacteria may actually be due to immunogenic milk peptides generated from the bacterial hydrolysis of milk constituents present in fermented milk products used to deliver the probiotic bacteria (Perdigon et al., 1988; Moineau and Goulet, 1991). This hypothesis warrants further examination as many of the studies in Table 3.4 administered probiotic bacteria in fermented milk products. If this contributes to the immune-stimulating effects of probiotic bacteria, it is unlikely to explain the immune effects of dietary prebiotic fibers.

Production of Short-Chain Fatty Acids from Fiber Fermentation

The gut microflora may modulate immune cells through the fermentation of dietary fibers to short-chain fatty acids (SCFAs). It is well established that the fermentation of IN and OF increases the production of SCFA, primarily acetate, butyrate, and propionate in the gut (Gibson and Roberfroid, 1995), which are absorbed into the bloodstream, where they are found in significant concentrations (Wolever et al., 1997). A number of studies support direct or indirect immunomodulatory properties of SCFAs.

Butyrate has been shown to decrease lymphocyte proliferation in vitro and to favor the secretion of Th-2- versus Th-1-type cytokines (Saemann et al., 2000; Cavaglieri et al., 2003). Butyrate also has been associated with immunomodulation of NK cells (Ishizaka et al., 1993; Ishizuka et al., 2004). We have demonstrated in a rat model that supplementing total parenteral nutrition (TPN) with SCFA results in increased NK cell activity (Pratt et al., 1996). Other studies have demonstrated anti-inflammatory properties of SCFA. Butyrate was reported to suppress both constitutive and cytokine-induced expression of the transcription factor NFκB in the colonic cell line HT-29 (Inan et al., 2000). Pharmacological doses of acetate administered intravenously to both healthy subjects and cancer patients increased peripheral blood antibody production, NK cell activity, and the allogeneic mixed lymphocyte reaction (Ishizaka et al., 1993). Whether these effects occur at concentrations seen after a high fermentable fiber meal is not known.

Additionally, SCFA production, particularly butyrate, in the colon may reduce the requirement of epithelial cells for glutamine, thereby sparing it for other cells, such as those of the immune system (Jenkins et al., 1999). This hypothesis is supported by the observation that lactulose administration can increase serum glutamine levels (Jenkins et al., 1997), and glutamine is an essential energy source for immune lymphocytes (Wu et al., 1991).

Modulation of Mucin Production

The layer of mucus overlying the gastrointestinal tract prevents the adherence and subsequent translocation of bacteria across the epithelial wall (Katayama et al., 1997). There is some evidence to indicate that the addition of fermentable fibers to the diet can increase mucin production (Satchithanandam et al., 1990). Greater mucin production might contribute to the lower incidence of bacterial translocation across the gut barrier reported in studies that fed dietary fibers (Deitch et al., 1993; Spaeth et al., 1994; Frankel et al., 1995; Xu et al., 1998). The increase in mucin production may occur in response to the decreased pH accompanying the production of SCFA (Bustos-Fernandez et al., 1978). Further support for SCFA stimulating mucin production comes from a perfused rat colon model where the production of acetate and butyrate from the fermentation of pectin, gum arabic, and cellulose stimulated mucin release, whereas the dietary fibers themselves did not (Barcelo et al., 2000). One study has reported that feeding IN increased sulfomucin production in both germ-free and heteroxenic rats (Fontaine et al., 1996). A more recent study showed that an active hexose correlated compound (AHCC), a product from the mycelium of Basidiomycete fungi that contains oligosaccharides increased the production of mucins 2 and 4 and trefoil factor 3 in rats with TNBS-induced colitis, significantly ameliorating inflammation (Daddaoua et al., 2007).

Conclusions

Although further work is needed to better define the changes, the mechanisms for immunomodulation, and the ultimate impact on immune health, there is convincing preliminary data to suggest that the consumption of prebiotic fibers can modulate immune parameters in GALT, secondary lymphoid tissues and peripheral circulation. Future studies on the physiological impact of consuming prebiotics should be designed to include assessments of the gut microflora, gut physiology, and the function and composition of the various regions of GALT.

Acknowledgements

The authors thank Mrs. Dawne Colwell for drawing the illustrations in this chapter.

References

Aattouri, N., Lemonnier, D. (1997) Production of interferon induced by *Streptococcus thermophilus*: role of CD4 and CD8 lymphocytes. *J. Nutr. Biochem.* 8, 25–31.

Abreu, M.T. (2002) The pathogenesis of inflammatory bowel disease: translational implications for clinicians. *Curr. Gastroenterol. Rep.* 4, 481–489.

Abreumartin, M.T., Targan, S.R. (1996) Regulation of immune responses of the intestinal mucosa. *Crit. Rev. Immunol.* 16, 277–309.

Akira, S., Takeda, K., Kaisho, T. (2001). Toll-like receptors: critical proteins linking innate and acquired immunity. *Nat. Immunol.* 2(8), 675–680.

Arslanoglu, S., Moro, G.E., Schmitt, J., Tandoi, L., Rizzardi, S., Boehm G. (2008) Early dietary intervention with a mixture of prebiotic oligosaccharides reduces the incidence of allergic manifestations and infections during the first two years of life. *J. Nutr.* 138, 1091–1095.

Arunachalam, K., Gill, H.S., Chandra, R.K. (2000) Enhancement of natural immune function by dietary consumption of *Bifidobacterium lactis* (HN019). *Eur. J. Clin. Nutr.* 54, 263–267.

Barcelo, A., Claustre, J., Moro, F., Chayvialle, J.A., Cuber, J.C., Plaisancié, P. (2000) Mucin secretion is modulated by luminal factors in the isolated vascularly perfused rat colon. *Gut* 46, 218–224.

Berg, R.D. (1985) Indigenous intestinal microflora and the host immune response. *EOS J. Immunol. Immunopharmacol.* 4, 161–168.

Bloom, B.R., Salgame, P., Diamond, B. (1992) Revisiting and revising suppressor T cells. *Immunol. Today* 13, 131–136.

Bouhnik, Y., Flourie, B., Riottot, M., et al. (1996) Effects of fructo-oligosaccharides ingestion on fecal bifidobacteria and selected metabolic indexes of colon carcinogenesis in healthy humans. *Nutr. Cancer* 26, 21–29.

Buddington, K.K., Donahoo, J.B., Buddington, R.K. (2002) Dietary oligofructose and inulin protect mice from enteric and systemic pathogens and tumor inducers. *J. Nutr.* 132, 472–477.

Buddington, R.K., Williams, C.H., Chen S-C, Witherly, S.A. (1996) Dietary supplement of neosugar alters the fecal flora and decreases activities of some reductive enzymes in human subjects. *Am. J. Clin. Nutr.* 63, 709–716.

Bustos-Fernandez, L., de P, I., Hamamura, S., et al. (1978) Does secretin influence rat colonic absorption and secretion? *Am. J. Gastroenterol.* 70, 265–269.

Calcinaro, F., Dionisi, S., Marinaro, M., et al. (2005) Oral probiotic administration induces interleukin-10 production and prevents spontaneous autoimmune diabetes in the non-obese diabetic mouse. *Diabetologia* 48, 1565–1575.

Cavaglieri, C.R., Nishiyama, A., Fernandes, L.C., Curi, R., Miles, E.A., Calder, P.C. (2003) Differential effects of short-chain fatty acids on proliferation and production of pro- and anti-inflammatory cytokines by cultured lymphocytes. *Life Sci.* 73, 1683–1690.

Chamaillard, M., Hashimoto, M., Horie, Y., et al. (2003) An essential role for NOD1 in host recognition of bacterial peptidoglycan containing diaminopimelic acid. *Nat. Immunol.* 4, 702–707.

Chen, Y., Kuchroo, V.K., Inobe, J., Hafler, D.A., Weiner, H.L. (1994) Regulatory T cell clones induced by oral tolerance: suppression of autoimmune encephalomyelitis. *Science* 265, 1237–1240.

Chiang, B.L., Sheih, Y.H., Wang, L.H., Liao, C.K., Gill, H.S. (2000) Enhancing immunity by dietary consumption of a probiotic lactic acid bacterium (*Bifidobacterium lactis* HN019): optimization and definition of cellular immune responses. *Eur. J. Clin. Nutr.* 54, 849–855.

Cross, M.L. (2002) Microbes versus microbes: immune signals generated by probiotic lactobacilli and their role in protection against microbial pathogens. *FEMS Immunol. Med. Microbiol.* 34, 245–253.

Cummings, J.H., Mann, J.I., Nishida, C., Vorster, H.H. (2009) Dietary fibre: an agreed definition. *Lancet* 373, 365–366.

Daddaoua, A., Martinez-Plata, E., Lopez-Posadas, R., et al. (2007) Active hexose correlated compound acts as a prebiotic and is antiinflammatory in rats with hapten-induced colitis. *J. Nutr.* 137, 1222–1228.

Daddaoua, A., Puerta, V., Requena, P., et al. (2006) Goat milk oligosaccharides are anti-inflammatory in rats with hapten-induced colitis. *J. Nutr.* 136, 672–676.

De Simone, C., Ciardi, A., Grassi, A., et al. (1992) Effect of *Bifidobacterium bifidum* and *Lactobacillus acidophilus* on gut mucosa and peripheral blood B lymphocytes. *Immunopharmacol. Immunotoxicol.* 14, 331–340.

De Simone, C., Ferrazzi, M., Di Seri, M., et al. (1987a) The immunoregulation of the intestinal flora: bifidobacteria and lactobacilli modulate the production of γ-IFN induced by pathogenic bacteria. *Int. J. Immunother.* 3, 151–158.

De Simone, C., Tzantzoglou, S., Baldinelli, L., et al. (1988) Enhancement of host resistance against *Salmonella typhimurium* infection by a diet supplemented with yogurt. *Immunopharmacol. Immunotoxicol.* 10, 399–415.

De Simone, C., Vesely, R., Negri, R., et al. (1987b) Enhancement of immune response of murine Peyer's patches by a diet supplemented with yogurt. *Immunopharmacol. Immunotoxicol.* 9, 87–100.

de Vrese, M., Schrezenmeir, J. (2008) Probiotics, prebiotics, and synbiotics. *Adv. Biochem. Eng. Biotechnol.* 111, 1–66.

de Vrese, M., Winkler, P., Rautenberg, P., et al. (2006) Probiotic bacteria reduced duration and severity but not the incidence of common cold episodes in a double blind, randomized, controlled trial. *Vaccine* 24, 6670–6674.

Deitch, E.A., Xu, D., Lu, Q., Berg, R. (1993) Elemental diet-induced immune suppression is caused by both bacterial and dietary factors. *J. Parenter. Enteral Nutr.* 17, 332–336.

Delves, P.J., Roitt, I.M. (2000a) The immune system: first of two parts. *N. Engl. J. Med.* 343, 37–49.

Delves, P.J., Roitt, I.M. (2000b) The immune system: second of two parts. *N. Engl. J. Med.* 343, 108–117.

DeWitt, R.C., Kudsk, K.A. (1999) The gut's role in metabolism, mucosal barrier function, and gut immunology. *Infect. Dis. Clin. North Am.* 13, 465–481.

Di Giancinto, C., Marinaro, M., Sanchez, M., Strober, W., Boirivant, M. (2005) Probiotics ameliorate recurrent Th1-mediated murine colitis by inducing IL-10 and IL-10-dependent TGF-beta-bearing regulatory cells. *J. Immunol.* 174, 3237–3246.

Donnet-Hughes, A., Rochat, F., Serrant, P., Aeschlimann, J.M., Schiffrin, E.J. (1999) Modulation of nonspecific mechanisms of defense by lactic acid bacteria: effective dose. *J. Dairy Sci.* 82, 863–869.

Ewaschuk, J.B., Backer, J.L., Churchill, T.A., Obermeier, F., Krause, D.O., Madsen, K.L. (2007) Surface expression of Toll-like receptor 9 is upregulated on intestinal epithelial cells in response to pathogenic bacterial DNA. *Infect. Immun.* 75, 2572–2579.

Femia, A.P., Luceri, C., Dolara, P., et al. (2002) Antitumorigenic activity of the prebiotic inulin enriched with oligofructose in combination with the probiotics *Lactobacillus rhamnosus* and *Bifidobacterium lactis* on azoxymethane-induced colon carcinogenesis in rats. *Carcinogenesis* 23, 1953–1960.

Field, C.J., McBurney, M.I., Massimino, S., Hayek, M.G., Sunvold, G.D. (1999) The fermentable fiber content of the diet alters the function and composition of canine gut associated lymphoid tissue. *Vet. Immunol. Immunopathol.* 72, 325–341.

Fitch, F.W., Stack, R., Fields, P., Lancki, D.W., Cronin, D.C. (1995) Regulation of T lymphocyte subsets. *Ciba Found. Symp.* 195, 68–80.

Fontaine, N., Meslin, J.C., Lory, S., Andrieux, C. (1996) Intestinal mucin distribution in the germ-free rat and in the heteroxenic rat harbouring a human bacterial flora: effect of inulin in the diet. *Br. J. Nutr.* 75, 881–892.

Frankel, W., Zhang, W., Singh, A., et al. (1995) Fiber: effect on bacterial translocation and intestinal mucin content. *World J. Surg.* 19, 144–149.

Fujimori, S., Gudis, K., Mitsui, K., et al. (2009) A randomized controlled trial on the efficacy of synbiotic versus probiotic or prebiotic treatment to improve the quality of life in patients with ulcerative colitis. *Nutrition* 25, 520–525.

Fukuda, M., Kanauchi, O., Araki, Y., et al. (2002) Prebiotic treatment of experimental colitis with germinated barley foodstuff: a comparison with probiotic or antibiotic treatment. *Int. J. Mol. Med.* 9, 65–70.

Fukushima, Y., Kawata, Y., Hara, H., Terada, A., Mitsuoka, T. (1998) Effect of a probiotic formula on intestinal immunoglobulin A production in healthy children. *Int. J. Food Microbiol.* 42, 39–44.

Fukushima, Y., Kawata, Y., Mizumachi, K., Kurisaki, J., Mitsuoka, T. (1999) Effect of bifidobacteria feeding on fecal flora and production of immunoglobulins in lactating mouse. *Int. J. Food Microbiol.* 46, 193–197.

Furrie, E., Macfarlane, S., Kennedy, A., et al. (2005) Synbiotic therapy (*Bifidobacterium longum*/Synergy 1) initiates resolution of inflammation in patients with active ulcerative colitis: a randomised controlled pilot trial. *Gut* 54, 242–249.

Gaskins, H.R., Mackie, R.I., May, T., Garleb, K.A. (1996) Dietary fructo-oligosaccharide modulates large intestinal inflammatory responses to *Clostridium difficile* in antibiotic-compromised mice. *Microb. Ecol. Health Dis.* 9, 157–166.

Gibson, G.R. (1998) Dietary modulation of the human gut microflora using prebiotics. *Br. J. Nutr.* 80(Suppl. 2), S209–S212.

Gibson, G.R., Beatty, E.R., Wang, X., Cummings, J.H. (1995) Selective stimulation of bifidobacteria in the human colon by oligofructose and inulin. *Gastroenterology* 108, 975–982.

Gibson, G.R., Roberfroid, M.B. (1995) Dietary modulation of the human colonic microbiota: introducing the concept of prebiotics. *J. Nutr.* 125, 1401–1412.

Gill, H.S. (2003) Probiotics to enhance anti-infective defences in the gastrointestinal tract. *Best Pract. Res. Clin. Gastroenterol.* 17, 755–773.

Gill, H.S., Rutherfurd, K.J., Cross, M.L. (2001) Dietary probiotic supplementation enhances natural killer cell activity in the elderly: an investigation of age-related immunological changes. *J. Clin. Immunol.* 21, 264–271.

Girardin, S.E., Tournebize, R., Mavris, M., et al. (2001) CARD4/Nod1 mediates NF-kappaB and JNK activation by invasive Shigella flexneri. *EMBO Rep.* 2, 736–742.

Goust, J.M., Bierer, B. (1993) Cell-mediated immunity. *Immunol. Ser.* 58, 187–212.

Green, D.R., Webb, D.R. (1993) Saying the 'S' word in public. *Immunol. Today* 14, 523–525.

Hafer, A., Kramer, S., Duncker, S., Krüger, M., Manns, M.P., Bischoff, S.C. (2007) Effect of oral lactulose on clinical and immunohistochemical parameters in patients with inflammatory bowel disease: a pilot study. *BMC Gastroenterol.* 7, 36.

Halpern, G.M., Vruwink, K.G., Van de Water, J., Keen, C.L., Gershwin, M.E. (1991) Influence of long-term yoghurt consumption in young adults. *Int. J. Immunother.* 7, 205–210.

Hatcher, G.E., Lambrecht, R.S. (1993) Augmentation of macrophage phagocytic activity by cell-free extracts of selected lactic acid-producing bacteria. *J. Dairy Sci.* 76, 2485–2492.

He, F., Tuomola, E., Arvilommi, H., Salminen, S. (2001) Modulation of human humoral immune response through orally administered bovine colostrum. *FEMS Immunol. Med. Microbiol.* 31, 93–96.

Hershberg, R.M., Cho, D.H., Youakim, A., et al. (1998) Highly polarized HLA class II antigen processing and presentation by human intestinal epithelial cells. *J. Clin. Invest.* 102, 792–803.

Hoentjen, F., Welling, G.W., Harmsen, H.J., et al. (2005) Reduction of colitis by prebiotics in HLA-B27 transgenic rats is associated with microflora changes and immunomodulation. *Inflamm. Bowel Dis.* 11(11), 977–985.

Hosono, A., Ozawa, A., Kato, R., et al. (2003) Dietary fructooligosaccharides induce immunoregulation of intestinal IgA secretion by murine Peyer's patch cells. *Biosci. Biotechnol. Biochem.* 67(4), 758–764.

Hurley, B.P., McCormick, B.A. (2004) Intestinal epithelial defense systems protect against bacterial threats. *Curr. Gastroenterol. Rep.* 6(5), 355–361.

Inan, M.S., Rasoulpour, R.J., Yin, L., Hubbard, A.K., Rosenberg, D.W., Giardina, C. (2000) The luminal short-chain fatty acid butyrate modulates NF- B activity in a human colonic epithelial cell line. *Gastroenterology* 118, 724–734.

Ishizaka, S., Kikuchi, E., Tsujii, T. (1993) Effects of acetate on human immune system. *Immunopharmacol. Immunotoxicol.* 15(2–3), 151–162.

Ishizuka, S., Tanaka, S., Xu, H., Hara, H. (2004) Fermentable dietary fiber potentiates the localization of immune cells in the rat large intestinal crypts. *Exp. Biol. Med. (Maywood)* 229(9), 876–884.

Jenkins, D.J.A., Kendall, C.W.C., Vuksan, V. (1999) Inulin, oligofructose and intestinal function. *J. Nutr.* 129, 1431S–1433S.

Jenkins, D.J.A., Popovich, D.G., Kendall, C.W.C., et al. (1997) Metabolic effects of non-absorbable carbohydrates. *Scand. J. Gastroenterol.* 32(Suppl. 222), 10–13.

Kagnoff, M.F. (1987) Immunology of the digestive system. In: Johnson, L.R. (editor), *Physiology of the Gastrointestinal Tract*. Raven Press, New York, 1699–1728.

Kagnoff, M.F. (1993) Immunology of the intestinal tract. *Gastroenterology* 105, 1275–1280.

Kaila, M., Isolauri, E., Soppi, E., Virtanen, E., Laine, S., Arvilommi, H. (1992) Enhancement of the circulating antibody secreting cell response in human diarrhea by a human Lactobacillus strain. *Pediatr. Res.* 32(2), 141–144.

Kanauchi, O., Iwanaga, T., Andoh, A., et al. (2001) Dietary fiber fraction of germinated barley foodstuff attenuated mucosal damage and diarrhea, and accelerated the repair of the colonic mucosa in an experimental colitis. *J. Gastroenterol. Hepatol.* 16(2), 160–168.

Katayama, M., Xu, D., Specian, R.D., Deitch, E.A. (1997) Role of bacterial adherence and the mucus barrier on bacterial translocation: effects of protein malnutrition and endotoxin in rats. *Ann. Surg.* 225(3), 317–326.

Kaufhold, J., Hammon, H.M., Blum, J.W. (2000) Fructo-oligosaccharide supplementation: effects on metabolic, endocrine and hematological traits in veal calves. *J. Vet. Med. Ser. A* 47(1), 17–29.

Kelly-Quagliana, K.A., Nelson, P.D., Buddington, R.K. (2003) Dietary oligofructose and inulin modulate immune functions in mice. *Nutr. Res.* 23, 257–267.

Kleesen, B., Sykura, B., Zunft, H-J., Blaut, M. (1997) Effects of inulin and lactose on fecal microflora, microbial activity, and bowel habit in elderly constipated persons. *Am. J. Clin. Nutr.* 65, 1397–1402.

Kudoh, K., Shimizu, J., Ishiyama, A., et al. (1999) Secretion and excretion of immunoglobulin A to cecum and feces differ with type of indigestible saccharides. *J. Nutr. Sci. Vitaminol.* 45, 173–181.

Kudoh, K., Shimizu, J., Wada, M., Takita, T., Kanke, Y., Innami, S. (1998) Effect of indigestible saccharides on B lymphocyte response of intestinal mucosa and cecal fermentation in rats. *J. Nutr. Sci. Vitaminol.* 44, 103–112.

Laissue, J.A., Gebbers, J-O. (1992) The intestinal barrier and the gut-associated lymphoid tissue. *Curr. Stud. Hematol. Blood Transfus.* 59, 19–43.

Lammers, K.M., Vergopoulos, A., Babel, N., et al. (2005) Probiotic therapy in the prevention of pouchitis onset: decreased interleukin-1beta, interleukin-8, and interferon-gamma gene expression. *Inflamm. Bowel Dis.* 11(5), 447–454.

Langkamp-Henken, B., Glezer, J.A., Kudsk, K.A. (1992) Immunologic structure and function of the gastrointestinal tract. *Nutr. Clin. Pract.* 7, 100–108.

Lara-Villoslada, F., Debras, E., Nieto, A., et al. (2006) Oligosaccharides isolated from goat milk reduce intestinal inflammation in a rat model of dextran sodium sulfate-induced colitis. *Clin. Nutr.* 25(3), 477–488.

Lim, B.O., Yamada, K., Nonaka, M., Kuramoto, Y., Hung, P., Sugano, M. (1997) Dietary fibers modulate indices of intestinal immune function in rats. *J. Nutr.* 127, 663–667.

Lindsay, J.O., Whelan, K., Stagg, A.J., et al. (2006) Clinical, microbiological, and immunological effects of fructo-oligosaccharide in patients with Crohn's disease. *Gut* 55(3), 348–355.

Link-Amster, H., Rochat, F., Saudan, K.Y., Mignot, O., Aeschlimann, J.M. (1994) Modulation of a specific humoral immune response and changes in intestinal flora mediated through fermented milk intake. *FEMS Immunol. Med. Microbiol.* 10, 55–64.

MacDonald, T.T. (1998) T cell immunity to oral antigens. *Curr. Opin. Immunol.* 10, 620–627.

Madar, Z., Gurevich, P., Ben-Hur, H., et al. (1998) Effects of dietary fiber on the rat intestinal mucosa exposed to low doses of a carcinogen. *Anticancer Res.* 18, 3521–3526.

Madsen, K.L., Doyle, J.S., Jewell, L.D., Tavernini, M.M., Fedorak, R.N. (1999) Lactobacillus species prevents colitis in interleukin 10 gene-deficient mice. *Gastroenterology* 116(5), 1107–1114.

Malin, M., Suomalainen, H., Saxelin, M., Isolauri, E. (1996) Promotion of IgA immune response in patients with Crohn's disease by oral bacteriotherapy with *Lactobacillus* GG. *Ann. Nutr. Metab.* 40, 137–145.

Manhart, N., Spittler, A., Bergmeister, H., Mittlböck, M., Roth, E. (2003) Influence of fructooligosaccharides on Peyer's patch lymphocyte numbers in healthy and endotoxemic mice. *Nutrition* 19(7–8), 657–660.

Marin, M.L., Lee, J.H., Murtha, J., Ustunol, Z., Pestka, J.J. (1997) Differential cytokine production in clonal macrophage and T-cell lines cultured with bifidobacteria. *J. Dairy Sci.* 80(11), 2713–2720.

Massimino, S., Field, C.J., Hayek, M.G., et al. (1998) Fermentable dietary fiber increases GLP-1 secretion and improves glucose homeostasis despite increased intestinal glucose transport capacity in healthy dogs. *J. Nutr.* 128, 1786–1793.

Matsuzaki, T. (1998) Immunomodulation by treatment with *Lactobacillus casei* strain Shirota. *Int. J. Food Microbiol.* 41, 133–140.

Matsuzaki, T., Yamazaki, R., Hashimoto, Y., Yokokura, T. (1998) The effect of oral feeding of *Lactobacillus casei* strain Shirota on immunoglobulin E production in mice. *J. Dairy Sci.* 81(1), 48–53.

McKay, D.M., Perdue, M.H. (1993) Intestinal epithelial function: the case for immunophysiological regulation. *Dig. Dis. Sci.* 38, 1377–1387.

Menne, E., Guggenbuhl, N., Roberfroid, M.B. (2000) Fn-type chicory inulin hydrolysate has a prebiotic effect in humans. *J. Nutr.* 130, 1197–1199.

Miettinen, M., Vuopio-Varkila, J., Varkila, K. (1996) Production of human tumor necrosis factor alpha, interleukin-6, and interleukin-10 is induced by lactic acid bacteria. *Infect. Immun.* 64(12), 5403–5405.

Mitsuoka, T., Hidaka, H., Eida, T. (1987) Effect of fructo-oligosaccharides on intestinal microflora. *Nahrung* 31(5–6), 427–436.

Moineau, S., Goulet, J. (1991) Effect of feeding fermented milks on the pulmonary macrophage activity in mice. *Milchwissenschaft* 46(9), 551–554.

Moro, G., Arslanoglu, S., Stahl, B., Jelinek, J., Wahn, U., Boehm, G. (2006) A mixture of prebiotic oligosaccharides reduces the incidence of atopic dermatitis during the first six months of age. *Arch. Dis. Child* 91(10), 814–819.

Mortensen, P.B., Clausen, M.R. (1996) Short-chain fatty acids in the human colon: relation to gastrointestinal health and disease. *Scand. J. Gastroenterol.* 216, 132–148.

Nagai, T., Ishizuka, S., Hara, H., Aoyama, Y. (2000) Dietary sugar beet fiber prevents the increase in aberrant crypt foci induced by γ irradiation in the colorectum of rats treated with an immunosuppressant. *J. Nutr.* 130, 1682–1687.

Nagendra, R., Venkat Rao, S. (1994) Effect of feeding infant formulations containing bifidus factors on *in vivo* proliferation of bifidobacteria and stimulation of intraperitoneal macrophage activity in rats. *J. Nutr. Immunol.* 2(4), 61–68.

Olivares, M., Paz Diaz-Ropero, M., Gomez, N., et al. (2006) Dietary deprivation of fermented foods causes a fall in innate immune response. Lactic acid bacteria can counteract the immunological effect of this deprivation. *J. Dairy Res.* 73(4), 492–498.

Palmer, C., Bik, E.M., DiGiulio, D.B., Relman, D.A., Brown, P.O. (2007) Development of the human infant intestinal microbiota. *PLoS Biol.* 5(7), e177.

Park, S.Y., Ji, G.E., Ko, Y.T., Jung, H.K., Ustunol, Z., Pestka, J.J. (1999) Potentiation of hydrogen peroxide, nitric oxide, and cytokine production in RAW 264.7 macrophage cells exposed to human and commercial isolates of *Bifidobacterium*. *Int. J. Food Microbiol.* 46, 231–241.

Pathmakanthan, S., Li, C.K., Cowie, J., Hawckey, C.J. (2004) *Lactobacillus plantarum* 299: beneficial in vitro immunomodulation in cells extracted from inflamed human colon. *J. Gastroenterol. Hepatol.* 19(2), 166–173.

Pelto, L., Isolauri, E., Lilius, E.M., Nuutila, J., Salminen, S. (1998) Probiotic bacteria down-regulate the milk-induced inflammatory response in milk-hypersensitive subjects but have an immunostimulatory effect in healthy subjects. *Clin. Exp. Allergy* 28(12), 1474–1479.

Perdigon, G., Alvarez, S., Nader de Macias, M.E., Medici, M. (1990) The oral administration of lactic acid bacteria increase the mucosal intestinal immunity in response to enteropathogens. *J. Food Prot.* 53(5), 404–410.

Perdigon, G., Alvarez, S., Nader de Macias, M.E., Savoy de Giori G., Medici, M., Nunez de K.M. (1991) Behaviour of natural and heated yogurt in the immune system and preventive capacity on enteric infections. *Milchwissenschaft* 46(7), 411–416.

Perdigon, G., Alvarez, S., Nader de Macias, M.E., Margni, R., Oliver, G., P. de Ruiz Holgado, A. (1986b) Lactobacilli administered orally induce release of enzymes from peritoneal macrophages in mice. *Milchwissenschaft* 41(6), 344–348.

Perdigon, G., Nader de Macias, M.E., Alvarez, S., Oliver, G., P. de Ruiz Holgado, A. (1986a) Effect of perorally administered lactobacilli on macrophage activation in mice. *Infect. Immun.* 53(2), 404–410.

Perdigon, G., Nader de Macias, M.E., Alvarez, S., Medici, M., Oliver, G., P. de Ruiz Holgado, A. (1986c) Effect of a mixture of *Lactobacillus casei* and *Lactobacillus acidophilus* administered orally on the immune system in mice. *J. Food Prot.* 49(12), 986–989.

Perdigon, G., Nader de Macias, M.E., Alvarez, S., Oliver, G., P. de Ruiz Holgado, A. (1987) Enhancement of immune response in mice fed with *Streptococcus thermophilus* and *Lactobacillus acidophilus*. *J. Dairy Sci.* 70(5), 919–926.

Perdigon, G., Nader de Macias, M.E., Alvarez, S., Oliver, G., and P. de Ruiz Holgado, A. (1988) Systemic augmentation of the immune response in mice by feeding fermented milks with *Lactobacillus casei* and *Lactobacillus acidophilus*. *Immunology* 63, 17–23.

Pierre, F., Perrin, P., Champ, M., Bornet, F., Meflah, K., Menanteau, J. (1997) Short-chain fructo-oligosaccharides reduce the occurrence of colon tumors and develop gut-associated lymphoid tissue in *min* mice. *Cancer Res.* 57, 225–228.

Poussier, P., Ning, T., Banerjee, D., Julius, M. (2002) A unique subset of self-specific intraintestinal T cells maintains gut integrity. *J. Exp. Med.* 195(11), 1491–1497.

Pratt, V.C., Tappenden, K.A., McBurney, M.I., Field, C.J. (1996) Short-chain fatty acid-supplemented total parenteral nutrition improves nonspecific immunity after intestinal resection in rats. *J. Parenter. Enteral Nutr.* 20, 264–271.

Rimoldi, M., Chieppa, M., Larghi, P., Vulcano, M., Allavena, P., Rescigno, M. (2005) Monocyte-derived dendritic cells activated by bacteria or by bacteria-stimulated epithelial cells are functionally different. *Blood* 106(8), 2818–2826.

Roller, M., Rechkemmer, G., Watzl, B., (2004) Prebiotic inulin enriched with oligofructose in combination with the probiotics *Lactobacillus rhamnosus* and *Bifidobacterium lactis* modulates intestinal immune functions in rats. *J. Nutr.* 134(1), 153–156.

Rumi, G., Tsubouchi, R., Okayama, M., Kato, S., Mózsik, G., Takeuchi, K. (2004) Protective effect of lactulose on dextran sulfate sodium-induced colonic inflammation in rats. *Dig. Dis. Sci.* 49(9), 1466–1472.

Saemann, M.D., Bohmig, G.A., Osterreicher, C.H., et al. (2000) Anti-inflammatory effects of sodium butyrate on human monocytes: potent inhibition of IL-12 and up-regulation of IL-10 production. *FASEB J.* 14(15), 2380–2382.

Satchithanandam, S., Vargofcak-Apker, M., Calvert, R.J., Leeds, A.R., Cassidy, M.M. (1990) Alteration of gastrointestinal mucin by fiber feeding in rats. *J. Nutr.* 120, 1179–1184.

Schiffrin, E.J., Rochat, F., Link-Amster, H., Aeschlimann, J.M., Donnet-Hughes, A. (1995) Immunomodulation of human blood cells following the ingestion of lactic acid bacteria. *J. Dairy Sci.* 78, 491–497.

Seifert, S., Watzl, B. (2007) Inulin and oligofructose: review of experimental data on immune modulation. *J. Nutr.* 137(11 Suppl), 2563S–2567S.

Sheih, Y.H., Chiang, B.L., Wang, L.H., Liao, C.K., Gill, H.S. (2001) Systemic immunity-enhancing effects in healthy subjects following dietary consumption of the lactic acid bacterium *Lactobacillus rhamnosus* HN001. *J. Am. Coll. Nutr.* 20(2 Suppl), 149–156.

Shu, Q., Lin, H., Rutherford, K.J., et al. (2000) Dietary *Bifidobacterium lactis* (HN019) enhances resistance to oral *Salmonella typhimurium* infection in mice. *Microbiol. Immunol.* 44(3), 213–222.

Solis, P.B., Lemonnier, D. (1991) Induction of 2′-5′ A synthetase activity and interferon in humans by bacteria used in dairy products. *Eur. Cytokine Netw.* 2(2), 137–140.

Solis Pereyra, B., Lemonnier, D. (1993) Induction of human cytokines by bacteria used in dairy foods. *Nutr. Res.* 13, 1127–1140.

Spaeth, G., Gottwald, T., Specian, R.D., Mainous, M.R., Berg, R.D., Deitch, E.A. (1994) Secretory immunoglobulin, A., intestinal mucin, and mucosal permeability in nutritionally induced bacterial translocation in rats. *Ann. Surg.* 220(6), 798–808.

Suzuki, K., Oida, T., Hamada, H., et al. (2000) Gut cryptopatches: direct evidence of extrathymic anatomical sites for intestinal T lymphopoiesis. *Immunity* 13(5), 691–702.

Takahashi, T., Nakagawa, E., Nara, T., Yajima, T., Kuwata, T. (1998) Effects of orally ingested *Bifidobacterium longum* on the mucosal IgA response of mice to dietary antigens. *Biosci., Biotechnol. Biochem.* 62(1), 10–15.

Takahashi, T., Oka, T., Iwana, H., Kuwata, T., Yamamoto, Y. (1993) Immune response of mice to orally administered lactic acid bacteria. *Biosci. Biotechnol. Biochem.* 57(9), 1557–1560.

Tannock, G.W. (2007) What immunologists should know about bacterial communities of the human bowel. *Semin. Immunol.* 19(2), 94–105.

Tejada-Simon, M.V., Lee, J.H., Ustunol, Z., Pestka, J.J. (1999a) Ingestion of yogurt containing *Lactobacillus acidophilus* and *Bifidobacterium* to potentiate immunoglobulin A responses to cholera toxin in mice. *J. Dairy Sci.* 82(4), 649–660.

Tejada-Simon, M.V., Ustunol, Z., Pestka, J.J. (1999b) Ex vivo effects of lactobacilli, streptococci, and bifidobacteria ingestion on cytokine and nitric oxide production in a murine model. *J. Food Prot.* 62(2), 162–169.

van Hoffen, E., Ruiter, B., Faber, J., et al. (2009) A specific mixture of short-chain galacto-oligosaccharides and long-chain fructo-oligosaccharides induces a beneficial immunoglobulin profile in infants at high risk for allergy. *Allergy* 64(3), 484–487.

Videla, S., Vilaseca, J., Antolin, M., et al. (2001) Dietary inulin improves distal colitis induced by dextran sodium sulfate in the rat. *Am. J. Gastroenterol.* 96(5), 1486–1493.

Weiner, H.L. (1997) Oral tolerance: immune mechanisms and treatment of autoimmune diseases. *Immunol. Today* 18, 335–343.

Wheeler, J.G., Shema, S.J., Bogle, M.L., et al. (1997) Immune and clinical impact of *Lactobacillus acidophilus* on asthma. *Ann. Allergy Asthma Immunol.* 79(3), 229–233.

Wolever, T.M., Josse, R.G., Leiter, L.A., Chiasson, J.L. (1997) Time of day and glucose tolerance status affect serum short-chain fatty acid concentrations in humans. *Metabolism* 46(7), 805–811.

Wu, G., Field, C.J., Marliss, E.B. (1991) Glutamine and glucose metabolism in rat splenocytes and mesenteric lymph node lymphocytes. *Am. J. Physiol.* 260, E141–E147.

Xu, D., Lu, Q., Deitch, E.A. (1998) Elemental diet-induced bacterial translocation associated with systemic and intestinal immune suppression. *J. Parent. Enteral Nutr.* 22(1), 37–41.

Yamada, K., Tokunaga, Y., Ikeda, A., et al. (1999) Dietary effect of guar gum and its partially hydrolyzed product on the lipid metabolism and immune function of Sprague-Dawley rats. *Biosci., Biotechnol. Biochem.* 63(12), 2163–2167.

Yasui, H., Mike, A., Ohwaki, M. (1989) Immunogenicity of *Bifidobacterium breve* and change in antibody production in Peyer's patches after oral administration. *J. Dairy Sci.* 72(1), 30–35.

Yasui, H., Ohwaki, M. (1991) Enhancement of immune response in Peyer's patch cells cultured with *Bifidobacterium breve*. *J. Dairy Sci.* 74(4), 1187–1195.

Yun, C-H, Estrada, A., Van Kessel, A., Gajadhar, A.A., Redmond, M.J., Laarveld, B. (1997) B-(1->3, 1->4) oat glucan enhances resistance to *Eimeria vermiformis* infection in immunosuppressed mice. *Int. J. Parasitol.* 27(3), 329–337.

Yun, C-H., Estrada, A., Van Kessel, A., Gajadhar, A., Redmond, M., Laarveld, B. (1998) Immunomodulatory effects of oat B-glucan administered intragastrically or parenterally on mice infected with *Eimeria vermiformis*. *Microbiol. Immunol.* 42(6), 457–465.

Zusman, I., Gurevich, P., Benhur, H., et al. (1998) The immune response of rat spleen to dietary fibers and to low doses of carcinogen: morphometric and immunohistochemical studies. *Oncol. Rep.* 5, 1577–1581.

Chapter 4

Lower Gut Hormones and Health Effects Associated with Consumption of Fermentable Fibers

Michael J Keenan, Jun Zhou, Reshani Senevirathne, Marlene Janes, and Roy J Martin

Introduction

Recently, the definition of fiber has been changed to include dietary, functional, and total fiber by a dietary reference intakes committee. The committee recommended using the terms, viscosity and fermentability, that represent physicochemical properties of fibers. Fructans and resistant starches (RS) are now included as fiber and are the best-known fermentable fibers. Several bacterial populations have been identified in the fermentation process. Consumption of these fermentable fibers is associated with increased production of short-chain fatty acids (SCFA) and the lower gut hormones, glucagon-like peptide 1 (GLP-1) and peptide YY (PYY), in the proximal large intestine. A direct effect of SCFAs on hormone production is unclear at this time. GLP-1 acts through GLP-1 receptors and is best known for its actions on the β-cells of the pancreas for glucose control. Its role in appetite and feeding behavior has also been investigated. PYY acts through PYY receptors and plays a role in energy balance and insulin sensitivity of peripheral tissues. Studies in rodents with dietary fermentable fibers demonstrated reduced body fat and improved glucose control. Human studies with RS or fructans showed improved insulin sensitivity, but only the fructan consumption was associated with increased plasma GLP-1. This may be the result of a difference in experimental design. Research with knockout mice

Nondigestible Carbohydrates and Digestive Health. Edited by T. M. Paeschke and W. R. Aimutis

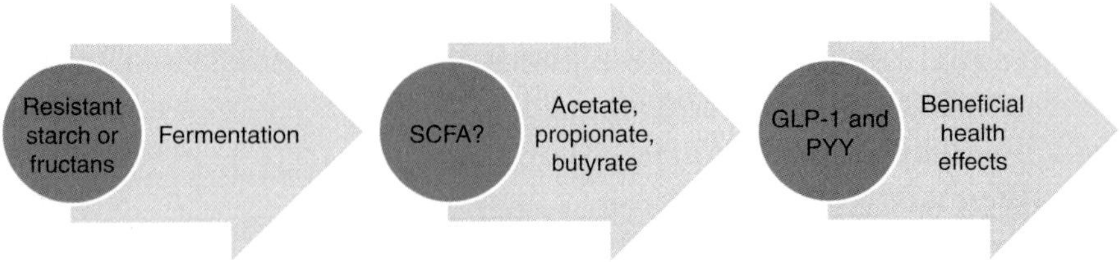

Figure 4.1. Fermentation of resistant starch and fructans is associated with increased plasma levels of lower gut hormones GLP-1 and PYY. The role of the fermentation products, acetate, propionate, and butyrate, in the production of the hormones is unclear at this time.

indicates that the beneficial health effects of fermentable fibers require lower gut hormone (GLP-1) production (Figure 4.1). Rodents fed large amounts of RS may be a good model for investigation of the "hindgut hypothesis" of Roux-en-Y gastric bypass surgery (RYGB) as the two have the result of elevated plasma PYY and GLP-1.

Definition of Fiber

A few years ago, the *Dietary Reference Intakes for Energy, Carbohydrate, Fiber, Fat, Fatty Acids, Cholesterol, Protein, and Amino Acids* was published. The chapter on fiber had the new definition of fiber that included dietary, functional, and total fiber. Dietary fiber includes "nondigestible carbohydrates and lignin that are intrinsic and intact in plants." The difference between the functional fiber category and dietary fiber category is that functional fiber includes "isolated, nondigestible carbohydrates that have beneficial physiological effects in humans." Total fiber is the combination of dietary and functional fiber. The term functional fiber replaced the term added fiber that was initially recommended by the Panel on the Definition of Dietary Fiber. The Panel on the Definition of Dietary Fiber also recommended the phasing out of the terms soluble and insoluble, and that they be replaced with other terms, viscosity and fermentability, that represent physicochemical properties. The opinion of the panel is that adoption of physicochemical properties of fibers will allow for better standardization of fiber.

Fructans and RS have been included as fiber and are two of the best-known fermentable fibers. Both are categorized as nonviscous

(Schneeman, 1999; Augustin et al., 2008). Inulins and oligofructoses (fructooligosaccharides) are fructans. There are basically four categories of RS. Category 1 (RS1) is found in whole grains with intact plant cell walls. The second category (RS2) consists of ungelatinized crystalline starch in high amylose cornstarches. The third (RS3) and fourth (RS4) types of RS are formed during food processing. RS3 is retrograded starch formed from cooking and cooling or extrusion. Potatoes in potato salad contain RS3. The last category of RS has chemically modified starch due to addition of esters, ethers, and cross-bonding. RS1 and RS2 are increasingly digested by α-amylase the longer they remain in the small intestine. However, RS3 and RS4 are not digested (Englyst, 1992; Cummings et al., 1996). The assigned category of dietary fiber is applied to the fiber source if the "plant cells and their three-dimensional interrelationships remain largely intact." Mechanical treatment does not eliminate the assignment to the dietary fiber category. Fructans found in plants would be dietary fiber, but synthesized or extracted fructans can be classified as functional fibers if there is sufficient data to show beneficial physiological effects in humans. RS1 and RS2 are dietary fibers, but RS3 and RS4 would be classified as functional fibers.

Fermentation of fructans and RS is primarily in the proximal large intestine (cecum), which is associated with the alterations in gut microflora, decreased pH, and increased SCFA and an enlarged cecum. Other associated effects of fermentation include production of lower or hindgut hormones (Reimer and McBurney, 1996; Keenan et al., 2006; Zhou et al., 2006; 2008; Delzenne et al., 2007; Shen et al., 2009).

Lower Gut Hormones

The lower gut or hindgut hormones are produced in greater amounts when increased food material passes through the small intestine to the large intestine. This occurs with increased levels of fiber or other indigestible macronutrients in the diet or in patients with gastric bypass surgery in which a significant portion of the small intestine is bypassed. PYY and GLP-1, are the two hormones, primarily reported to increase (Reimer et al., 1997; Cani et al., 2005; Borg et al., 2006; Keenan et al., 2006; Morinigo et al., 2006; Zhou et al., 2008; Shen et al., 2009). One of the reasons for this focus may be that commercial kits for these two hormones have been available for several years. PYY is produced by

L endocrine cells with the order of highest production—rectum, ileum, and large intestine (McGowan and Bloom, 2004). GLP-1 is also produced by L endocrine cells in the ileum and large intestine (Kreymann et al., 1988). Comparison of the gene expression in the terminal ileum and the proximal colon of the two hormones revealed the greater expression in the terminal ileum, but the effect of RS was only evident in the proximal colon (Zhou et al., 2006). GLP-1 is actually the product of posttranslational modification of the proglucagon peptide (Kieffer and Habener, 1999). The gene, referred to as either the preproglucagon or proglucagon gene, is expressed in the pancreas, intestine, and brain. In the pancreatic α-cells, the product of posttranslational modification is glucagon. GLP-1, GLP-2, and oxyntomodulin are the products in the intestine and brain.

Roles of PYY and GLP-1

GLP-1 acts through binding to the GLP-1 receptors. These receptors are found on many cell types including beta cells of the pancreas, neuronal cells in the brain, and in adrenal, pituitary, kidney, and throughout the gastrointestinal tract (Dunphy et al., 1998; Gotthardt et al., 2006). Thus far, most research has focused on the action of GLP-1 on pancreatic beta cells. GLP-1 enhances both early and late phases of insulin secretion stimulated by glucose. This action of GLP-1 has led to the development of drugs aimed at mimicking or enhancing this function in the treatment of diabetes mellitus (Lovshin and Drucker, 2009). Because of GLP-1's short half-life, Exendin-4, a specific GLP-1-receptor agonist with a longer half-life has been developed. Exendin-4 decreases the blood glucose level, reduces body weight, and improves insulin sensitivity in obese diabetic (ob/ob, db/db) mice, diabetic fatty Zucker rats, and diabetic rhesus monkeys (Young et al., 1999).

GLP-1 has been shown to have multiple actions in the brain including control of feeding behavior, sympathetic output, and neuronal survival. Its role in appetite and feeding behavior has received extensive investigation (Hameed et al., 2009; Moran and Dailey, 2009). The mechanisms appear to involve GLP-1 suppression of agouti-related protein (AgRP) and neuropeptide Y (NPY) and stimulation of proopiomelanocortin (POMC) expression in the arcuate nucleus of the hypothalamus. AgRP and NPY are the most powerful peptides known to be involved in the stimulation of feeding behavior. The increased gene expression of NPY and AgRP mRNA induced by 48-hour fasting are significantly attenuated by the intracerebroventricular (icv) injection of GLP-1 in rats (Seo et al., 2008). In addition, GLP-1 increases the spontaneous action-potential firing of POMC neurons (Ma et al., 2007). Excitation of POMC cells is well established to reduce food intake (Lee and Wardlaw, 2007) and activate the sympathetic nervous system (Usui et al., 2009). This later action of GLP-1 is likely to result in increased energy expenditure (Osaka et al., 2005; Pannacciulli et al., 2006). These multiple actions of GLP-1 on energy intake and energy expenditure make it a logical candidate for weight control.

In addition to having a role in control of diabetes and weight management, exciting new evidence for a role of GLP-1 in healthy aging has been found. For example, activation of GLP-1 receptors provides beneficial effects on deterioration of glucose homeostasis and insulin secretion in aging mice (Irwin et al., 2007) and improves bone resorption through a calcitonin-dependent pathway (Yamada et al., 2008). Furthermore, GLP-1 provides neuroprotection in an animal model of sensory neuropathy (Perry et al., 2007) and GLP-1 receptor stimulation preserves primary cortical and dopamineric neurons in cellular and rodent models of stroke and Parkinsonism (Li et al., 2009). Similarly, enhancing central nervous system endogenous GLP-1 receptor pathways has been suggested for intervention in Alzheimer's disease (Perry and Greig, 2005). Further study of the role of GLP-1 in healthy aging is critically needed.

PYY is another gut peptide that may play a role in modulating energy balance and adiposity through control of food intake and nutrient partitioning. In plasma, PYY appears as full-length PYY1–36 and truncated PYY3–36. PYY3–36 seems to inhibit both food intake and gut motility more effectively than PYY 1–36 (Witte et al., 2009). PYY acts through the G-protein coupled receptors Y1, Y2, Y4, Y5, and Y6. Of these receptors, Y2 seems to be important for the control of feeding behavior. Mice with a targeted deletion of the Y2-receptor are resistant to the anorectic effects of PYY3–36 (Batterham et al., 2002). Administration of PYY increases fat oxidation (Adams et al., 2006) and transgenic mice that overexpress PYY have increased basal temperature indicative of increased thermogenesis (Boey et al., 2008). The mechanism by which PYY alters energy partitioning and fat oxidation is still undetermined.

PYY is also important in the control of glucose homeostasis. This control appears to be achieved though the effects of PYY in regulating insulin output and insulin sensitivity (Boey et al., 2007). PYY stimulates beta cell insulin secretion and increases insulin sensitivity in peripheral tissues.

Fermentation to SCFA, Production of Lower Gut Hormones, and Physiological Effects

The microflora in the large intestine feed upon the fermentable fibers. The major products of fermentation are the SCFA. Acetate, propionate, and butyrate are the most abundant products of fermentation in the large intestine. Butyrate has been a major focus of interest for health of the gut. It is a preferred source of energy for colonocytes (Bergman, 1990) and through its role as a histone deacetylase inhibitor (Gao et al., 2009), it may actually promote apoptosis of precancerous cells (Le Leu et al., 2005). A lot of the research with fermentable fibers has focused on butyrate production and the bacteria that produce butyrate. Figure 4.1 shows bacteria that break down RS and fructans and their products, and then the bacteria that produce butyrate.

Fermentation of prebiotics such as RS in the large intestine, and production of butyrate is a complex process. Production of butyrate from RS requires the involvement of several groups of bacteria. The study of the fermentation of RS has demonstrated initial bacterial adherence to starch molecules especially by *Bacteroides thetaiotaomicron* in *Bacteroides* spp., *Bifidobacterium longum* in *Bifidobacterium* spp. and some *Lactobacillus* spp. (Bird et al., 2000; Xu et al., 2003; Louis et al., 2007). However, *B. thetaiotaomicron*, *Lactobacillus* spp., or *Bifidobacterium* spp. do not produce butyrate. The main acidic fermentation products from *B. thetaiotaomicron* are acetate, propionate, and succinate. Lactate and acetate are produced by *Bifidobacterium* spp. and *Lactobacillus* spp. Thus, these bacterial species adhere to the surface of starch molecules and ferment RS into intermediate products that are converted by other species to butyrate (Duncan et al., 2002; Duncan et al., 2004a, 2004b; Louis et al., 2007).

The butyrate-producing bacteria fall under the Clostridium clusters I, III, IV, XI, XIVa, XV, and XVI. The majority of the bacterial species that are butyrate producers are included in two *Clostridium* clusters—*Clostridium* cluster IV and *Clostridium* cluster XIV. Duncan and his collaborators initially discovered *Coprococcus* spp. and *Roseburia* spp. classified in *Clostridium* cluster XIV and the species *Faecalibacterium prausnitzii* that is classified in *Clostridium* cluster IV. These species are associated with production of butyric acid from acetate. However, these bacteria do not utilize lactate for production of butyrate (Duncan et al., 2002; Cani and Delzenne, 2007). Later, this same research group discovered lactate utilizing bacteria from human feces that produce butyrate as their main fermentation end product. *Eubacterium hallii* and *Anaerostipes caccae* from *Clostridium* cluster XIV were able to utilize both the D and L isomers of lactic acid while *Clostridium indolis* from *Clostridium* cluster XIVa was only able to use D lactic acid (Duncan et al., 2004b). *Eubacterium limosum* from *Clostridium* cluster XIVb also was able to convert lactate into acetate and butyrate in the presence of *B. longum* (Sato et al., 2008). These *Clostridium* clusters are based on more recent genotypic methods of taxonomy (16S rRNA gene sequence) rather than on the older process based on phenotypic characteristics (Clarridge, 2004).

The process for fermentation of fructans is similar with most of the research with fructans focused on Bifido bacteria and Lactobacilli, but other groups of bacteria are involved in fermentation of fructans to SCFA including butyrate (Louis et al., 2007). Several species in Clostridial cluster XIVa found in human fecal samples, increase with in vitro culture with fructans (Manderson et al., 2005).

It appears that fermentation of RS and fructans is required for production of lower gut hormones, PYY and GLP-1. The fermentation to SCFA is associated with production of the hormones. Primary culture of cells from the cecum of rats revealed that addition of SCFA to the culture medium increased production of PYY or GLP-1 gene expression (Zhou et al., 2006). Butyrate was the most effective compared to acetate and propionate. Cannulation studies in rodents, pigs, and humans give somewhat mixed results with infusion of SCFA into the ileum or cecum and production of lower gut hormones. Cuche et al. (2000) surgically produced innervated and denervated ileal loops and demonstrated increased plasma PYY, but not GLP-1 levels in pigs infused with a 0.5 M solution of 60% acetate, 30% propionate, and 10% butyrate at a rate of 2.4 mL/minute for 1 hour. Plasma PYY was increased in rats that had 2 mmol/hour of a mix of 65% acetate, 20% propionate, and 15% butyrate infused into the colon over 2 hours (Cherbut et al., 1998).

Infusion of 135 mmol of a mix of SCFA (70% acetic, 20% propionate, and 10% butyrate) into the colon of human subjects over 90 minutes at 3 mL/minute resulted in no difference in plasma PYY or oxyntomodulin compared to infusions of lactose, hypertonic saline, or isotonic saline (Piche et al., 2000). Kien et al. (2007) observed no differences in plasma GLP-2 among four groups of piglets with cecal cannulas: (1) control infusion of phosphate-buffered saline; (2) butyrate infusion (1 mL/hour of 12. 7 μmol/minute = 2.3μmol/kg/minute); (3) control infusion, but fed inulin (3 g/L of sow's milk replacer); and (4) butyrate infusion and fed inulin. Longo et al. (1991) observed a stepwise release of PYY in rabbits infused with graded concentrations of butyrate in the colon. Additionally, the SCFA receptor GPR43 was shown to be expressed in the L enteroendocrine cells of the lower gut (Karaki et al., 2006). This indicates that SCFA may play a role in some function of these endocrine cells. Thus, the results from cell culture combined with the cannulation studies are inconclusive about the role of SCFA and production of lower gut hormones. Infusion appears to favor production of PYY, but not the peptides from the proglucagon peptide. Other fatty acids, such as formate, succinate, or lactate (Louis et al., 2007), other products of fermentation, or the presence of various bacterial populations may be necessary for production of lower gut hormones. Alternatively, the infusion studies are short-term and there may be an adaptation period of a day or more with consumption of RS for production of GLP-1.

Increased production of lower gut hormones with increased fermentable fibers in the diet has been associated with several healthful physiological effects. In rodents, these include reduced body fat (Keenan et al., 2006; Shen et al., 2009; Zhou et al., 2009), increased fat oxidation (Zhou et al., 2009), reduced fat synthesis (Delzenne et al., 2007), improved glucose control (Cani et al., 2005; Delzenne et al., 2007; Zhou et al., 2008; Aziz et al., 2009), and reduced body weight gain (Cani et al., 2005; Delzenne et al., 2007; Aziz et al., 2009). In another rodent study, consumption of RS decreased postprandial lipogenesis, but lower gut hormones were not measured in this study (Higgins et al., 2006).

Human studies have also demonstrated healthful effects with feeding of fermentable fibers. In one study, RS consumption increased fat oxidation (Higgins et al., 2004). Robertson et al. (2005) demonstrated that insulin sensitivity was higher and forearm muscle glucose clearance was greater for subjects given 30 g/day of a RS supplementation.

The measurements of Robertson et al. (2005) were not associated with increased plasma GLP-1, but were associated with increased plasma acetate and propionate levels. However, PYY and GLP-1 were elevated in a study by Cani et al. (2009) in which they supplemented subjects with 16 g/day of a fructan and demonstrated improved glucose control and increased satiety. The reason for the differences in observance of increased production of lower gut hormones is not clear. Timing of blood sampling nearer to the fermentable fiber reaching the large intestine may be a factor or there may be differences inherent in the type of fermentable fiber: fructan versus RS. The difference between the human study by Robertson et al. (2005) and many rodent studies is the level of RS used. In many rodent studies, the level of RS approaches 30% of the diet (w/w) (Keenan et al., 2006; Zhou et al., 2008; Aziz et al., 2009; Shen et al., 2009; Zhou et al., 2009). A result from one study, and presumably the others, showed consistently elevated plasma levels of PYY and GLP-1 over a 24-hour period (Zhou et al., 2008). In rodent studies, the diet is consistent with no variety and always contains RS. Thus, fermentation is fairly constant over the 24-hour period. In support of this, rodents killed in the fasted state demonstrated less evidence of fermentation than rodents killed in the nonfasted state (Keenan et al., unpublished data). The issue of fasting or amount of food intake before the blood draw may be important for the differences in lower gut hormone production between the Robertson et al. (2005) and Cani et al. (2009) studies. Robertson et al. (2005) reported no effect of RS supplementation on GLP-1 concentrations, but did not show the data. From the report, it is not clear if the blood draw for GLP-1 analysis was after a 12-hour overnight fast prior to the euglycemic-hyperinsulinemic clamp or on a different occasion after a fast or after consumption of a 500 kcal meal (60 g carbohydrate and 21 g fat) for the meal tolerance test. However, Cani et al. (2009), after having their subjects fast overnight, fed their subjects a free-choice buffet breakfast that had to be consumed within 15 minutes. The blood draws were before and 10, 30, 60, and 120 minutes after consumption of the breakfast. Apparently in agreement with the Robertson et al. (2005) study, the values for GLP-1 for fructan versus control subjects in the fasted state were not different. At 10 minutes postprandially, GLP-1, relative to baseline, was significantly greater for fructan versus control subjects and there was a significant time × treatment effect with repeated measures analysis.

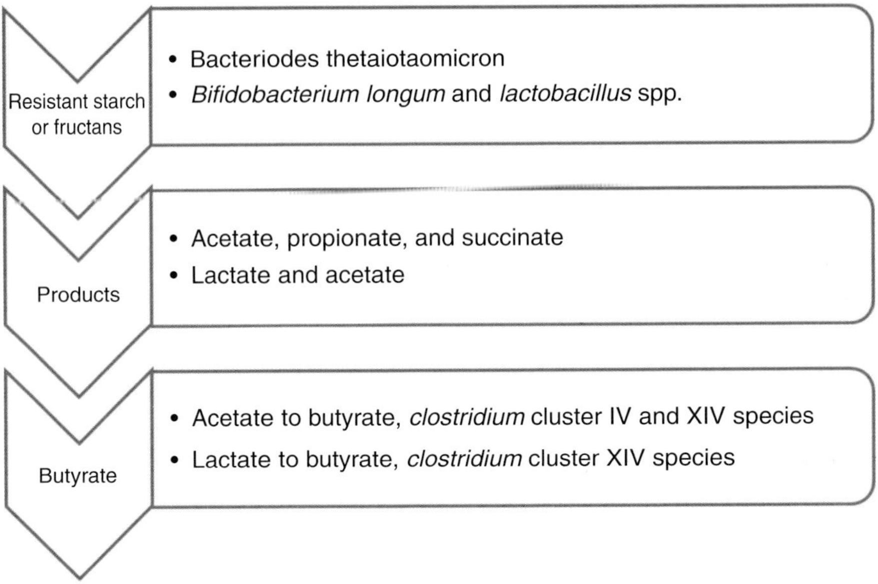

Figure 4.2. Fermentation of resistant starch and fructans to butyrate.

Thus, a lack of increased production of lower gut hormones with feeding of RS in the human study by Robertson et al. (2005) may be due to a fasting blood sample in the experimental design. This is a more plausible argument than a difference in the type of fermentable fiber (Figure 4.2).

Mechanism of Healthful Effects of Lower Gut Hormones

There is evidence that the healthful effects, associated with consumption of fermentable fibers in rodents, are a result of the production of the lower gut hormones. Cani et al. (2006) demonstrated the importance of the GLP-1 receptor for observance of the healthful effects of oligofructose. This fructan in the diet of rodents "improved glucose tolerance, fasting blood glucose, glucose-stimulated insulin secretion, and insulin-sensitive hepatic glucose production and reduced body weight gain." Use of the GLP-1 receptor antagonist exendin 9–39 and use of GLP-1 receptor knockout mice totally prevented the healthful benefits of the dietary fructan. Preliminary data by our research group (Zhou et al., unpublished data) also demonstrated that an intact GLP-1 receptor was required for reduced body fat with feeding of RS. Thus, data support an endocrine mechanism for the healthful effects associated with the feeding of fermentable fibers.

The effects of GLP-1 in promoting improved glucose control are likely a direct effect on the pancreas and promotion of increased β-cell mass (Service et al., 2005). It is reported that GLP-1 increases β-cell mass in rodents by production of new cells (Brubaker and Drucker, 2004) and prevents apoptosis of β-cells in humans (Farilla et al., 2003).

Reduction in fat mass as a result of increased fat oxidation or reduced fat synthesis in peripheral tissues may also be due to a direct effect of the lower gut hormones. Ding et al. (2006) treated primary cultures of hepatocytes with GLP-1. Treatment resulted in an increase in cyclic adenosine monophosphate (cAMP) and reductions in mRNA for genes involved in fatty acid synthesis and an increase in mRNA for genes involved in fatty acid oxidation including peroxisome proliferators-activated receptor α (PPARα). Generation of cAMP is known to lead to activation of gene expression of peroxisome proliferators-activated receptor-gamma coactivator 1 α (PGC1α), which regulates components of adaptive thermogenesis and mitochondrial biogenesis (Puigserver and Spiegelman, 2003; Tritos et al., 2003). However, there may also be an indirect loop involving the brain as increased plasma GLP-1 and PYY with feeding of RS were associated with increased POMC in the arcuate nucleus of the hypothalamus (Shen et al., 2009). Treatment with capsaicin to knock out unmyelinated neurons did not eliminate the effects of RS in reducing body fat and increased production of POMC.

Similarity and Differences to Roux-en-Y Gastric Bypass Surgery

Rodent models fed large amounts of RS have consistently elevated plasma levels of PYY and GLP-1(Zhou et al., 2008). Human patients (Borg et al., 2006; le Roux et al., 2006) and rat models (Suzuki et al., 2005; le Roux et al., 2006) undergoing RYGB also have elevated plasma levels of PYY and GLP-1. Many obese human patients with type 2 diabetes or poor glucose control that undergo RYGB develop healthy glucose control before appreciable weight loss (Rubino, 2008). Two hypotheses are currently proposed for this health improvement with

RYGB: (1) "hypothesis of the distal bowel;" and (2) "hypothesis of the proximal intestine." According to the first hypothesis, more food and nutrients reach the distal small intestine and large intestine because of the bypass of a large portion of the proximal small intestine. The obvious signals for this mechanism are GLP-1 and other hind gut hormones. The argument of the second hypothesis is that subjects with type 2 diabetes have a dysfunctional duodenum that releases a signal that interferes with healthy glucose control. At this time, no candidate signal has been identified (Rubino, 2008).

Unlike human subjects and rodent models undergoing RYGB that generally exhibit reduced food intake (Suzuki et al., 2005; Borg et al., 2006; le Roux et al., 2006), rodent models fed large amounts of RS have equal or greater energy intakes (Keenan et al., 2006; Shen et al., 2009; Zhou et al., 2009). At one time it was believed that reduced ghrelin after RYGB may account for the reduced food intake. Although ghrelin was reported reduced in rats that underwent RYGB (Suzuki et al., 2005), further studies followed by a reanalysis indicate that reduced ghrelin secretion is not a result of RYGB for human patients (le Roux et al., 2006).

Thus, use of a rodent model fed large amounts of RS may serve as a means to study the effects of the "hypothesis of the distal bowel." This model provides consistently elevated levels of hindgut hormones without the surgery of a reduced stomach size and bypass of a large portion of the small intestine. The effects in this model are reduced body fat (Keenan et al., 2006; Shen et al., 2009; Zhou et al., 2009), increased fat oxidation (Zhou et al., 2009), improved glucose control (Zhou et al., 2008; Aziz et al., 2009), and no reduction in energy intake (Keenan et al., 2006; Shen et al., 2009; Zhou et al., 2009) compared to controls with diets with equal energy densities (Tulley et al., 2009). Results using this rodent model indicate that increased plasma levels of PYY and GLP-1 in RYGB patients are important for health benefits of the surgery. The surgical changes of RYGB are likely responsible for an enhancement of the effect of increased hindgut hormones, such as improved glucose control and reduced body fat. Thus, it appears that there is sufficient evidence for the hind gut hypothesis.

Summary

Well-known fermentable fibers, RS and fructans, are fermented to SCFA by several characterized bacterial populations. Consumption of RS and fructans is associated with increased lower gut hormones PYY and GLP-1. The effect of SCFA on production of the hormones is unclear. The healthful effects of dietary RS and fructans in rodents are reduced with GLP-1 antagonists or in GLP-1 receptor knockout mice. Human studies demonstrate beneficial effects of supplemental RS or a fructan, but the study with RS that measured GLP-1 observed no increase compared to control subjects. The study using the fructan observed increased plasma levels of PYY and GLP-1. Differences in lower gut hormone production may be the result of experimental design—fasting versus nonfasting blood drawing. Finally, the rodent studies with feeding of high levels of RS with consistently elevated plasma levels of PYY and GLP-1 may provide a model for testing the two hypotheses for the effects of RYGB.

References

Adams, S.H., Lei, C., Jodka, C.M., et al. (2006) PYY[3–36] administration decreases the respiratory quotient and reduces adiposity in diet-induced obese mice. *J. Nutr.* 136, 195–201.

Augustin, M.A., Sanguansri, P., Htoon A. (2008) Functional performance of a resistant starch ingredient modified using a microfluidiser. *Innov. Food Sci. Emerg. Technol.* 9, 224–231.

Aziz, A.A., Kenney, L.S., Goulet, B., Abdel-Aal el, S. (2009) Dietary starch type affects body weight and glycemic control in freely fed but not energy-restricted obese rats. *J. Nutr.* 139, 1881–1889.

Batterham, R.L., Cowley, M.A., Small, C.J., et al. (2002) Gut hormone PYY(3–36) physiologically inhibits food intake. *Nature* 418, 650–654.

Bergman, E.N. (1990) Energy contributions of volatile fatty acids from the gastrointestinal tract in various species. *Physiol. Rev.* 70, 567–590.

Bird, A.R., Brown, I.L., Topping, D.L. (2000) Starches, resistant starches, the gut microflora and human health. *Curr. Issues Intest. Microbiol.* 1, 25–37.

Boey, D., Lin, S., Enriquez, R.F., et al. (2008) PYY transgenic mice are protected against diet-induced and genetic obesity. *Neuropeptides* 42, 19–30.

Boey, D., Sainsbury, A., Herzog H. (2007) The role of peptide YY in regulating glucose homeostasis. *Peptides* 28, 390–395.

Borg, C.M., le Roux, C.W., Ghatei, M.A., Bloom, S.R., Patel, A.G., Aylwin, S.J. (2006) Progressive rise in gut hormone levels after Roux-en-Y gastric bypass suggests gut adaptation and explains altered satiety. *Br. J. Surg.* 93, 210–215.

Brubaker, P.L., Drucker, D.J. (2004) Minireview: glucagon-like peptides regulate cell proliferation and apoptosis in the pancreas, gut, and central nervous system. *Endocrinology* 145, 2653–2659.

Cani, P.D., Delzenne, N.M. (2007) Gut microflora as a target for energy and metabolic homeostasis. *Curr. Opin. Clin. Nutr. Metab. Care* 10, 729–734.

Cani, P.D., Knauf, C., Iglesias, M.A., Drucker, D.J., Delzenne, N.M., Burcelin R. (2006) Improvement of glucose tolerance and hepatic insulin sensitivity by oligofructose requires a functional glucagon-like peptide 1 receptor. *Diabetes* 55, 1484–1490.

Cani, P.D., Lecourt, E., Dewulf, E.M., et al. (2009) Gut microbiota fermentation of prebiotics increases satietogenic and incretin gut peptide production with consequences for appetite sensation and glucose response after a meal. *Am. J. Clin. Nutr.* 90, 1236–1243.

Cani, P.D., Neyrinck, A.M., Maton, N., Delzenne, N.M. (2005) Oligofructose promotes satiety in rats fed a high-fat diet: involvement of glucagon-like Peptide-1. *Obes. Res.* 13, 1000–1007.

Cherbut, C., Ferrier, L., Roze, C., et al. (1998) Short-chain fatty acids modify colonic motility through nerves and polypeptide YY release in the rat. *Am. J. Physiol.* 275, G1415–G1422.

Clarridge, J.E., 3rd. (2004) Impact of 16S rRNA gene sequence analysis for identification of bacteria on clinical microbiology and infectious diseases. *Clin. Microbiol. Rev.* 17, 840–862, table of contents.

Cuche, G., Cuber, J.C., Malbert, C.H. (2000) Ileal short-chain fatty acids inhibit gastric motility by a humoral pathway. *Am. J. Physiol. Gastrointest. Liver Physiol.* 279, G925–G930.

Cummings, J.H., Beatty, E.R., Kingman, S.M., Bingham, S.A., Englyst, H.N. (1996) Digestion and physiological properties of resistant starch in the human large bowel. *Br. J. Nutr.* 75, 733–747.

Delzenne, N.M., Cani, P.D., Neyrinck, A.M. (2007) Modulation of glucagon-like peptide 1 and energy metabolism by inulin and oligofructose: experimental data. *J. Nutr.* 137; 2547S–2551S.

Ding, X., Saxena, N.K., Lin, S., Gupta, N.A., Anania, F.A. (2006) Exendin-4, a glucagon-like protein-1 (GLP-1) receptor agonist, reverses hepatic steatosis in ob/ob mice. *Hepatology* 43, 173–181.

Duncan, S.H., Barcenilla, A., Stewart, C.S., Pryde, S.E., Flint, H.J. (2002) Acetate utilization and butyryl coenzyme A (CoA):acetate-CoA transferase in butyrate-producing bacteria from the human large intestine. *Appl. Environ. Microbiol.* 68, 5186–5190.

Duncan, S.H., Holtrop, G., Lobley, G.E., Calder, A.G., Stewart, C.S., Flint, H.J. (2004a) Contribution of acetate to butyrate formation by human faecal bacteria. *Br. J. Nutr.* 91, 915–923.

Duncan, S.H., Louis, P., Flint, H.J. (2004b) Lactate-utilizing bacteria, isolated from human feces, that produce butyrate as a major fermentation product. *Appl. Environ. Microbiol.* 70, 5810–5817.

Dunphy, J.L., Taylor, R.G., Fuller, P.J. (1998) Tissue distribution of rat glucagon receptor and GLP-1 receptor gene expression. *Mol. Cell Endocrinol.* 141, 179–186.

Englyst H. (1992) Classification and measurement of nutritionally important starch fractions. *Eur. J. Clin. Nutr.* 46 Suppl 2, S33–S50.

Farilla, L., Bulotta, A., Hirshberg, B., et al. (2003) Glucagon-like peptide 1 inhibits cell apoptosis and improves glucose responsiveness of freshly isolated human islets. *Endocrinology* 144, 5149–5158.

Gao, Z., Yin, J., Zhang, J., et al. (2009) Butyrate improves insulin sensitivity and increases energy expenditure in mice. Diabetes *58,* 1509–1517.

Gotthardt, M., Lalyko, G., van Eerd-Vismale, J., et al. (2006) A new technique for in vivo imaging of specific GLP-1 binding sites: first results in small rodents. *Regul. Pept.* 137, 162–167.

Hameed, S., Dhillo, W.S., Bloom, S.R. (2009) Gut hormones and appetite control. *Oral Dis.* 15, 18–26.

Higgins, J.A., Brown, M.A., Storlien, L.H. (2006) Consumption of resistant starch decreases postprandial lipogenesis in white adipose tissue of the rat. *Nutr. J.* 5, 25.

Higgins, J.A., Higbee, D.R., Donahoo, W.T., Brown, I.L., Bell, M.L., Bessesen, D.H. (2004) Resistant starch consumption promotes lipid oxidation. *Nutr. Metab. (Lond).* 1, 8.

Irwin, N., McClean, P.L., Harriott, P., Flatt, P.R. (2007) Beneficial effects of sub-chronic activation of glucagon-like peptide-1 (GLP-1) receptors on deterioration of glucose homeostasis and insulin secretion in aging mice. *Exp. Gerontol.* 42, 296–300.

Karaki, S., Mitsui, R., Hayashi, H., et al. (2006) Short-chain fatty acid receptor, GPR43, is expressed by enteroendocrine cells and mucosal mast cells in rat intestine. *Cell Tissue Res.* 324, 353–360.

Keenan, M.J., Zhou, J., McCutcheon, K.L., et al. (2006) Effects of resistant starch, a non-digestible fermentable fiber, on reducing body fat. *Obesity (Silver Spring).* 14, 1523–1534.

Kieffer, T.J., Habener, J.F. (1999) The glucagon-like peptides. *Endocr. Rev.* 20, 876–913.

Kien, C.L., Blauwiekel, R., Bunn, J.Y., Jetton, T.L., Frankel, W.L., Holst, J.J. (2007) Cecal infusion of butyrate increases intestinal cell proliferation in piglets. *J. Nutr.* 137, 916–922.

Kreymann, B., Yiangou, Y., Kanse, S., Williams, G., Ghatei, M.A., Bloom, S.R. (1988) Isolation and characterisation of GLP-1 7–36 amide from rat intestine. Elevated levels in diabetic rats. *FEBS Lett.* 242, 167–170.

Le Leu, R.K., Brown, I.L., Hu, Y., et al. (2005) A synbiotic combination of resistant starch and *Bifidobacterium lactis* facilitates apoptotic deletion of carcinogen-damaged cells in rat colon. *J. Nutr.* 135, 996–1001.

le Roux, C.W., Aylwin, S.J., Batterham, R.L., et al. (2006) Gut hormone profiles following bariatric surgery favor an anorectic state, facilitate weight loss, and improve metabolic parameters. *Ann. Surg.* 243, 108–114.

Lee, M., Wardlaw, S.L. (2007) The central melanocortin system and the regulation of energy balance. *Front. Biosci.* 12, 3994–4010.

Li, Y., Perry, T., Kindy, M.S., et al. (2009) GLP-1 receptor stimulation preserves primary cortical and dopaminergic neurons in cellular and rodent models of stroke and Parkinsonism. *Proc. Natl. Acad. Sci. USA.* 106, 1285–1290.

Longo, W.E., Ballantyne, G.H., Savoca, P.E., Adrian, T.E., Bilchik, A.J., Modlin, I.M. (1991) Short-chain fatty acid release of peptide YY in the isolated rabbit distal colon. *Scand. J. Gastroenterol.* 26, 442–448.

Louis, P., Scott, K.P., Duncan, S.H., Flint, H.J. (2007) Understanding the effects of diet on bacterial metabolism in the large intestine. *J. Appl. Microbiol.* 102, 1197–1208.

Lovshin, J.A., Drucker, D.J. (2009) Incretin-based therapies for type 2 diabetes mellitus. *Nat. Rev. Endocrinol.* 5, 262–269.

Ma, X., Bruning, J., Ashcroft, F.M. (2007) Glucagon-like peptide 1 stimulates hypothalamic proopiomelanocortin neurons. *J. Neurosci.* 27, 7125–7129.

Manderson, K., Pinart, M., Tuohy, K.M., et al. (2005) In vitro determination of prebiotic properties of oligosaccharides derived from an orange juice manufacturing by-product stream. *Appl. Environ. Microbiol.* 71, 8383–8389.

McGowan, B.M., Bloom, S.R. (2004) Peptide YY and appetite control. *Curr. Opin. Pharmacol.* 4, 583–588.

Moran, T.H., Dailey, M.J. (2009) Minireview: gut peptides: targets for antiobesity drug development? Endocrinology. 150, 2526–2530.

Morinigo, R., Moize, V., Musri, M., et al. (2006) Glucagon-like peptide-1, peptide, Y.Y., hunger, and satiety after gastric bypass surgery in morbidly obese subjects. *J. Clin. Endocrinol. Metab.* 91, 1735–1740.

Osaka, T., Endo, M., Yamakawa, M., Inoue S. (2005) Energy expenditure by intravenous administration of glucagon-like peptide-1 mediated by the lower brainstem and sympathoadrenal system. *Peptides* 26, 1623–1631.

Panel on the Definition of Dietary Fiber. (2001) *Dietary Reference Intakes Proposed Definition of Dietary Fiber*. National Academy Press, Washington, DC.

Panel on Macronutrients. (2002/2005) *Dietary Reference Intakes for Energy, Carbohydrates, Fiber, Fat, Protein, and Amino Acids (Macronutrients)*. National Academy Press, Washington, DC.

Pannacciulli, N., Bunt, J.C., Koska, J., Bogardus, C., Krakoff J. (2006) Higher fasting plasma concentrations of glucagon-like peptide 1 are associated with higher resting energy expenditure and fat oxidation rates in humans. *Am. J. Clin. Nutr.* 84, 556–560.

Perry, T., Greig, N.H. (2005) Enhancing central nervous system endogenous GLP-1 receptor pathways for intervention in Alzheimer's disease. *Curr. Alzheimer Res.* 2, 377–385.

Perry, T., Holloway, H.W., Weerasuriya, A., et al. (2007) Evidence of GLP-1-mediated neuroprotection in an animal model of pyridoxine-induced peripheral sensory neuropathy. *Exp. Neurol.* 203, 293–301.

Piche, T., Zerbib, F., Varannes, S.B., et al. (2000) Modulation by colonic fermentation of LES function in humans. *Am. J. Physiol. Gastrointest. Liver Physiol.* 278, G578–G584.

Puigserver, P., Spiegelman, B.M. (2003) Peroxisome proliferator-activated receptor-gamma coactivator 1 alpha (PGC-1 alpha): transcriptional coactivator and metabolic regulator. *Endocr. Rev.* 24, 78–90.

Reimer, R.A., McBurney, M.I. (1996) Dietary fiber modulates intestinal proglucagon messenger ribonucleic acid and postprandial secretion of glucagon-like peptide-1 and insulin in rats. *Endocrinology* 137, 3948–3956.

Reimer, R.A., Thomson, A.B., Rajotte, R.V., Basu, T.K., Ooraikul, B., McBurney, M.I. (1997) A physiological level of rhubarb fiber increases proglucagon gene expression and modulates intestinal glucose uptake in rats. *J. Nutr.* 127, 1923–1928.

Robertson, M.D., Bickerton, A.S., Dennis, A.L., Vidal, H., Frayn, K.N. (2005) Insulin-sensitizing effects of dietary resistant starch and effects on skeletal muscle and adipose tissue metabolism. *Am. J. Clin. Nutr.* 82, 559–567.

Rubino F. (2008) Is type 2 diabetes an operable intestinal disease? A provocative yet reasonable hypothesis. *Diabetes Care.* 31 Suppl 2, S290–S296.

Sato, T., Matsumoto, K., Okumura, T., et al. (2008) Isolation of lactate-utilizing butyrate-producing bacteria from human feces and in vivo administration of *Anaerostipes caccae* strain L2 and galacto-oligosaccharides in a rat model. *FEMS Microbiol. Ecol.* 66, 528–536.

Schneeman, B.O. (1999) Fiber, inulin and oligofructose: similarities and differences. *J. Nutr.* 129, 1424S–1427S.

Seo, S., Ju, S., Chung, H., Lee, D., Park S. (2008) Acute effects of glucagon-like peptide-1 on hypothalamic neuropeptide and AMP activated kinase expression in fasted rats. *Endocr. J.* 55, 867–874.

Service, G.J., Thompson, G.B., Service, F.J., Andrews, J.C., Collazo-Clavell, M.L., Lloyd, R.V. (2005) Hyperinsulinemic hypoglycemia with nesidioblastosis after gastric-bypass surgery. *N. Engl. J. Med.* 353, 249–254.

Shen, L., Keenan, M.J., Martin, R.J., et al. (2009) Dietary resistant starch increases hypothalamic POMC expression in rats. *Obesity (Silver Spring).* 17, 40–45.

Suzuki, S., Ramos, E.J., Goncalves, C.G., Chen, C., Meguid, M.M. (2005) Changes in GI hormones and their effect on gastric emptying and transit times after Roux-en-Y gastric bypass in rat model. *Surgery.* 138, 283–290.

Tritos, N.A., Mastaitis, J.W., Kokkotou, E.G., Puigserver, P., Spiegelman, B.M., Maratos-Flier E. (2003) Characterization of the peroxisome proliferator activated receptor coactivator 1 alpha (PGC 1alpha) expression in the murine brain. *Brain Res.* 961, 255–260.

Tulley, R.T., Appel, M.J., Enos, T.G., et al. (2009) Comparative methodologies for measuring metabolizable energy of various types of resistant high amylose corn starch. *J. Agric. Food Chem.* 57, 8474–8479.

Usui, D., Yamaguchi-Shima, N., Okada, S., Shimizu, T., Wakiguchi, H., Yokotani K. (2009) Selective activation of the sympathetic ganglia by centrally administered corticotropin-releasing factor in rats. *Auton. Neurosci.* 146, 111–114.

Witte, A.B., Gryback, P., Holst, J.J., et al. (2009) Differential effect of PYY1–36 and PYY3–36 on gastric emptying in man. *Regul. Pept.* 158, 57–62.

Xu, J., Bjursell, M.K., Himrod, J., et al. (2003) A genomic view of the human–*Bacteroides thetaiotaomicron* symbiosis. *Science* 299, 2074–2076.

Yamada, C., Yamada, Y., Tsukiyama, K., et al. (2008) The murine glucagon-like peptide-1 receptor is essential for control of bone resorption. *Endocrinology* 149, 574–579.

Young, A.A., Gedulin, B.R., Bhavsar, S., et al. (1999) Glucose-lowering and insulin-sensitizing actions of exendin-4: studies in obese diabetic (ob/ob, db/db) mice, diabetic fatty Zucker rats, and diabetic rhesus monkeys (*Macaca mulatta*). *Diabetes* 48, 1026–1034.

Zhou, J., Hegsted, M., McCutcheon, K.L., et al. (2006) Peptide YY and proglucagon mRNA expression patterns and regulation in the gut. *Obesity (Silver Spring)* 14, 683–689.

Zhou, J., Martin, R.J., Tulley, R.T., et al. (2008) Dietary resistant starch upregulates total GLP-1 and PYY in a sustained day-long manner through fermentation in rodents. *Am. J. Physiol. Endocrinol. Metab.* 295, E1160–E1166.

Zhou, J., Martin, R.J., Tulley, R.T., et al. (2009) Failure to ferment dietary resistant starch in specific mouse models of obesity results in no body fat loss. *J. Agric. Food Chem.* 57, 8844–8851.

Chapter 5

Animal, In Vitro, and Cell Culture Models to Study the Role of Dietary Fibers in the Gastrointestinal Tract of Humans

Trevor A. Faber and George C. Fahey, Jr.

Introduction

The beneficial physiological roles of plant fiber in the human diet are presently undisputed. Dietary fibers are proven to affect laxation and attenuate blood cholesterol and glucose concentrations (Dietary Reference Intakes, 2006). In addition, dietary fibers play a preventative role in the causation of cardiovascular disease (King, 2005) and, perhaps, colon cancer (Young et al., 2005) and inflammatory bowel disease incidences (Galvez et al., 2005). Many novel dietary fiber sources and analogous substrates are being discovered that could potentially benefit human health. Human subjects are the gold standard in the evaluation of such substrates; however, human experiments have negative aspects including expense, ethical considerations, safety concerns, and dietary compliance issues. Evaluating substrates using either animal, in vitro, and (or) cell culture models results in less expense and is safer in determining efficacy of fiber substrate inclusion in human diets or supplements.

Studies using animal models provide insight into how the body might be affected by the inclusion of fermentable carbohydrate in the diet. The host–microbiota–substrate interaction is a key determinant of the efficacy of a fermentable ingredient. This interaction essentially controls how the substrate might benefit the host. Dietary fiber influences

Nondigestible Carbohydrates and Digestive Health. Edited by T. M. Paeschke and W. R. Aimutis

digestion by altering microbial enzyme activities and gut anatomical characteristics (Boisen et al., 1985; Eggum, 1991; Jørgensen et al., 1996).

Animal evaluations of test substrates offer a cost-effective method to prescreen the efficacy of a substrate before conducting a human trial. The rodent is a common animal model due to its small size, moderate initial and maintenance costs, and relatively simple housing requirements. Pigs and dogs also serve as important models due to their relative size and digestive similarities to the human. In addition, the ileal-cannulated pig and dog provide means of evaluating both hydrolytic and fermentative digestion processes independently. Animal models demonstrate adverse effects such as diarrhea, vomiting, and other signs of gastrointestinal upset that may be due to the test substrate.

In vitro models are inexpensive, rapid methods to predict fiber digestion in the human gastrointestinal tract. Several types of in vitro models exist including batch cultures, chemostat simulators, and computer-controlled systems. These systems range widely in complexity and capability. Some systems are capable of simulating either hydrolytic or fermentative digestion, while others can simulate both processes.

Cell culture models allow evaluation of nutrient absorption and interaction at the cellular level. Cell lines can mimic the physiological responses of epithelial cells in digestive organs. These models also may indicate how test substrates affect nutrient transporter expression in epithelial cells, similar to that noted in vivo. In addition, method of absorption may be determined, whether it is active transport or passive diffusion.

Animal Models

Use of animal models allows a level of experimental control by reducing variance due to diet and environment. Animals are fed experimental diets in a fixed quantity and at the same times each day. Dietary control allows accurate digestibility calculations and specific ingredient effects to be determined. Control is difficult with human subjects due to compliance issues and their desire for diet variety. Animals are housed in environmentally controlled facilities in relatively steady-state conditions. Environmental control allows long-term studies to be conducted.

Although use of animal models is a less expensive alternative to human experimentation, animals nevertheless require specialized, costly facilities. Variation in animal handling and feeding protocols results in differences among research studies. Conclusions from studies may vary considerably due to differences in dietary fiber quantity, basal diet composition, study length, and length of diet adaptation periods (Edwards et al., 1992). Certain animal models simulate the human gastrointestinal tract very well. Nevertheless, differences exist and must be considered when interpreting results. In addition, the microbial population in the large bowel dictates the nature of the fermentative process and varies among host species. Animal welfare is a major concern due to opposition to animal use from special interest groups. Animal care and welfare must be evaluated carefully before, during, and after each experiment.

Rodent

The rodent is commonly used in dietary fiber/fermentable substrate studies. Numerous rodents may be housed in one facility, allowing studies to be well replicated. In addition, food consumption is considerably less than for other animal models, which is important if test substrate availability is limited. Rodent models are used to evaluate substrate–gene interactions and fermentative effects through intestinal tissue collection and genetic knock-out models. Genetic knockout animals are bred to express desired disease states (i.e., colon cancer, colitis) to determine dietary fiber effects on the disease state.

The rodent gastrointestinal tract is proportionally larger than that of the human; however, the rate of passage is shorter due to the need of the rodent to meet their high energy requirement (Van Soest et al., 1982). The rat stomach has regions of nonsecreting epithelium, which is unlike that of the human. The rat jejunum, making up 75% of the total intestinal tract, is proportionally larger than that of the human, which is only 37% of the total intestinal tract. The rat ileum is proportionally smaller, only 2% of the total intestinal tract, while a comparable value for the human is 37% (Hebel and Stromberg, 1976; Ritschel, 1991). The rat colon is approximately one-third the proportional size of the human colon. This is likely due to the fact that rats are cecal fermenters, while humans are colonic fermenters. Fermentative capacity is the percentage of the digestive tract or particular organ devoted to microbial fermentation.

The cecum contributes 32% of the total 61% of fermentative capacity in the rat whereas the human has no cecal fermentation (Parra, 1978). Rodents practice coprophagy that may alter food intake and digestibility measurements. Coprophagy may be prevented, but with some difficulty.

Intestinal microflora differences occur among host species. Techniques to alter the microflora have been developed with some success. One method is to inoculate the cecum of germ-free rodents with a human fecal suspension to establish a microbial population similar to the human (Hazenberg et al., 1981). This model is better known as the human flora-associated (HFA) model. Djouzi and Andrieux (1997), Silvi et al. (1999), Kleessen et al. (2001), Kleessen et al. (2003), and Tamura and Saitoh (2006) successfully evaluated fermentable carbohydrates using HFA rats. Kleessen et al. (2001) found that bacterial populations similar to those of the human can be maintained in HFA rats. Kleessen et al. (2003) compared normal rats with HFA rats, both fed a fructan diet, and noted greater bacterial concentrations and intestinal villus height in the HFA rats. Use of this model has been limited due to the expense and difficulty of maintaining these animals.

The rat, compared to the pig or dog, is not considered as an effective model for the human. When the rat is compared to the pig, digestibility differences are noted. Digestibility of nonstarch polysaccharides (Goodlad and Mathers, 1991; Canibe et al., 1997; Wisker and Bach Knudsen, 2003) and arabinoxylans (Wisker and Bach Knudsen, 2003) was lower in rats compared to pigs, while digestibility of β-glucans was similar between species (Wisker and Bach Knudsen, 2003). This suggests that anatomical and physiological differences between the rat and pig play a role in determining the apparent digestibility of fermentable carbohydrates.

The rodent is an advantageous model for the human as related to dietary fiber due to its relatively low cost and need for simple facilities compared to larger animal models. However, few direct correlations may be made between the rodent and human because of the anatomical and physiological differences that exist. Intestinal length, cecal fermentation, and coprophagy are the main differences. Nonetheless, the rodent model is commonly used and accepted.

Pig

The pig is a popular animal model in dietary fiber research due, in part, to digestive physiology similarities to that of the human. Pigs are omnivorous monogastrics like humans. Ileal-cannulated pigs allow separate determination of hydrolytic (i.e., small intestine) and fermentative (i.e., large bowel) digestion. Hydrolytic digestion effects can be measured due to the ability to access and remove ileal digesta before it passes to the large bowel. Sample quantities that may be removed from pigs are sufficiently large to allow for precise measurement of a wide array of ileal constituents, both exogenous and endogenous.

Despite the similarities in body size and weight, the pig has a larger digestive tract compared to humans. The pig stomach has a greater volume compared to humans (6–8 L compared to 1–1.5 L, respectively) and contains a greater region of nonglandular and cardiac tissue (Dressman and Yamada, 1991). The pig small intestine is three times the length of the human intestine (18.3 m compared with 6.3 m, respectively). The large intestine is also longer than that of the human (4.25 m compared with 1.75 m, respectively; Van Soest et al., 1982). Pigs have a large cecum that contributes 15% of the total 48% of fermentative capacity (Parra, 1978), whereas humans do not exhibit any cecal fermentation. The colon is the main site of fermentation and short-chain fatty acid (SCFA) absorption. The microbial population found in the large bowel is similar to that of the human (Wang et al., 2005). As is the case for rodents, germ-free piglets may be inoculated with a human fecal suspension to establish a human-type microbial population within the large bowel (Pang et al., 2007).

Although the stomach and small intestine of the pig are proportionally larger, the passage rate of liquids and particles is similar in these organs of the pig and human. However, the large intestine has a longer retention time (45 hours compared with 40 hours in man) (Arhan and Devroede, 1981; Gorard et al., 1996; Glistø et al., 1998; Leeuwen and Jansman, 2007). This longer retention time allows for greater fermentation, thus increasing SCFA production.

The diet, size, and similarities in digestive physiology of the pig and the human make the pig an attractive fermentative model for the human. The ileal-cannulated pig is an important model since it enables hydrolytic and fermentative digestion to be evaluated separately. Pigs are not as cost-effective as the rodent as they require large, specialized facilities. In addition, feed and animal care costs are greater when using pigs as compared to small animal models. The anatomical differences, tract length, and increased fermentation capacity are additional disadvantages of the pig.

Dog

The dog is an omnivorous monogastric capable of consuming diets similar in ingredient and chemical composition to that of humans. Certain breeds of dog also have anatomical similarities due to their comparable body size. As with pigs, cannulated dogs are used to measure ileal nutrient digestibility. An advantage of the cannulated dog compared with the cannulated pig is longevity. Most cannulated pigs are functional for approximately 6 months, while cannulated dogs may live their entire lives (12+ years) in a colony. This feature reduces the number of animals needed in research and surgical expense. Similar to pigs, the housing and maintenance of dogs is more expensive than is the case for small animal models. Since dogs are considered companion animals, much emotion is attached to their use and care. Thus, welfare of the dog always is a high priority.

The size and structure of the dog stomach is very similar to that of the human, with both species having a stomach capacity of approximately 1–1.5 liters (Dressman and Yamada, 1991). The cardiac, fundic, and pyloric stomach regions are similar in size to that of the human. The dog small intestine is shorter in length than is the case for humans (4.1 m and 6.3 m, respectively; Stevens, 1977), resulting in shorter transit times (15 minutes–3.5 hours compared to 3–5 hours, respectively; Youngberg et al., 1985). The large intestine of the dog also is shorter (0.7 m compared to 1.7 m) and unsacculated. Despite this, dogs and humans have similar fermentative capacities (14% for dogs as compared to 17% for humans; Parra, 1978) and relatively similar microbial populations in the large intestine (Balish et al., 1977; Davis et al., 1977; Bell et al., 2009). Dog size impacts gastric emptying time, digestive efficiency, and gastrointestinal tract size (Meyer et al., 1993; Weber et al., 2003; Bourreau et al., 2004). This should be considered when using the beagle (8–9 kg) compared to larger breeds such as hounds (15–25 kg).

The dog, like the pig, is an effective model for the human. Similarities such as digestive system, large bowel fermentation capacity, and overall body size allow data to be collected that are relatively similar to that of the human. The ileal-cannulated dog allows differentiation between hydrolytic and fermentative digestion events. The dog has a shorter intestinal tract, thus increasing digesta transit rate. In addition, animal care costs and welfare issues have limited the use of the dog as a common animal model for human fiber utilization research.

In Vitro Systems

In vitro systems can screen test substrates for hydrolytic and/or fermentative digestion capacity and kinetics. Fermentative in vitro systems are rapid methods to determine the microbial digestion of a substrate. In addition, the pure substrate may be evaluated individually or in a diet matrix. In vitro systems also provide a means of evaluating the efficacy of potential prebiotics and probiotics (Gibson and Fuller, 2000).

According to Longland (1991), an adequate gastrointestinal in vitro model should have the following characteristics: physiological quantities of enzymes should be used in sequence; the pH should allow the activation of enzymes and other cofactors (e.g., bile salts and coenzymes); digestive end products should be removed from the system; at each stage, digesta should be adequately mixed; and finally, the transit rate should simulate the representative host animal species. The appropriate system to evaluate a substrate is dependent on the chemical composition of the substrate. If it contains hydrolytic and fermentatively digestible components, a system capable of both types of digestion should be used. However, if evaluating an ingredient that is fermentatively digested only, (e.g., fiber-rich ingredients, resistant starches, oligosaccharides), then a fermentation model alone is appropriate. No matter which system is used, a strict anaerobic environment is required for survival and growth of the gut microbiota used in fermentation experiments.

Hydrolytic and Fermentative Digestion Systems

Batch Systems

The simplest in vitro system to evaluate hydrolytically and fermentatively digestible substrates is the batch system. A batch system consists of a single, self-contained vessel. Samples are first hydrolytically digested using simulated gastric juice and intestinal enzymes such as pepsin, pancreatin, trypsin, and renin. Duodenal or jejunal fluid from a human or animal may be substituted to supply the necessary enzymes. The sample is then allowed to incubate and digest. Following hydrolytic digestion, fermentative digestion may be evaluated by filtering the sample and adding a fecal suspension to act as a microbial inoculum (Metz and Van Der Meer, 1985). Microbial enzymes, such as cellulase, hemicellulase, arabinase, xylanase, and pectinase, may be added as alternatives to the fecal suspension (Boisen and Fernández, 1991). The batch system is a rapid, inexpensive method to evaluate the complete digestive process. This method does not allow addition or removal of nutrients from the vessel. Over time, this causes the microbes to enter the death phase, resulting in a cessation of fermentation.

The three-stage method described above has been validated with in vivo data (Beames et al., 1996; Boisen and Fernández, 1997; Hervera et al., 2009). Boisen and Fernández (1997) evaluated the organic matter digestibility of 31 different feedstuffs using the three-stage method and intact pigs. Authors noted that organic matter digestibility was similar between models with a correlation coefficient of 0.94. The authors state that this in vitro method accurately quantifies dietary fiber, particularly nonstarch polysaccharide, digestion. Beames et al. (1996) evaluated organic matter and energy digestibilities of select barley products and noted high correlations (0.97) between in vitro disappearance and pig in vivo digestibility coefficients. Hervera et al. (2009) evaluated total tract apparent digestibility of crude protein of 28 dry dog foods using beagles and the three-stage in vitro method. The in vitro system significantly ($P < 0.05$) overestimated digestion of crude protein (average 4.35 percentage units) in most samples; nevertheless, a correlation coefficient of 0.81 was noted.

Hydrolytic digestion of starch, specifically resistant starch, may be evaluated by using an in vitro technique developed by Muir and O'Dea (1992). This method involves chewing of a test substrate without swallowing, then removing the substrate and suspending it in a pepsin and acid solution for 30 minutes to simulate gastric digestion. To simulate small intestinal digestion, the solution is neutralized with sodium hydroxide and incubated with a pancreatic enzyme mixture and amyloglucosidase for 6 hours. All remaining starch is classified as resistant starch, which provides a fermentative substrate for microbiota in the large bowel. Åkerberg et al. (1998) measured resistant starch concentrations in select foods and compared data with values obtained in ileostomy studies. A 0.97 correlation coefficient was calculated, indicating excellent agreement between in vitro and in vivo methodologies.

Simulated Human Intestinal Microbial Ecosystem

The simulated human intestinal microbial ecosystem (SHIME) is a chemostatic system that evaluates the complete digestive process from

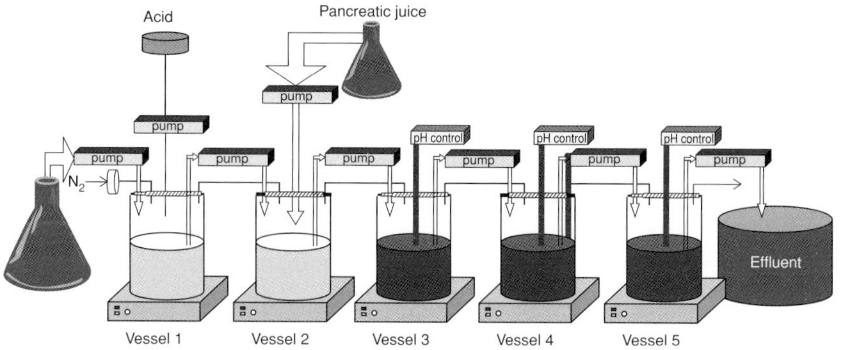

Figure 5.1. Schematic of the SHIME. Courtesy of Dr. Sam Possemiers, Ghent University, Ghent, Belgium.

stomach to colon in one closed system (Molly et al., 1993). This computer-controlled system consists of six vessels simulating the digestive tract: stomach (vessel 1), duodenum and jejunum (vessel 2), ileum (vessel 3), proximal colon (vessel 4), transverse colon (vessel 5), and distal colon (vessel 6). A vessel placed after the distal colon collects the effluent from the system. The contents of vessels 1 and 2 are pumped semicontinuously (every 4 hours) while the contents of vessels 3, 4, and 5 are pumped continuously. The system is kept anaerobic at a temperature of 37°C. In addition, pH of each vessel is computer-controlled to simulate each representative segment of the gastrointestinal tract (Figures 5.1 and 5.2).

Figure 5.2. The simulated human intestinal microbial ecosystem. Courtesy of Dr. Sam Possemiers, Ghent University, Ghent, Belgium.

The first three vessels are inoculated with 10 mL of supernatant from a diet suspension (e.g., 15% protein, 20% fat, and 45% carbohydrate), which serves as a nutrient source for the microbes, for 8 consecutive days, while the last three vessels are inoculated with 50 mL of fecal suspension for 10 consecutive days. After inoculation, 200 mL of fresh SHIME media, containing arabinogalactan, pectin, xylan, and starch (Molly et al., 1993), are added to vessel 1 three times per day. After 4 hours, the contents of vessel 1 are pumped into vessel 2 along with 100 mL of pancreatic juice mixed with bile. After 4 hours, the contents of vessel 2 are pumped into vessel 3. After a 20-hour-residence time, the contents of vessel 3 are pumped into vessel 4. There, the contents have a 32-hour-residence time, after which the contents are pumped to vessel 5. Once in vessel 5, contents have a 24-hour-residence before being pumped out of the system. The total transit time of the media is 84 hours. After 8 days of SHIME-only media being pumped through the system, the media is blended with the test substrate and added to the system for 12 days. After 12 days, SHIME-only media is again pumped through the system for 8–10 days. Use of the different vessels allows predictions of the hydrolytic and fermentative processes occurring in each region of the intestinal tract. However, despite the consistent movement of media through the vessels, this system lacks the ability to remove digested nutrients, all of which enter the simulated colon vessels. This provides additional nutrients (i.e., nutrients that would have been hydrolytically digested in vivo) for the microbiota, altering the microbial populations and, thus, any inferences that can be made about the resulting data.

During development of the SHIME system, Molly et al. (1993) compared four different media solutions and found that the solution containing arbinogalactan, pectin, xylan, and starch substrates closely correlated with in vivo data from published literature sources. Molly et al. (1994) compared data from the SHIME system with that from eight human volunteers and noted similar fermentative patterns. Other studies (Kontula et al., 1998; Van de Wiele et al., 2007; Vanhaecke et al., 2009) have evaluated inulins and probiotics and noted that the SHIME system is capable of detecting prebiotic and probiotic responses. Except for Molly et al. (1994), no experiment has evaluated the relationship between SHIME data and animal responses in the same study. Further validation is necessary to determine the efficacy of this system.

TNO Intestinal Model

The TNO Intestinal Model (TIM-1) is a complex in vitro system that measures hydrolytic digestion responses (Minekus et al., 1995). It was developed at the TNO Nutrition and Food Research Institute in Zeist, the Netherlands. This computer-controlled system is capable of simulating characteristics such as meal size, meal duration, peristalsis, pH, gastric and intestinal secretions, gastrointestinal transit time, and the absorption of water and nutrients. TIM-1 consists of four successive compartments simulating the stomach, duodenum, jejunum, and ileum as is shown in Figure 5.3 at points A, C, E, and G, respectively.

Each compartment consists of a glass tube with an inner flexible wall. Through the use of circulating water outside of the flexible wall,

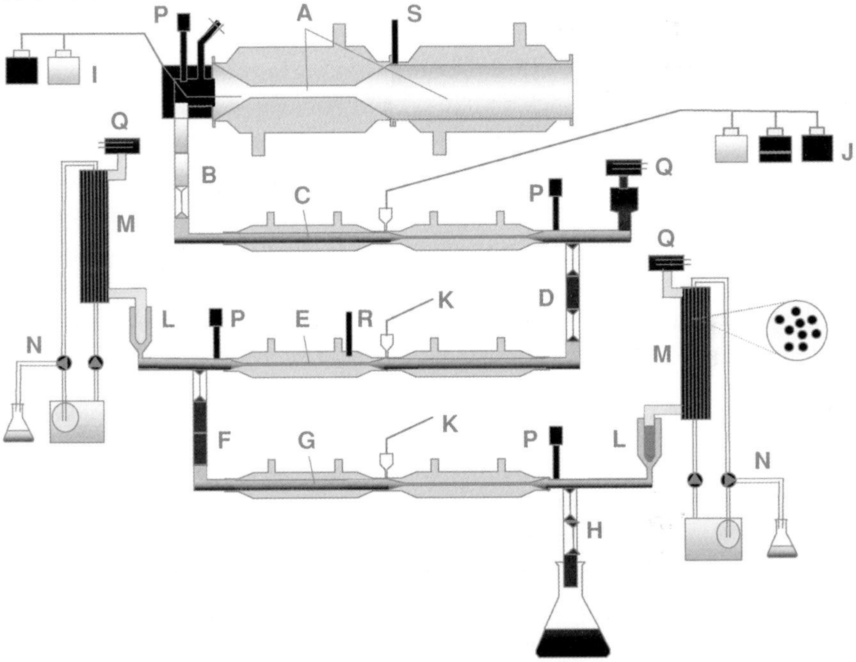

Figure 5.3. TIM-1 system: (a) stomach compartment; (b) pyloric sphincter; (c) duodenum compartment; (d) peristaltic valve; (e) jejunum compartment; (f) peristaltic valve; (g) ileum compartment; (h) ileo-cecal sphincter; (i) stomach secretion; (j) duodenum secretion; (k) jejunum/ileum secretion; (l) prefilter; (m) semipermeable membrane; (n) water absorption; (p) pH electrodes; (q) level sensors; (r) temperature sensor; (s) pressure sensor. Courtesy of Dr. Koen Venema, TNO, Zeist, The Netherlands.

a mixing and peristaltic effect is achieved by pumping water at varying pressures. Digesta transit rate is controlled by controlling water flow. Computer-regulated syringe pumps are used in each compartment to maintain the pH at physiological levels and also to provide gastric juices, enzymes, bile, and pancreatic juices. Hollow fiber membranes in the jejunal and ileal compartments absorb digested nutrients and water from the chyme. The system is dosed by adding substrates to the gastric vessel at fixed intervals. The final effluent may be used in the TIM-2 model, a system that simulates the large bowel (see below).

TIM-1 is designed to mimic physiological characteristics (i.e., gastrointestinal transit rate and gastric and intestinal secretions) based on in vivo data (Minekus et al., 1995); however, few comparisons of digestibility have actually been made to in vivo data. Meunier et al. (2008) compared TIM-1 data with ileal digestibility data from pigs. The TIM-1 model was modified to simulate the physiological conditions of the growing pig intestine and to digest a complete swine diet varying in fiber concentration as opposed to evaluation of a single ingredient. Authors noted a lower digestibility of starch and organic matter (average of 20 and 19 percentage units, respectively) in TIM-1 than for the ileal-cannulated pig. Low-fiber diets had a higher crude protein digestibility (15 percentage units) in the TIM-1 model, while high-fiber diets had a lower in vitro crude protein digestibility (average of 5 percentage units).

Due to the unique characteristics of this model system, normal biological variation found in animals and humans is removed, thus resulting in a less variable dataset than occurs normally. In addition, it provides a method to prescreen ingredients for their ability to alter hydrolytic and enzymatic digestion. The ability of this system to remove digestive end products is unique, providing an important advantage over other in vitro systems. Further validation using in vivo data is required to fully accept TIM-1 as an accurate in vivo intestinal model, but the potential of the system is great.

The functional gastrointestinal dog model (FIDO) is similar to TIM-1 but has been modified to mimic the gastrointestinal tract of the dog. The system setup is similar to TIM-1; however, physiological function is based on data obtained from the dog as opposed to the human. Smeets-Peeters (2000) validated the FIDO model using previously published data obtained from ileal-cannulated dogs at the University of Illinois. First, diets containing one of five carbohydrate sources were tested using the FIDO system. Similar ileal protein digestibility coefficients were obtained between models, except for potato starch, which had a greater protein digestibility in vitro than in vivo. Poultry by-product meal and fresh poultry diets also were evaluated using FIDO. The author noted similar protein digestibility coefficients between diets when evaluated in vitro. However, in the ileal-cannulated dog, fresh poultry was noted to have a significantly higher protein digestibility than poultry by-product meal. This difference was not detected in FIDO. Digestibility coefficient standard deviations were much lower using the FIDO system as compared to the ileal-cannulated dog. The author noted that the FIDO system was an excellent model to rank substrates in terms of digestibility, but not to predict actual digestibility coefficients that would be noted in vivo.

In another study, Smeets-Peeters (2000) evaluated the effect of gelling agents commonly used in canned dog foods on glycemic response in the FIDO model. In previously published in vivo studies, a lowered glycemic response was found among gelling agents; however, Smeets-Peeters (2000) found no differences among gelling agents when evaluated using FIDO. The author noted that possible differences in viscosity among gelling agents could explain the differences between in vitro and in vivo data.

Fermentative In Vitro Systems

Some in vitro systems are capable of simulating the entire gastrointestinal tract, whereas others simulate only large intestinal events. These systems rely on a microbial inoculum to ferment the test substrate. This inoculum is commonly feces from the target species, but pure microbial media, to include probiotic microorganisms, may be substituted as can cecal contents and digesta from any of the three sections of the large bowel.

Batch

The batch system is the simplest, involving a single anaerobic vessel. The vessel contains a nutrient media mixed with the test substrate. The media mixture contains concentrations of nutrients and compounds (e.g., bile salts) that are representative of ileal fluid. This media solution may be replaced by actual intestinal fluid. The vessel then is inoculated with a fecal suspension containing microbiota and incubated at 37°C for 2–24 hours. The fermentability of a substrate is quantified by measurement of disappearance of substrate dry matter, pH change, microbial differences, and production of SCFA and gases.

Batch systems have been used in numerous studies (Wang and Gibson, 1993; Barry et al., 1995; Tzortzis et al., 2004) to evaluate the fermentability of a wide variety of substrates. Batch culture system data have been compared with in vivo data. Flickinger et al. (2000) used batch culture fermentation to evaluate the fermentability of α-glucooligosaccharide and maltodextrin-like oligosaccharide. They noted that the oligosaccharides were resistant to hydrolytic digestion, but were fermentable. These results agreed with those collected using the ileal-cannulated dog. Sunvold et al. (1995) evaluated carbohydrates varying in fermentability using the batch culture method and also in the dog model and found good agreement between the two. Daniel et al. (1997) evaluated fermentability of nonstarch polysaccharides in the human and using batch culture fermentation. The correlation coefficient between the two methods was 0.72.

Reading Simulator

A more complex fermentation model is the reading simulator, a three-stage continuous culture device consisting of three vessels mimicking the proximal, transverse, and distal colon (Macfarlane et al., 1998). Each vessel is set at a different pH (5.8, 6.2, and 6.8) to mimic the physiological pH of each segment of the colon. First, each vessel is inoculated with 100 mL of 20% (wt/vol) feces and allowed to incubate overnight. Fresh simulator fluid, containing carbohydrates, vitamins, and minerals, then is continuously pumped into the first vessel. Each vessel is allowed to overflow into the next, and the third vessel overflows into a waste container. The total retention time of the system can be controlled to range between 27 and 67 hours. After 14 days, a steady-state condition is achieved. The test substrate then is mixed with the simulation fluid and pumped into the first vessel. Once a new steady state is achieved, the system is run for 22 days. After 22 days, the test substrate is removed from the simulation fluid and the washout phase begins. The washout period involves pumping substrate-free media through the system for 50 days. During this time, aliquots from each vessel are removed to determine how long substrate-induced changes continue in the absence of the fresh test substrate.

The reading simulator was validated by comparing chemical and microbial data obtained from sudden death victims to data collected from the simulator (Macfarlane et al., 1998). They noted many similarities between datasets. Gibson and Wang (1994) and Tzortzis et al. (2005) further concluded that the reading simulator was capable of demonstrating prebiotic effects when evaluating oligofructose and galactooligosaccharides, respectively. Tzortzis et al. (2005) compared bacterial changes measured in vitro and in vivo using pigs and noted many similarities.

EnteroMix® Colon Simulator

The EnteroMix® Colon Simulator is a semicontinuous device that consists of four vessels connected consecutively (Mäkivuokko et al., 2005). Fresh media is added to vessel one and effluent is collected after passage through the fourth vessel. Each vessel (1–4) represents a different region of the large intestine: cecum + ascending, transverse, descending, and distal colon, respectively. Each vessel is filled with increasing volumes (3, 5, 7, 9 mL) of media and, then, vessel 1 is inoculated with 10 mL of fecal slurry consisting of feces and media. After 2 minutes, fluid from vessel 1 is transferred to vessel 2. The same procedure is repeated for vessels 3 and 4. After each vessel is inoculated with the fecal suspension, the test substrate is mixed with fresh media and pumped into the first vessel. After 3 hours, 3 mL of fluid from vessel 1 is transferred to vessel 2. After 6 hours from initial inoculation, 3 mL of fluid from vessel 2 is transferred to vessel 3, and so forth. After vessel 4 empties into the effluent chamber, vessel 1 receives another 3 mL of test substrate media mix and the process continues for 48 hours, after which the simulation is stopped and samples for analysis are taken from each vessel. Due to the small working volumes in each vessel, it is not possible to pull samples during the run unlike the case for other in vitro systems.

Mäkivuokko et al. (2005) first evaluated this system using polydextrose as a fermentative substrate. They noted similar results to other published data, reporting increases in SCFA concentrations after ingestion of polydextrose. Mäkivuokko et al. (2006) evaluated the effect of lactose in the colon and noted similar results to studies using human colonoscopic matter. The Enteromix model has been used to evaluate the fermentative effects of cocoa, polydextrose, and xylitol in the large bowel (Mäkeläinen et al., 2007; Mäkivuokko et al., 2007). Both studies indicated prebiotic effects and results comparable to previously published in vivo literature.

TNO Intestinal Model

The TNO Intestinal Model (TIM-2) model is a complex, computer-controlled in vitro model designed similarly to TIM-1; however, TIM-2 simulates the proximal colon only (Minekus et al., 1999) as is shown in Figure 5.4.

The system consists of a looped glass vessel with a flexible inner wall. Water is pumped on the outside of this wall to simulate peristaltic movement that affects colonic transit rate. The system is inoculated with fresh fecal inoculum and allowed to incubate for 16 hours. After incubation, the simulation is started by semicontinuously adding ileal medium or simulated ileal fluid from TIM-1. The pH of the system is kept constant at 5.8, the pH of the proximal colon. During fermentation, hollow-fiber membranes connected to a dialysis machine remove

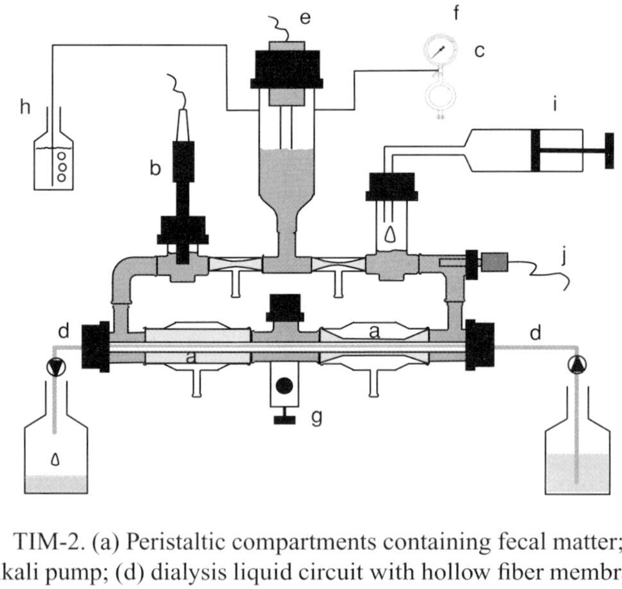

Figure 5.4. TIM-2. (a) Peristaltic compartments containing fecal matter; (b) pH electrode; (c) alkali pump; (d) dialysis liquid circuit with hollow fiber membrane; (e) level sensor; (f) N2 gas inlet; (g) sampling port; (h) gas outlet; (i) "ileal efflux" container; (j) temperature sensor. Courtesy of Dr. Koen Venema, TNO, Zeist, The Netherlands.

fermentative end products, such as SCFA and water, to prevent build-up of metabolites that hinder bacterial growth and survival. The ability of TIM-2 to remove microbial metabolites is a unique advantage compared with other fermentative in vitro systems.

Several studies have used the TIM-2 model to evaluate fermentable substrates. Physiological characteristics of the system were established from data of sudden death victims. Minekus et al. (1999) compared the fermentative end products of pectin, lactulose, lactitol, and fructooligosaccharide with published in vivo literature. They noted similar molar ratios of SCFA and found that microbial population concentrations were within physiological ranges. Three studies (Van der Werf and Venema, 2001; Van Nuenen et al., 2003; Venema et al., 2003) used the TIM-2 model to evaluate the fermentability of lactulose and inulin. Venema et al. (2003) evaluated lactulose and noted that SCFA ratios differed from ratios evaluated in fresh fecal samples obtained from women. The ratio of acetate, propionate, and butyrate obtained in vitro were 79.9, 16.8, and 3.2, respectively, as opposed to 63.8, 20.8, and 15.4, respectively, obtained in vivo. The authors stated that these differences could be due to sampling site. In vivo SCFA measurements were obtained from feces, while the TIM-2 system samples were obtained from the digesta inside the system, where fatty acids are yet to be absorbed. Van der Werf and Venema (2001) evaluated the prebiotic potential of lactulose and noted similar bifidobacteria concentrations to previously published literature using human fecal samples. Van Nuenen et al. (2003) evaluated inulin in the TIM-2 model and noted shifts of microbial species comparable to those in humans as reported in previously published literature.

Method Comparison

Batch culture systems provide rapid, simple, and inexpensive methods to evaluate the hydrolytic and/or fermentative digestion of substrates. They allow substrate disappearance to be determined in addition to gas production, SCFA production, and microbial concentrations. However, they mimic only one segment of the intestine and steady-state conditions are not achieved, limiting their ability to completely mimic in vivo conditions. Chemostats and the TNO systems mimic fermentation in the human large intestine. The movement of fluid from vessel to vessel simulates gastrointestinal tract digesta retention times. Multivessel systems are able to mimic different regions of the intestinal tract. From each vessel, samples can be taken to evaluate the dynamics in each region of the tract. Due to the anaerobic environment and ability to remove fermentative end products, bacterial populations are able to survive in the log phase for extended periods of time. In addition, specific bacterial populations are able to be grown and maintained in these systems. These systems are labor-intensive and expensive to maintain. In addition, a perfect correlation to the human is impossible due to limitations of the systems. Few studies have evaluated test substrates both in vitro and in vivo. Most validation studies compare results with previously published literature as compared to running a simultaneous in vivo experiment. Because there are very few direct comparisons available, in vitro systems are best used as screening tools to reduce the number of substrates that must be evaluated in vivo.

Cell Culture Models

Intestinal cell culture models provide the ability to evaluate test substrates in vitro at the cellular level. Cell culture models are epithelial cells commonly derived from fetal rodent intestine or cancerous tumors. Intestinal cells from fetal rodents retain both morphological and functional characteristics for an extended period of time. This is difficult to accomplish with intestinal cells from adult rodents. Cancer cells express many of the same transporters found in the small intestinal epithelium, and cells are able to differentiate quickly, an advantage compared with other cell lines. Cell cultures are commonly used by the pharmaceutical industry as a means to evaluate the mechanism(s) of absorption of potential drugs. Nutritionists and cell biologists have adapted this model to evaluate cellular transporters of nutrients and the impact of novel compounds on disease states such as inflammatory bowel disease.

Cells grow and differentiate on microporous culture supports found in 6- or 24-well cluster plates. Cell growth commences until a monolayer of cells across the microporous filter develops. The microporous filter divides the plate well into two compartments. The first compartment, the apical compartment, mimics the intestinal lumen and contains the endogenous medium. The second compartment, the basal compartment, mimics the interstitial space and both vascular and lymphatic systems and contains the basolateral medium. This arrangement mimics the structural organization of intestinal mucosa and promotes development of cell polarity and transport of nutrients (Sambruy et al., 2001).

Evaluation of cell continuity across the support structure is important to determine if molecules are able to "leak" across the filter. Transepithelial electrical resistance (TEER) is a technique to measure both the integrity and closure of the tight junctions among cells. This test is performed by placing a pair of Ag/AgCl electrodes into the apical layer and basolateral layer and measuring the electrical resistance of the cells and filter. The TEER value is determined by subtracting the resistance of the filter alone (no cells) from the resistance of the filter with cells. This value then is multiplied by the growth area of the filter. The TEER value is expressed as $\Omega \times cm^2$ (Sambruy et al., 2001). Tightness of cell junctions is based on the TEER value and provides an indication of the passive permeability of the cell line.

Caco-2 Cells

Caco-2 cells are the most commonly used cell line and are derived from human colon adenocarcinoma cells. Despite being cancer cells, Caco-2 cells are capable of producing physiological amounts of hydrolase, esterase, and brush border enzymes to simulate those present in the human small intestine (Bailey et al., 1996); however, Caco-2 cells lack goblet cells that produce mucin (Rubas et al., 1993). Ion transport properties and carrier systems are similar to those of intestinal epithelium (Bailey et al., 1996). Caco-2 cells have a reported TEER value of 500 Ω cm^2, which is different from the reported values for the rat ileum (88 Ω cm^2) and large intestine (150 Ω cm^2) (Duizer et al., 1997; Sutton et al., 1992). Rubas et al. (1993) suggested the Caco-2 cell line may adequately predict drug absorption in the human after comparing permeability of several compounds such as PEG 4000, mannitol, and naproxen. Yee (1997) evaluated numerous drugs and calculated an overall correlation of 0.90 between human absorption and Caco-2 cell permeability. The inference was made that the cell line could be applied to evaluate the major mechanisms of gastrointestinal absorption.

Two studies have evaluated the effect of carbohydrates on cellular expression using Caco-2 cells. Mesonero et al. (1995) evaluated GLUT5 expression after exposing PD7-cloned Caco-2 cells to a medium containing glucose and fructose. Changes in GLUT5 expression were noticed using the cell line. Expression of COX-1, 2, and 3 in Caco-2 cells was evaluated after cells had been exposed to the fermentative end products of polydextrose. Polydextrose had been fermented in the EnteroMix colon simulator. Fermentative end products from the simulator then were added to a media well containing Caco-2 cells. COX-1 and 3 expressions increased, while COX-2 expression decreased after polydextrose fermentation (Mäkivuokko et al., 2005).

Intestinal Epithelial Cell 18 Cells

Intestinal epithelial cell 18 (IEC-18) cells are obtained from the small intestine of the fetal rat. Tight cellular junctions are formed among cells, which closely resemble those noted in the ileum (Duizer et al., 1997). Duizer et al. (1997) suggested this cell line has a higher paracellular permeability and better discrimination potential between hydrophilic compounds than Caco-2 cells. This was confirmed by Versantvoort et al. (2002) who compared the permeability of methylglucose across IEC-18 and Caco-2 cells. Methylglucose permeated IEC-18 cells at a faster rate due to a decreased number of developed tight junctional complexes between cells allowing smaller molecular weight molecules to permeate the cells. The decreased number of tight junctions is indicated by a difference in the TEER value between IEC-18 cells and Caco-2 cells (50 Ω cm^2 and 500 Ω cm^2, respectively) (Versantvoort et al., 2002). No studies could be found comparing the permeability of IEC-18 cells with in vivo absorption data.

Madin Darby Canine Kidney Cells

The Madin Darby Canine Kidney (MDCK) cell line is commonly used to evaluate permeability of nutrients and drugs. MDCK cells are classified as columnar epithelial cells derived from canine kidney (Cho et al., 1989). This difference in species and organ may result in differences in levels of transporter expression than noted in the human intestine. MDCK cells are popular due to a shorter cultivation time compared to Caco-2 cells (3 days compared to 3 weeks, respectively) (Balimane and Chong, 2005). This shorter cultivation time decreases the labor requirement and downtime if cell contamination is noticed. MDCK cells have a reported TEER value of 200 Ω cm^2, similar to that of the rat large intestine (150 Ω cm^2) (Cho et al., 1989). When MDCK cells were first proposed as a model for cellular transport, Cho et al. (1989) evaluated the permeability of different carbohydrates, inulin, sucrose, and dextran. Authors found that the cell line provided the ability to differentiate between paracellular and transcellular vesicular transport. However, further studies are required to determine the correlation between MDCK cell permeability and human nutrient absorption. Species and organ differences should be considered when evaluating results obtained from this cell line.

Numerous cell lines exist representing different digestive organs in the body. Each cell line is unique in rate of differentiation, molecule permeability, and transporter expression. The proper cell line to use is dependent on the organ of interest and the nature of the test substrate.

Cells lines may serve as important models to evaluate the permeability of carbohydrate and fiber constituents. The ability to express nutrient transporters and simulate passive and active transport allows for a better understanding of the mechanism of nutrient transport. Cell culture models rank the permeability of nutrients as opposed to predicting true in vivo absorption. The growth and maintenance of a cell line may be very tedious and expensive. Contamination of cell lines is of the greatest concern. Due to the fact that most cells originate from either nonhuman tissue or cancer cells, physiological differences may exist, leading to an inability to accurately mimic in vivo conditions. In addition, cell lines lack the ability to demonstrate how a test substrate will impact the organ or system as a whole.

Conclusion

As the discovery of new, novel carbohydrates continues, there will be a need to evaluate these compounds for efficacy as dietary fibers or fiber proxies for humans. Animal models provide an effective method to evaluate substrates in vivo. The ability to control the diet and environmental conditions provide a robust method to evaluate the efficacy and potential adverse effects of a substrate. However, results may vary depending on the animal model used. The anatomy and physiology of the animal model must be considered carefully.

In vitro models also provide methods to evaluate substrates. In vitro systems range from very simple to complex and are able to simulate either a part of or the entire digestive tract. In vitro systems are unable to mimic the host–microbial–substrate interaction that would occur with epithelial tissue and immune system components of an animal model. In addition, batch systems do not allow fermentative end products to be removed, which does not allow for steady-state conditions to be achieved long term.

Use of cell culture models indicates how test substrates interact with epithelial cells at the cellular level. Also, permeability of the test substrate through the epithelial layer may predict how the substrate might be absorbed in vivo. Cell culture models do not indicate how the test substrate will affect the organ or tissue as a whole, but only the epithelial cells, thus hindering its effectiveness in simulating in vivo conditions. Cell origin must be considered when evaluating results due to differences in species and physiological responses.

Animal, in vitro, and cell culture models are powerful screening tools for studying metabolism and utilization of dietary fiber and their fermentative end products. The in vitro system may be used first to prescreen test substrates. This is a reliable approach that prevents the unnecessary use of animals if no substrate effect exists. An animal model then may be used to determine if a similar effect can be detected in vivo. However, cost and ethical considerations must be accounted for before conducting an animal experiment. If positive effects are noted, cell culture models then may be used to determine mechanism of action of substrate at the cellular level. In vitro and animal data must be carefully interpreted to best predict how the test substrate might function in the human. No model exists to allow a direct comparison to the human, but animal, in vitro, and cell culture models allow correlations to be made to predict the potential efficacy of test substrates used in human diets or supplements.

References

Åkerberg, A.K.E., Liljeberg, H.G.M., Granfeldt, Y.E., Drews, A.W., Björck, I.M.E. (1998) An *in vitro* method, based on chewing, to predict resistant starch content in foods allows parallel determination of potentially available starch and dietary fiber. *J. Nutr.* 128, 651–660.

Arhan, P., Devroede. (1981) Segmental colonic transit time. *Dis. Colon Rectum*. 24, 625–629.

Bailey, C.A., Bryla, P., Malick, A.W. (1996) The use of intestinal epithelial cell culture model, Caco-2, in pharmaceutical development. *Adv. Drug Del. Rev.* 22, 85–103.

Balimane, P.V., Chong, S. (2005) Cell culture-based models for intestinal permeability: a critique. *Drug Discovery Today* 10, 335–343.

Balish, E., Cleven, D., Brown, J., Yale, C.E. (1977) Nose, throat, and fecal flora of beagle dogs housed in "locked" or "open" environments. *Appl. Environ. Microbiol.* 34, 207–221.

Barry, J-L., Hoebler, C., Macfarlane, G.T., et al. (1995) Estimation of the fermentability of dietary fibre *in vitro:* a European interlaboratory study. *Br. J. Nutr.* 74, 303–322.

Beames, R.M., Helm, J.H., Eggum, B.O., Boisen, S., Bach Knudsen, K.E., Swift, M.L. (1996) A comparison of methods for measuring the nutritive value for pigs of a range of hulled and hulless barley cultivars. *Anim. Feed Sci. Technol.* 62, 189–201.

Bell, J.A., Kopper, J.J., Turnbull, J.A., Barbu, N.I., Murphy, A.J., Mansfield, L.S. (2009) Ecological characterization of the colonic microbiota of normal and diarrheic dogs. *Interdiscip. Perspect. Infect. Dis.* 2008, 1–17.

Boisen, S., Agergaard, N., Rotenberg, S., Kragelund, Z. (1985) Effects of gut flora on intestinal activities of trypsin, chymotrypsin, elastase and amylase in growing rats fed purified diets with cellulose, pectin or sand. *J. Anim. Physiol. Anim. Nutr.* 53, 245–254.

Boisen, S., Fernández, J.A. (1991) In vitro digestibility of energy and amino acids in pig feeds. In: Verstegen, M.W.A., Huisman, J., Hartog, L.A. (editors), *Digestive Physiology in Pigs*, pp. 231–236. PuDoc, Wageningen, The Netherlands.

Boisen, S., Fernández, J.A. (1997) Prediction of the total tract digestibility of energy in feedstuffs and pig diets by in vitro analyses. *Anim. Feed Sci. Technol.* 68, 277–286.

Bourreau, J., Hernot, D., Bailhache, E., et al. (2004) Gastric emptying rate is inversely related to body weight in dog breeds of different sizes. *J. Nutr.* 134, 2039S–2041S.

Canibe, N., Bach Knudsen, K.E., Eggum, B.O. (1997) Digestibility and nitrogen balance in rats given dried or toasted peas (*Pisum sativum*) of different years of harvest. *J. Sci. Food Agric.* 73, 21–33.

Cho, M.J., Thompson, D.P., Cramer, C.T., Vidmar, T.J., Scieszka, J.F. (1989) The Madin Darby Canine Kidney (MDCK) epithelial cell monolayer as a model cellular transport barrier. *Pharm. Res.* 6, 71–77.

Daniel, M., Wisker, E., Rave, G., Feldheim, W. (1997) Fermentation in human subjects of nonstarch polysaccharides in mixed diets, but not in a barley fiber concentrate, could be predicted by in vitro fermentation using human fecal inocula. *J. Nutr.* 127, 1981–1988.

Davis, C.P., Cleven, D., Balish, E., Yale, C.E. (1977) Bacterial association in the gastrointestinal tract of beagle dogs. *Appl. Environ. Microbiol.* 34, 194–206.

Dietary Reference Intakes (2006) Fiber. In: Otten, J.J., Hellwig, J.P., Meyers, L.D. (editors), *Dietary References Intakes: The Essential Guide to Nutrient Requirements*, pp. 111–121. The National Academies Press, Washington, DC.

Djouzi, Z., Andrieux, C. (1997) Compared effects of three oligosaccharides on metabolism of intestinal microflora in rats inoculated with a human faecal flora. *Br. J. Nutr.* 78, 313–324.

Dressman, J.B., Yamada, K. (1991) Animal models for oral drug absorption. In: Welling, P.G., Tse, F.L.S., Dighe, S.V. (editors), *Pharmacological Bioequivalence*, pp. 235–266. Marcel Dekker, New York.

Duizer, E., Penninks, A.H., Steinhuis, W.H., Groten, J.P. (1997) Comparison of permeability characteristics of the human colonic Caco-2 and rat small intestinal IEC-18 cell lines. *J. Control. Release* 49, 39–49.

Edwards, C.A., Adiotomre, J., Eastwood, M.A. (1992) Dietary fibre: the use of in vitro and rat models to predict action on stool output in man. *J. Sci. Food Agric.* 59, 257–260.

Eggum, B.O. (1991) The influence of dietary fibre on protein digestion and utilization. In: Schweizer, T.F., Edwards, C.A. (editors), *Dietary Fibre—a Component of Food—Nutritional Function in Health and Disease*, pp. 181–196. Springer-Verlag, ILSI Human Nutrition Reviews, London, UK.

Flickinger, E.A., Wolf, B.W., Garleb, K.A., et al. (2000) Glucose-based oligosaccharides exhibit different in vitro fermentation patterns and affect in vivo apparent nutrient digestibility and microbial populations in dogs. *J. Nutr.* 130, 1267–1273.

Galvez, J., Rodriguez-Cabezas, M.E., Zarzuelo, A. (2005) Effects of dietary fiber on inflammatory bowel disease. *Mol. Nutr. Food Res.* 49, 601–608.

Gibson, G.R., Fuller, R. (2000) Aspects of in vitro and in vivo research approaches directed toward identifying probiotics and prebiotics for human use. *J. Nutr.* 130, 391S–395S.

Gibson, G.R., Wang, X. (1994) Enrichment of bifidobacteria from human gut contents by oligofructose using continuous culture. *FEMS Microbiol. Lett.* 118, 121–128.

Glistø, L.V., Brunsgaard, G., Højsgaard, S., Sandström, B., Bach Knudsen, K.E. (1998) Intestinal degradation in pigs of rye dietary fibre with different structural characteristics. *Br. J. Nutr.* 80, 457–468.

Goodlad, J.S., Mathers, J.C. (1991) Digestion by pigs of non-starch polysaccharides in wheat and raw peas (*Pisum sativum*) fed in mixed diets. *Br. J. Nutr.* 65, 259–270.

Gorard, D.A., Gomborone, J.E., Libby, G.W., Farthing, M.J.G. (1996) Intestinal transit in anxiety and depression. *Gut* 39, 551–555.

Hazenberg, M.P., Bakker, M., Verschoor-Burggraaf, A. (1981) Effects of the human intestinal microflora on germ-free mice. *J. Appl. Bacteriol.* 50, 95–106.

Hebel, R., Stromberg, M.W. (1976) Digestive system. In: Hebel, R., Stromberg, M.W. (editors), *Anatomy of the Laboratory Rat*, pp. 43–52. Williams and Wilkins, Baltimore, MD.

Hervera, M., Baucells, M.D., González, G., Pérez, E., Castrillo, C. (2009) Prediction of digestible protein content of dry extruded dog foods: comparison of methods. *J. Anim. Physiol. Anim. Nutr.* 93, 366–372.

Jorgensen, H., Zhaoy, X-Q., Eggum, B. (1996) The influence of dietary fibre and environmental temperature on the development of the gastrointestinal tract, digestibility, degree of fermentation in the hind-gut and energy metabolism in pigs. *Br. J. Nutr.* 75, 365–378.

King, D.E. (2005) Dietary fiber, inflammation, and cardiovascular disease. *Mol. Nutr. Food Res.* 49, 594–600.

Kleessen, B., Hartmann, L., Blaut, M. (2001) Oligofructose and long-chain inulin: Influence on the gut microbial ecology of rats associated with a human faecal flora. *Br. J. Nutr.* 86, 291–300.

Kleessen, B., Hartmann, L., Blaut, M. (2003) Fructans in the diet cause alterations of intestinal mucosal architecture, released mucins and mucosa-associated bifidobacteria in gnotobiotic rats. *Br. J. Nutr.* 89, 597–606.

Kontula, P., Jaskari, J., Nollet, L., et al. (1998) The colonization of a simulator of the human intestinal microbial ecosystem by a probiotic strain fed on a fermented oat bran product: Effects on the gastrointestinal microbiota. *Appl. Microbiol. Biotechnol.* 50, 246–252.

Leeuwen, P.V., Jansman, A.J.M. (2007) Effects of dietary water holding capacity and level of fermentable organic matter on digesta passage in various parts of the digestive tract in growing pigs. *Livestock Science* 109, 77–80.

Longland, A.C. (1991) Digestive enzyme activities in pigs and poultry. In: Fuller, M.F. (editor), *In Vitro Digestion for Pigs and Poultry*, pp. 3–18. CAB International, Wallingford, UK.

Macfarlane, G.T., Macfarlane, S., Gibson, G.R. (1998) Validation of a three-stage compound continuous culture system for investigating the effect of retention time on the ecology and metabolism of bacteria in the human colon. *Microb. Ecol.* 35, 180–187.

Mäkeläinen, H.S., Mäkivuokko, H.A., Salminen, S.J., Rautonen, N.E., Ouwehand, A.C. (2007) The effects of polydextrose and xylitol on microbial community and activity in a 4-stage colon simulator. *J. Food Sci.* 72, 153–159.

Mäkivuokko, H., Kettunen, H., Saarinen, M., et al. (2007) The effect of cocoa and polydextrose on bacterial fermentation in gastrointestinal tract simulations. *Biosci. Biotechnol. Biochem.* 71, 1834–1843.

Mäkivuokko, H., Nurmi, J., Nurminen, P., Stowell, J., Rautonen, N. (2005) In vitro effects on polydextrose by colonic bacteria and caco-2 cell cyclooxygenase gene expression. *Nutr. Cancer* 52, 94–104.

Mäkivuokko, H., Saarinen, M.T., Ouwehand, A.C., Rautonen, N.E. (2006) Effects of lactose on colon microbial community structure and function in a four-stage semi-continuous culture system. *Biosci. Biotechnol. Biochem.* 70, 2056–2063.

Mesonero, J., Matosin, M., Cambier, D., Rodriguez-Yoldi, M-J., Brot-Laroache, E. (1995) Sugar-dependent expression of the fructose transporter GLUT5 in Caco-2 cells. *J. Biochem.* 312, 757–762.

Metz, S.H.M., Van Der Meer, J.M. (1985) Nylon bag and in vitro techniques to predict in vivo digestibility of organic matter in feedstuffs for pigs. In: Just, A., Jørgensen, H., Fernández, J.A. (editors), *Digestive Physiology in the Pig*, pp. 373–376. National Institute of Animal Science, Copenhagen.

Meunier, J.P., Manzanilla, E.G., Anguita, M., et al. (2008) Evaluation of a dynamic in vitro model to simulate the porcine ileal digestion of diets differing in carbohydrate composition. *J. Anim. Sci.* 86, 1156–1163.

Meyer, H., Kienzle, E., Zentek, J. (1993) Body size and relative weights of gastrointestinal tract and liver in dogs. *J. Vet. Nutr.* 2, 31–35.

Minekus, M., Marteau, P., Havenaar, R., Huis in't Veld, J.H.J. (1995) A multicompartmental dynamic computer-controlled model simulating the stomach and small intestine. *Alternatives to Laboratory Animals.* 23, 197–209.

Minekus, M., Smeets-Peeters, M., Bernalier, A., et al. (1999) A computer-controlled model to simulate conditions of the large intestine with peristaltic mixing, water absorption and absorption of fermentation products. *Appl. Microb. Biotechnol.* 53, 108–114.

Molly, K., Vande Woestyne, M., De Smet, I., Verstraete, W. (1994) Validation of the Simulator of the Human Intestinal Microbial Ecosystem (SHIME) reactor using microorganism-associated activities. *Microb. Ecol. Health Dis.* 7, 191–200.

Molly, K., Vande Woestyne, M., Verstraete, W. (1993) Development of a 5-step multi-chamber reactor as a simulation of the human intestinal microbial ecosystem. *Appl. Microb. Biotechnol.* 39, 254–258.

Muir, J.G., O'Dea, K. (1992) Measurement of resistant starch: Factors affecting the amount of starch escaping digestion in vitro. *Am. J. Clin. Nutr.* 56, 123–127.

Pang, X., Hua, X., Yang, Q., et al. (2007) Inter-species transplantation of gut microbiota from human to pigs. *Int. Soc. Microb. Ecol. J.* 1, 156–162.

Parra, R. (1978) Comparison of foregut and hindgut fermentation in herbivores. In: Montgomery, G.G. (editor), *The Ecology of Arboreal Folivores*, pp. 205–229. Smithsonian Institution Press, Washington, DC.

Ritschel, W.A. (1991) Targeting in the gastrointestinal tract: new approaches. *Methods Find. Exp. Clin. Pharmacol.* 13, 313–336.

Rubas, W., Jezyk, N., Grass, G.M. (1993) Comparison of the permeability characteristics of a human colonic epithelial (Caco-2) cell line to colon of rabbit, monkey, and dog intestine and human drug absorption. *Pharm. Res.* 10, 113–118.

Sambruy, Y., Ferruzza, S., Ranaldi, G., De Angelis, I. (2001) Intestinal cell culture models – applications in toxicology and pharmacology. *Cell Biol. Toxicol.* 17, 301–317.

Silvi, S., Rumney, C.J., Cresci, A., Rowland, I.R. (1999) Resistant starch modifies gut microflora and microbial metabolism in human flora-associated rats inoculated with faeces from Italian and UK donors. *J. Appl. Microbiol.* 86, 521–530.

Smeets-Peeters, M.J.E. (2000) Feeding FIDO: development, validation and application of a dynamic in vitro model of the gastrointestinal tract of the dog. PhD Thesis Wageningen University. Universal Press, Veenendaal, The Netherlands.

Stevens, C.E. (1977) Comparative physiology of the digestive system. In: Swenson, M.J. (editor), *Dukes Physiology of Domestic Animals*, pp. 216–232. Comstock, Ithaca, NY.

Sunvold, G.D., Fahey, G.C., Jr, Merchen, N.R., et al. (1995) Dietary fiber for dogs: IV. In vitro fermentation of selected fiber sources by dog fecal inoculum and in vivo digestion and metabolism of fiber-supplemented diets. *J. Anim. Sci.* 73, 1099–1109.

Sutton, S.C., Forbes, A.E., Cargill, R., Hochman, J.H., LeCluyse, E.L. (1992) Simultaneous in vitro measurement of intestinal tissue permeability and transepithelial electrical resistance (TEER) using sweetana-grass diffusion cells. *Pharm. Res.* 9, 316–319.

Tamura, M., Saitoh, H. (2006) Comparison of the in vitro metabolism of isoflavones by fecal flora from human flora-associated mice and human. *J. Sci. Food Agric.* 86, 1567–1570.

Tzortzis, G., Goulas, A.K., Baillon, M.-L.A., Gibson, G.R., Rastall, R.A. (2004) In vitro evaluation of the fermentation properties of galactooligosaccharides synthesised by α-galactosidase from *Lactobacillus reuteri*. *Appl. Microbiol. Biotechnol.* 64, 106–111.

Tzortzis, G., Goulas, A.K., Gee, J.M., Gibson, G.R. (2005) A novel galactooligosaccharide mixture increases the bifidobacterial population numbers in a continuous in vitro fermentation system and in the proximal colonic contents of pigs in vivo. *J. Nutr.* 135, 1726–1731.

Van de Wiele, T., Boon, N., Possemiers, S., Jacobs, H., Verstraete, W. (2007) Inulin-type fructans of longer degree of polymerization exert more pronounced in vitro prebiotic effects. *J. Appl. Microbiol.* 102, 452–460.

Van Der Werf, M.J., Venema, K. (2001) Bifidobacteria: genetic modification and the study of their role in the colon. *J. Agric. Food Chem.* 49, 378–383.

Van Soest, P.J., Jeraci, J., Fosse, T., Wrick, K., Ehle, F. (1982) Comparative fermentation of fibre in man and other animals. In: Wallace, G., Bell, L. (editors), *Fibre in Human and Animal Nutrition*, pp. 75–80. The Royal Society of New Zealand, Wellington.

Vanhaecke, L., Grootaert, C., Verstraete, W., Van de Wiele, T. 2009. Chemopreventive effects from prebiotic inulin towards microbial 2-amino-1-methyl-6-phenylimidazo [4,5-b] pyridine bioactivation. *J. Appl. Microbiol.* 106, 474–485.

Van Nuenen, M., Meyer, D., Venema, K. 2003. The effect of various inulins and *Clostridium difficile* on the metabolic activity of the human colonic microbiota in vitro. *Microb. Ecol. Health Dis.* 15, 137–144.

Venema, K., van Nuenen, M.H.M.C., Van Den Heuvel, E.G., Pool, W., Van Der Vossen, J.M.B.M. (2003) The effect of lactulose on the composition of the intestinal microbiota and short-chain fatty acid production in human volunteers and a computer-controlled model of the proximal large intestine. *Microb. Ecol. Health Dis.* 15, 94–105.

Versantvoort, C.H.M., Ondrewater, R.C.A., Duizer, E., Van de Sandt, J.J.M., Gilde, A.J., Groten, J.P. (2002) Monolayers of IEC-18 cells as an in vitro model for screening the passive transcellular and paracellular transport across the intestinal barrier: comparison of active and passive transport with the human colon carcinoma Caco-2 cell line. *Environ. Toxicol. Pharmacol.* 11, 335–344.

Wang, M., Ahrné, S., Jeppsson, B., Molin, G. (2005) Comparison of bacterial diversity along the human intestinal tract by direct cloning and sequencing of 16S rRNA genes. *FEMS Microbiol. Ecol.* 54, 219–231.

Wang, X., Gibson, G.R. (1993) Effects of the in vitro fermentation of oligofructose and inulin by bacteria growing in the human large intestine. *J. Appl. Bacteriol.* 75, 373–380.

Weber, M., Martin, L., Biourge, V., Nguyen, P., Dumon, H. (2003) Influence of age and body size on the digestibility of a dry expanded diet in dogs. *J. Anim. Physiol. Anim. Nutr.* 87, 21–31.

Wisker, E., Bach Knudsen, K.E. (2003) The rat as a model for pigs: comparative values for the digestibility of NSP and other macronutrients. *Br. J. Nutr.*. 90, 373–383.

Yee, S. 1997. In vitro permeability across Caco-2 cells (colonic) can predict in vivo (small intestinal) absorption in man – fact or myth. *Pharm. Res.* 14, 763–766.

Young, G.P., Hu, Y., Le Leu, R.K., Nyskohus, L. (2005) Dietary fiber and colorectal cancer: a model for environment-gene interactions. *Mol. Nutr. Food Res.* 49, 571–584.

Youngberg, C.A., Wlodyga, J., Schmaltz, S., Dressman, J.B. (1985) Radiotelemetric determination of gastrointestinal pH in four healthy beagles. *Am. J. Vet. Res.* 46, 1516–1521.

Chapter 6

Impact of Fiber on Gastrointestinal Microbiota

Koen Venema

Introduction

Extensive studies have been conducted to examine impact of fiber ingestion on microbiota populations in the gastrointestinal (GI) tract. This chapter reviews methods utilized to conduct these studies and results of pivotal work.

Why Study the Composition of the Gastrointestinal Microbiota?

What happens in our intestines? The human GI tract is the primary site of food intake, food conversion, and food uptake. In the small intestine, highly active enzymatic hydrolysis of the major dietary components (carbohydrates, fats, and proteins) takes place and the resultant digestive products are absorbed. Thus, the bulk of our food intake is processed by the small intestine. What then is the function of the second part of the intestine, the large intestine (or colon)? Until recently, the colon was considered just a storage place for undigested food components after the removal of water. However, the past 10–15 years have changed this view drastically. Nowadays, the large intestine is called the "forgotten organ": the cells of microorganisms (totaling approximately 10^{14} cells and most of which are bacteria) present in the colon outnumber the cells of the host by a factor of $\sim$10 and all these bacteria

Nondigestible Carbohydrates and Digestive Health. Edited by T. M. Paeschke and W. R. Aimutis

contribute to food processing by fermentation of undigested/indigestible material. The biochemical (metabolic) potential of this complex assemblage of different microorganisms is considered to equal that of the liver. This community of (mostly anaerobic) bacteria influences human gut physiology and health by exerting a number of activities including fermentation of dietary compounds (including fibers) which escape digestion in the small intestine, processing of mucosal cells shed in the small intestine, and breakdown of intestinally secreted mucus. Thus, polymers of sugars are degraded by the colonic microbes into gasses such as hydrogen, carbon dioxide, and methane as well as short-chain fatty acids (SCFA) (Bergman, 1990; Cummings, 1991), notably acetate, propionate, and butyrate (Figure 6.1). These SCFA are taken up by the host which contributes to its energy and health status. In addition, the microbial community produces a variety of other health-related compounds including vitamins (Hill, 1995) and other growth-promoting compounds.

However, also toxic, mutagenic, and carcinogenic substances may be formed that negatively affect the health of the host (Cummings and Macfarlane, 1997). It is also known nowadays that the microbiota in the

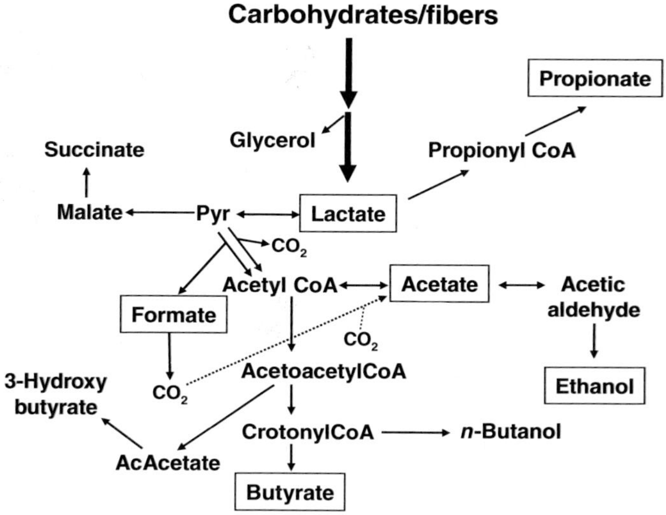

Figure 6.1. Overview of anaerobic bacterial metabolic pathways involved in carbohydrate metabolism. Boxes show the important short-chain fatty acids from formate (C1-acid) up to butyrate (C4-acid), lactate and ethanol.

colon plays a role in the modulation of the immune system (Chadwick and Anderson, 1995), the transformation of bile acids (Aries and Hill, 1970), and the provision of a barrier against pathogenic bacteria (Hill, 1995).

The above-mentioned general facts already indicate that the processes going on inside the colon are important for us. There is indeed evidence that the bacterial metabolic processes in the colon are becoming an increasingly important issue for man. Numerous indications of a correlation between the intestinal health status and the occurrence of various intestinal diseases, such as colon cancer, inflammatory bowel disease (IBD) and irritable bowel syndrome (IBS) (Chadwick and Anderson, 1995) have been reported. Also, disorders, at first, seemingly unrelated to the GI tract such as cirrhosis, autism, and allergy may be affected by the composition and activity of the microbiota. The general hypothesis is that there is a correlation between proteolytic fermentation in the (distal) colon and the occurrence of colon cancer and IBD (Macfarlane et al., 1992a; Levine et al., 1998). Protein fermentation leads to the production of microbial metabolites that can be toxic to the host (Macfarlane et al., 1992b; MacFarlane and MacFarlane 1995). These may especially include sulfur-containing metabolites (Hill, 1995; Rowland, 1995). Carbohydrate fermentation on the other hand leads amongst others to the production of SCFA, which are considered to be health promoting (Cummings, 1991, 1995). During carbohydrate fermentation, protein is incorporated into microbial biomass (Birkett et al., 1996), preventing fermentation of protein. However, most carbohydrates are completely fermented in the proximal colon, which leads to the depletion of these substrates in the distal colon. Here, the microorganisms switch to fermenting protein. It is hypothesized that it would therefore be of importance to prolong fermentation of carbohydrates. This can be accomplished for instance by including more slowly fermentable carbohydrates or fibers in the diet (e.g., certain types of resistant starch). More general, the idea developed that changing the diet in such a way that harmful bacteria (or their harmful activities) are suppressed and beneficial bacteria (or their beneficial activities) are stimulated, may contribute to improving gut health. The exploitation of this has been largely restricted to the use or addition of fiber to the diet. However, the microbial processes occurring inside the colon are hitherto largely unknown, because the accessibility of the lumen of the human colon is severely limited in practice.

Purpose of This Review

The studies discussed in the foregoing all served to demonstrate that there is a strong yet intricate link between intestinal bacterial metabolism and gut health. There is also little doubt that nutrition or nutritional components such as fiber have an impact on the composition and activity of the intestinal microbiota and thereby influences human health and well-being. However, the mechanisms behind all these processes are still largely unknown. If we want to be able to improve human health by optimizing nutrition, we must know what bacteria are stimulated by the different dietary components such as fibers, what are the regulatory mechanisms that govern how our intestinal bacteria react to various food ingredients, and also what will be the determinant processes for the reaction of the host (i.e., our own body) in return. While this may seem a far-away goal, a key step in this research, i.e., the characterization and analysis of the microbiota composition and physiological behavior, has become possible using all the powerful analytical methods available today. Studying the dynamics of an organism's metabolic fluxes in response to various stimuli (such as different nutritional components) provides the unique and key insights to understand physiological regulation. As pointed out above, it is precisely this regulation which is important to uncover, so as to finally enable a targeted modulation of intestinal bacterial composition and more importantly bacterial metabolism to achieve beneficial effects on human health. Therefore, the present review puts a special focus on the application and perspectives of metabolite analysis and its main enabling technology, stable isotope labeling, to study the effect of fibers on microbiota composition and activity.

Methods to Study Bacterial Physiology: Many Fields of Science Come Together

Not surprisingly, recent years have seen the development of specially adapted experimental techniques to be able to study the colon and its inhabitants. These include in vitro model systems, cell culture models, animal models, microdialysis, and breath tests (Table 6.1).

The challenge for the coming years is to establish a truly integrative research approach that will enable the analysis and correlation of

Table 6.1. Experimental techniques to study the activity of the microbiota in the colon.

Experimental Technique	Some Characteristics
In vitro model systems	From simple anaerobic cultivation tubes to full-fledged sophisticated computer-controlled in vitro model systems Enable quantitative studies under defined conditions using an inoculum isolated either from feces or, in an invasive manner, from specific sites along the GI tract The effect of, e.g., prebiotics on the composition and activity of the microbiota has been studied in such a model
Cell culture models	From Caco-2 or HT29 cell lines to intact intestinal mucosal strips Enable to selectively study the properties of different types of (transformed) colon cells in isolation The advantage of mucosal strips over cell cultures is that conditions probably better approach the in vivo situation because the strips include an intact basolateral lining of cells, as, e.g., reflected in increased values for butyrate uptake that differ markedly from isolated colonocytes
(Monogastric) animal models	Especially dogs and pigs, with several methods of sampling: either (1) dissection after sacrificing the animals; (2) use of stomas implanted in the living animal to probe specific luminal sites in the GI tract; or (3) multicatheterization of blood vessels so as to gain access to arterial, portal, and hepatic venous blood all at the same time in the living, conscious animal Upon chemical analysis of blood samples, the techniques described in the line above allow for detailed quantitative studies of net splanchnic absorption and/or intestinal production of metabolites, but such studies generally bear a (strong) invasive character

Table 6.1. (*Continued*)

Experimental Technique	Some Characteristics
Microdialysis	Used safely with low-grade invasiveness in humans with small catheters placed in specific tissue beds of interest Allowing continuous sampling of the interstitial space over prolonged periods of time without taking any biopsies Quantitative measures for metabolic activity are not easily obtained because the exact amount of tissue involved in the dialysis is not known. Results of intraperitoneal microdialyses were shown to strongly depend on catheter position
Breath tests and gas analyses	Enable noninvasive assessments of especially intestinal (carbohydrate) metabolism Information is limited and quantitative aspects are not trivial since measured values represent overall metabolism of the complete organism

Modified from de Graaf and Venema 2008.

the many diverse aspects involved in intestinal microbial metabolism and its interaction with the human host in a meaningful manner. This will require cooperation between scientists from various different scientific disciplines such as microbial physiology, human physiology, and gastroenterology, as well as highly technical disciplines including genomics, proteomics, metabolomics, analytical sciences, and bioinformatics, and possibly nanotechnology.

In Vitro Models of the Human Colon

Because the colon is largely inaccessible, and analysis of fecal samples far from reflects the situation in the human (proximal) colon itself, several in vitro colon models have been developed. The first colon models, developed in the 1980s, were simple fermenters comprising a single vessel with a stirring device and pH-control (Miller and Wolin 1981;

Manning et al., 1987). Models developed later (Macfarlane et al., 1998; De Boever et al., 2000) had several vessels in which the physiological transit through the colon could be better simulated. However, as in the earlier models, microorganisms did not reach physiological densities due to inhibition of microbially produced metabolites that accumulated with time. In contrast, in the TNO in vitro model of the large intestine (TIM-2; Minekus et al., 1999) metabolites and water are constantly removed from the system. Moreover, peristaltic mixing is simulated, and in this system microorganisms reach physiological densities (ca. 10^{10}/mL). In this computer-controlled model, parameters such as transit time and pH are controlled, and age-dependent colon simulations can be run. The system has been frequently used for fiber research (e.g., Venema et al., 2004).

In general, in vitro models allow fast, reproducible experiments under standardized conditions. Frequent sampling at various locations of the model and the possibility to vary several parameters influencing the fermentative processes in the colon (pH, food, transit time, etc.) permit detailed mechanistic insight into these processes, e.g., the fermentation of nondigestible carbohydrates, fibers, and prebiotics. In addition, experiments in models allow complete mass balances. A sound validation of these models is difficult, since this requires comparison with the hardly accessible colon itself. However, a comparison with data obtained from sudden-death victims (Cummings et al., 1987; Macfarlane et al., 1992a) suggests that models can functionally approach the in situ situation in the colon, regarding microbial enzyme activities and production and concentration of SCFAs (Macfarlane et al., 1998; Minekus et al., 1999), especially in the TIM-2 system, where inhibiting metabolites are removed, as it occurs in vivo.

Thus, despite several limitations (e.g., absence of host cells, no spatial gradients in microbial colonization, inoculation with fecal microbiota) in vitro models can be valuable tools to investigate and develop hypotheses about microbe-mediated processes occurring in the colonic lumen during the passage of food. Studies involving the TIM-2 model frequently addressed microbial degradation of dietary relevant, nondigestible carbohydrates, fibers, or prebiotics [recent examples not only include inulin (van Nuenen et al., 2003; van de Wiele et al., 2004), tagatose (Venema et al., 2005), lactulose (Venema et al., 2003) but also mucin-degradation (Macfarlane et al., 2005)]. So far, fermentation processes were mostly interpreted on the basis of ratio and amounts of produced SCFAs and changes in microbial community structure; the use of stable isotopes for nucleic acid-based stable isotope probing (SIP) analyses and metabolic flux analyses (see below) promises a deeper insight into microbial functionality in these model systems.

Stable Isotopes Offer Unique Insights

Perhaps one of the greatest difficulties in research of intestinal metabolism and function is the requirement for adequate minimally invasive experimental techniques that allow for in vivo research in humans. Such techniques will have to cope with severely limited possibilities for manipulation of experimental conditions as well as for material sampling due to ethical constraints. The use of stable isotopes may prove to be a key factor to success here. Stable isotope-labeled molecules follow the same metabolic routes, and function identically in physiological processes as their natural, unlabeled counterparts. The isotopic label however allows for their specific detection at any desired stage after their administration, allowing to indirectly monitor the processes in which they are involved. Over the past 20 years, stable isotope-labeling techniques have proven to be most powerful tools to get quantitative as well as qualitative information about the metabolic processes going on in living organisms in general, including microorganisms, plants, animals, and humans, and also in the colon in vivo (Moran and Jackson, 1990; Wolfe, 1992; de Graaf, 2000; Shulman and Rothman, 2001; Pouteau et al., 2003; Kelleher, 2004; McCabe and Previs, 2004; Dolnikowski et al., 2005; Ratcliffe and Shachar-Hill, 2006). The use of isotopically labeled compounds enables to selectively study that part of metabolism in which the isotopic tracer is involved, offering ample possibilities to probe microbial as well as host metabolism, or both at the same time. Isotopic labeling in compounds can be conveniently and specifically detected by mass spectrometry- and/or nuclear magnetic resonance (NMR)-based analytical techniques.

What Is the Sort of Information You Can Retrieve from Stable Isotope Experiments?

Basically, stable isotopes can convey three sorts of information: (1) how fast a specific metabolic process is running; (2) which are the products of the process; and (3) (in a complex mixture of microbes such as the intestinal microbiota) which players are involved in use of the substrate (Figure 6.2). That is, the speed of incorporation of an isotopic label in a molecule gives information on the synthesis rate of that metabolite, and the position of the label in the molecule gives information on the biosynthetic pathway of the molecule. In special cases, namely, experimental protocols where a steady state is established, it is the degree of isotopic enrichment at different positions in the concerned molecules that gives information on the synthesis rate (or, equivalently, the biosynthetic pathway flux) of the molecule.

In addition, in situ isotope tracking techniques aim to unravel the substrate utilization of microbes in their natural habitat by linking the isotopic signature of biomarker molecules to their inherent phylogenetic information (Manefield et al., 2004). They are useful for a broad and unprejudiced activity-screen in complex communities, but also to verify whether selected groups of microbes utilize a certain substrate or not. In experiments with animal models, in vitro models simulating the human GI tract or human feeding trials, methods using stable isotopes appear preferable over those involving radioactive isotopes for practical and ethical reasons. Since little is known about the actual in situ functions of human colonic microorganisms, SIP techniques appear particularly promising.

SIP approaches, recently reviewed (Dumont and Murrell 2005; de Graaf and Venema, 2008), involve incubation of microorganisms with a stable isotope labeled substrate under conditions resembling the environmental situation (Figure 6.2). Up to date, predominantly ^{13}C-labeled substrates have been used. Only the actively metabolizing microbes incorporate the label into their biomass, including phylogenetic biomarkers such as DNA and RNA. After incubation, identification of active microbes is achieved by molecular analysis of labeled versus unlabeled biomarkers. Here, nucleic acids clearly have the highest phylogenetic resolution and even allow identification of active, but not-yet-cultured species, through use of the "fingerprint" or "signature" in the 16S rDNA gene.

Microbial Functionality: Focus on Colonic Fermentations

The members of the human intestinal microbiota are considered key players in maintaining human health and well-being, because they are

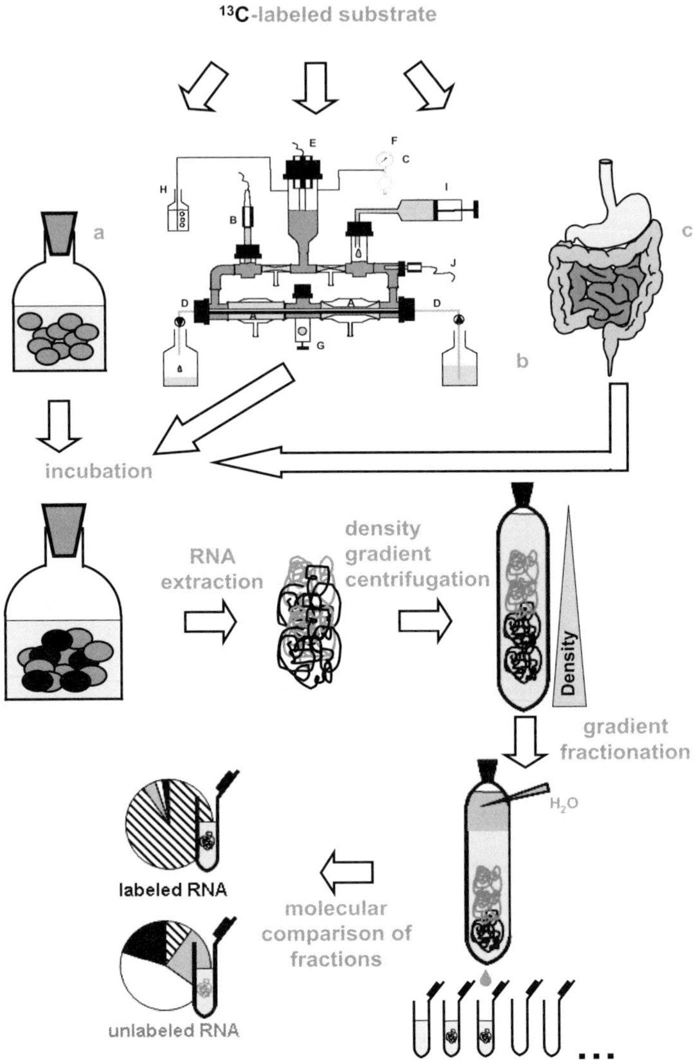

Figure 6.2. Principle of RNA-based stable isotope probing (SIP) for detection and characterization of microbes that actively metabolize the labeled substrate. ^{13}C-labeled substrates are incubated in (a) simple in vitro models (test-tube or flask); (b) sophisticated in vitro systems such as the TIM-2 system; or (c) in vivo (animal or even human individuals). Samples obtained from these experiments [in the figure only exemplified for samples from (a)], are subjected to RNA isolation and density gradient centrifugation. After separation of the gradient in fractions, molecular fingerprinting techniques, such as terminal restriction fragment length polymorphism (T-RFLP) (Kovatcheva-Datchary et al., 2009) can be used to determine the enrichment in the heavier fractions of those microorganisms that specifically fermented the substrate. Adapted from de Graaf and Venema (2008).

implicated in developmental (Xu and Gordon, 2003), immunological (Noverr and Huffnagle, 2004), and nutritional (Bäckhed et al., 2004; Ley et al., 2005) host functions. In turn, distortions in gut microbial community structure and/or functionality are thought to be causative for several intestinal diseases (Guarner and Malagelada, 2003; Lim et al., 2005). However, due to the limited accessibility of the human GI tract and the complexity of the resident microbiota, the underlying mechanisms are far from being understood. This also holds true for the believed beneficial effects of some microbial fermentation products in the human colon.

Carbohydrates and proteins available for microbial fermentation in the human colon are mostly of dietary origin, but can also be host-derived (Table 6.2). Mucus is such a host-derived carbohydrate source for intestinal microorganisms, of which some even prefer mucins to other sugars (Derrien et al., 2004). Endogenous proteins are either of secretory origin or enter the gut lumen as desquamated epithelial cells. Most microorganisms in the colon prefer to ferment carbohydrates and switch to protein fermentation when fermentable carbohydrates have been used up (Ouwehand et al., 2005). While carbohydrate fermentation

Table 6.2. Fermentable substrates that reach the human colon, based on Egert et al. (2006).

Substrate		Amount [g/day]
Carbohydrates	Resistant starch; type RS1–RS4	5–35
	Nondigestible polysaccharides	10–25
	Oligosaccharides (e.g., fructo-, galacto-, or glucooligosaccharides, inulin)	2–8
	Mono- and disaccharides (e.g., lactose, fructose, lactitol)	2–5
	Mucins	3–5
	Synthetic carbohydrates (e.g., lactulose, polydextrose, modified cellulose)	Variable
Protein	Of dietary origin	1–12
	Of endogenous origin (e.g., pancreatic enzymes and other secretions)	4–8
	Desquamated epithelial cells	30–50
Other	Nonprotein nitrogen (e.g., urea, nitrate)	~0.5
	Organic acids, lipids, bacterial recycling	Unknown

leads to perceived health-promoting metabolites (SCFAs, lactate) (Scheppach, 1994), protein fermentation yields toxic metabolites (e.g., the sulfur-containing compounds hydrogen sulfide and methanethiol, ammonia, and several phenolic and indolic compounds). In a simplified view, carbohydrate fermentation takes place in the proximal part of the colon, while protein fermentation occurs rather in the distal colon (Guarner and Malagelada, 2003). Since it is primarily this latter part of the colon where colon cancer occurs and IBD starts, it is tempting to hypothesize the causative nature of such toxic metabolites (Van Nuenen et al., 2004).

From the SCFAs (primarily acetate, propionate, and butyrate) produced by the colonic microbiota, butyrate attracted specific attention, as it is the major energy source of colonocytes (Roediger, 1980). A deficient supply of butyrate or a distortion in butyrate metabolism may contribute to the etiology of intestinal inflammations and carcinogenesis (Scheppach and Weiler, 2004). Butyrate shortage can result from a general lack of fermentable substrates (e.g., fiber) or from an excess of substrates that are exclusively fermented by non-butyrate producers (e.g., lactulose; Venema et al., 2003). A deficient metabolism of butyrate can also be caused by high concentrations of S-containing metabolites (e.g., H_2S and methanethiol), originating from S-containing amino acids in fermentable proteins or from sulfate reduction by sulfate-reducing bacteria. These S-containing metabolites block butyrate oxidation by the colonocytes, which may lead to epithelial atrophy (Babidge et al., 1998). On the other hand, high luminal concentration of SCFAs may also cause mucosal injury (Argenzio and Meuten, 1991). The uptake level of butyrate by colonocytes may determine its actual effects, i.e., whether it serves a potentially vital nutritional role or causes toxicity when present in excess.

Clearly, a deeper understanding of their role in health and disease requires unraveling the in situ functionality of the microorganisms colonizing the human GI tract. For this, researchers have to cope with the individuality and complexity of a community in a hardly accessible habitat (Zoetendal et al., 1998). On the other hand, since there are only a limited number of biochemical routes leading from a given substrate (e.g., glucose) to a metabolite (e.g., acetate), there probably is an enormous functional redundancy in this ecosystem. Nevertheless, many of the conversions carried out by individual species of the GI tract microbiota and the resulting metabolic fluxes have been largely anecdotal and are only beginning to become known, due to the use of stable isotope-labeled substrates.

Ways to Study the (Shift in) Composition of the Microbiota

Phylogenetic Microarrays

The study of the composition of the (intestinal) microbiota has recently gained enormous momentum through the development of DNA microarray methods to study the presence of hundreds of species at the same time (Palmer et al., 2007; Rajilić-Stojanović et al., 2009; Rose et al., 2010). Like other molecular approaches that have been used in the recent past, the microarray platform also relies on the species-specific DNA signatures in the 16S rDNA. Due to high throughput DNA sequencing possibilities of this decennium, these microarrays can be customized for every conceivable complex microbiota. Shifts in the composition of the microbiota—for instance, before and after intake of fiber—can be quickly addressed using this technology. The HITChip (Human intestinal tract Chip; Rajilić-Stojanović et al., 2009) contains over 4800 dedicated tiling oligonucleotide probes of 1140 unique microbial phylotypes and covers most bacteria that have been detected in the intestine up to date. The arrays developed by Palmer and colleagues (Palmer et al., 2006) contain 7,167 unique sequences selected to provide quantitative information on the taxonomic composition of diverse microbial populations. The DNA-microarray platform developed by TNO (Rose et al., 2010; Maathuis et al., 2009), named the Intestinal-Chip (or I-Chip), contains over 350 probes that detect in a semiquantitative way different bacteria at the level of individual species, genera, and larger phylogenetic groups, and is suitable for high throughput use.

Examples of the use of phylogenetic microarrays

We are born sterile, but soon after birth a new microbial ecosystem establishes in our GI tract. Although it is a universal and integral part of human biology, the temporal progression of this process, the sources of the microbes that make up the ecosystem, how and why it varies from one infant to another, and how the composition of this ecosystem influences human physiology, development, and disease are still poorly understood. Palmer et al. (2007) used a microarray, along with sequencing of cloned libraries of PCR-amplified 16S rDNA, to profile the microbial communities in stool samples from 14 healthy, full-term human infants, including a pair of dizygotic twins. To investigate possible origins of the infant microbiota, they also profiled vaginal and breast milk samples from most of the mothers, and stool samples from all of the mothers, most of the fathers, and two siblings. Using the array they showed that the composition and temporal patterns of the microbial communities varied widely from baby to baby. Of course, this had previously been established with denaturing gradient gel electrophoresis (DGGE) (Favier et al., 2002), but not with the amount of detail that could be obtained with the microarray. Despite considerable temporal variation, the distinct features of each baby's microbial community were recognizable for intervals of weeks to months. The strikingly parallel temporal patterns of the twins suggested that incidental environmental exposures play a major role in determining the distinctive characteristics of the microbial community in each baby. By the end of the first year of life, the microbial ecosystems in each baby, although still distinct from one baby to the other, had converged toward a profile characteristic of the complex adult GI tract.

In their studies with the HITChip, Rajilić-Stojanović et al. investigated the dynamics of the human adult microbiota, studied the specific microbiota composition in patients suffering from IBS or ulcerative colitis (UC), and evaluated the microbial diversity in TIM-2, TNO's in vitro model of the colon (Rajilic-Stojanovic, 2007). The studies with the human adults showed that the temporal variations of the microbiota composition are more pronounced over an extended time span than was believed before, showing that environmental stimuli (such as diet) are an important factor in shaping the microbial community. Despite this, the microbiota of each individual has a unique profile which makes it distinct from other individuals. A 9-year follow-up study showed that the microbiota consists of a permanent core with other members being present transiently, occasionally, or incidentally. The HITChip analyses of the fecal microbiota of IBS patients showed that it is disturbed compared to healthy individuals and on top of that, specific for each type of IBS (diarrhea-predominant or constipation-predominant). In the diarrhea-predominant patients, several subgroups of *Clostridium* cluster XIVa were more abundant than in healthy controls, whereas in constipation-predominant patients, the major bacterial group *Clostridium* cluster IV was significantly higher. In UC patients, the microbiota was found to be significantly different from that of healthy controls, with a positive correlation with relapse of *Peptostreptococcus* species, *Clostridium difficile*, and various sulfate-reducing bacterial species such as *Desulfovibrio*.

The HITChip was also used to study the inoculum preparation and composition of the microbiota in TIM-2. The fecal inoculum used in this study was derived from fecal donations of seven healthy volunteers. For the purpose of this study, two slightly different procedures of inoculum preparation were used. For procedure IN1, 80 g of fresh mixed fecal material was mixed in a 2-liter fermenter with 670 mL of a complex medium representing terminal ileal chyme (modified from (Gibson et al., 1988) as described by (van Nuenen et al., 2003)) and incubated at pH 5.8 and 37°C. Additional medium was added after an initial 4 hours batch incubation during 40 hours to reach a final volume of 2000 mL. For procedure IN2, 80 g of fecal material was mixed with 670 mL of dialysis liquid used in TIM-2 (van Nuenen et al., 2003). Medium as described above, but concentrated 1.5-fold and amended with 1.7% of pig gastric mucin, was added during 44 hours of incubation, yielding the same final volume as described for IN1. Samples were snap-frozen in liquid nitrogen and stored in the freezer. For HITChip, samples were taken at 0, 4, 20, 24, 28, and 44 hours of incubation. Approximately 3,700 probes gave a signal on the array, similar to that observed for fresh fecal control samples. After the 44 hours incubation, the similarity index based on Pearson product moment correlation coefficient was 85% for IN1 and 79% for IN2, indicating that very few changes occurred during the incubation. Subsequently, the inoculum was introduced into TIM-2, and over the course of 3 days the composition was followed. Specifically during freezing, thawing, and introduction of the microbiota into the in vitro model, the similarity score dropped, indicating that these are crucial steps for the preparation of a standardized inoculum for such systems. Specifically, members of the *Bacteroides* group became overrepresented, while members of the *Clositridium* clusters XIVa and IV were underrepresented (but were still present).

Using the TNO I-Chip, recently, the effect of the undigested fraction of maize products on the activity and composition of the microbiota in TIM-2 were determined (Maathuis et al., 2009). All fibers showed increases of bifidobacterial and some *Lactobacillus* species, although different species were stimulated by different fibers. Pullulan showed the largest increase of bifidobacteria. Despite the fact that the I-Chip is semiquantitative (as are the other platforms), by comparing multiple hybridizations, specifically the one at T_0 and after 72 hours of fermentation in TIM-2, a factor of increase or decrease could be determined for the different species. This allowed for a quantitative determination of the shifts occurring after addition of the different fibers. Also, the technology was developed to allow for 96 samples to be processed simultaneously (Rose et al., 2010).

Stable Isotope Probing: Clues to Metabolic Function from Genomics Data

As discussed above, molecular DNA technologies allow for a comprehensive and integrated approach to assessing the structure of microbial communities, providing a perspective in GI tract microbiology. Although the application of these tools has significantly advanced our understanding of the gut microbial diversity, it does not provide functional insight on which microbes are relevant for specific dietary conversions (de Vos, 2001; Egert et al., 2006). The real challenge here is to develop and apply methodologies for analyzing the functionality of the microbiome (full genome of all members of the microbiota), to predicts its effect on human health. For this, it is important to know which microorganisms are responsible for the observed activities, elucidate dominant microbial functionalities in the human GI-tract, the impact of specific dietary components such as fiber on these functionalities, and ultimately the effect on gut health. Stable isotopes can play an important role in answering these questions.

To couple the microbial diversity to metabolic function in the colon, in situ SIP approaches appear very promising (Egert et al., 2006). Typically, in nucleic acid-based SIP studies ^{13}C-labeled compounds (usually, carbohydrates or fibers) that act as substrates in the food chain, are delivered to cultures of (intestinal) bacteria (Figure 6.2). This can be in test tubes, sophisticated in vitro models such as TIM-2, or in vivo in human volunteers. Subsequently, the ribosomal DNA or RNA of the microbial community is isolated and subjected to density gradient centrifugation to isolate the heavy, labeled fraction of nucleic acids (Figure 6.2). These heavier fractions stem from bacteria that have consumed the substrate, and incorporated the ^{13}C (which is heavier than the more abundant natural isotope: ^{12}C) in their nucleic acids. Either general or group-specific PCR amplification allows 16S rDNA fragments to be amplified

(Satokari et al., 2001; Heilig et al., 2002). These can then be characterized by high throughput rDNA sequence analysis providing insight into the microbial diversity of these fractions. Alternatively, phylogenetic microarrays, such as the ones discussed above, can be used for community profiling in the different fractions. By following the development of the rDNA sequence diversity in time, the specific groups of microbes involved in the food chain from, e.g., carbohydrates to SCFA can be reconstructed. This approach has also been used in the gut, where recently we have taken this SIP strategy and applied it to a human gut microbial community (Egert et al., 2007; Kovatcheva-Datchary et al., 2009). In the study by Egert et al. (a proof-of-principal study), 16S rRNA-based SIP profiling was used to identify bacteria-fermenting glucose (as a model substrate) under conditions simulating the human small intestine. As expected, phylogenetic analysis of the ^{13}C-labeled 16S rRNA revealed a microbial community dominated by microorganisms closely related to lactic acid bacteria and *Clostridium perfringens*, not unlike the microbiota in the terminal ileum where glucose is prevalent. Fingerprints of the most-labeled rRNA fraction identified *Streptococcus bovis* and *C. perfringens* as the most active glucose fermenters in the model. We used RNA-SIP, since RNA is the most responsive (sensitive) biomarker for SIP analyses because it occurs in greater cellular copy numbers, has a higher turnover rate than does DNA, and is produced more or less independent of cellular replication (Manefield et al., 2002). Owing to fewer variations in its guanine-cytosine (GC)-content compared to DNA, ribosomal 16S-based RNA-SIP might also be less susceptible to GC-effects that interfere with the separation of labeled 16S rRNA. It should be realized though that it is exactly this difference in GC-content between different members of the microbiota which results in the fact that a density gradient of unlabeled RNA already contains RNA in the heavier fractions. For instance, bifidobacteria can be found in these fractions (our own unpublished results).

Dietary fibers and carbohydrates, such as resistant starch, represent a substantial part of the human diet (Annison and Topping, 1994), and are known to affect the functioning of the human colon. Data from several animal studies showed increased large bowel butyrate levels and counts of *Bifidobacterium*, *Lactobacillus*, *Eubacterium*, and *Bacteroides* spp. after consumption of resistant starch (Kleessen et al., 1997; Jacobasch et al., 1999; Bird et al., 2000; Le Blay et al., 2003; Wong et al., 1995). However, there is still lack of knowledge with respect to the key microbes involved in colonic breakdown of these carbohydrates. In a combined SIP, HITChip, and NMR spectroscopy-based metabolic profiling approach, we recently studied the fermentation of starch in TIM-2 (Kovatcheva-Datchary et al., 2009). In this case, the inoculum of TIM-2 was a fresh fecal sample (mixture from seven healthy human adults), rather than a frozen inoculum. One gram of 98% uniformly [^{13}C]-labeled starch, isolated from potatoes grown in the presence of $^{13}CO_2$, was administered to the in vitro model, corresponding to a concentration of 40 mM glucose-units. Samples were taken at various time points after starch addition and subjected to density gradient centrifugation to separate the heavy from the light RNA. Four hours after the addition of the [^{13}C]-labeled starch, RNA was detected in the high-density fractions of the density gradient centrifugations, indicating the incorporation of the label in the RNA of starch-utilizing populations. To compare the microbial community in the gradient fractions at different time points, terminal restriction fragment length polymorphism (T-RFLP) fingerprint analysis was applied (Figure 6.3a). Species most closely related to *Ruminococcus bromii*, an uncultured Bacteroidetes group species related to *Prevotella*, and *Eubacterium rectale* corresponded to the major peaks enriched in fractions with highest density, indicating that these three organisms are involved in the degradation of starch in the in vitro model of the large intestine. A peak corresponding to *Bifidobacterium* was present also in the control (T_0) fraction, and did

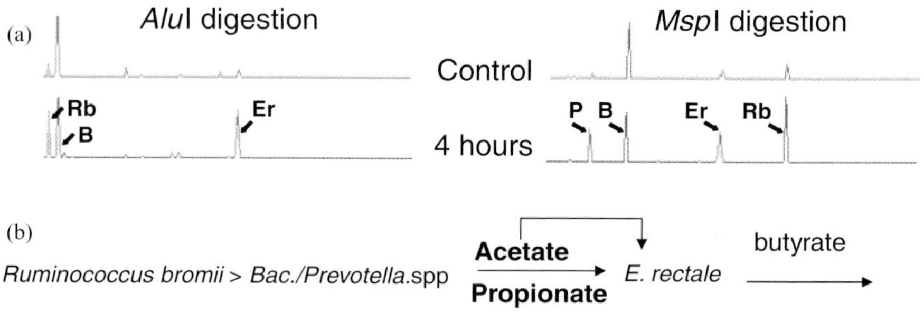

Figure 6.3. (a) Terminal restriction fragment length polymorphism (T-RFLP) profile of the most dense fractions of a control (unlabeled) sample and a sample taken after 4 hours of incubation with ^{13}C-labeled starch. PCR-fragments were digested with *Alu*I (left) or *Msp*I (right). Rb: *Ruminococcus bromii*; B: *Bifidobacterium*; P: species closely related to *Prevotella*; Er: *Eubacterium rectale*. (b) Proposed cross-feeding within the complex microbiota on starch after 4–8-hour fermentation of the labeled substrate.

not change in the 4 hours fraction. As indicated above, the presence in this fraction is likely due to the high GC-content of bifidobacteria, and illustrates that it is important to take the proper controls in these type of analyses.

Apart from the SIP analyses, the centrifugation fractions were also analyzed using the HITChip platform. This showed that this in vitro colon model is in fact well representing the fecal microbiota present in vivo in human volunteers, with all of the most abundant bacterial groups normally found in human fecal samples present. Furthermore, HITChip analysis also revealed that a strong numerical increase in the *Clostridium* cluster IV was detected in the "heavy" (^{13}C-labeled) fractions, especially concerning clones related to *R. bromii*. Significant changes in the "heavy" fractions were also detected in the Bacteroidetes phylum, mostly due to an increase in phylotypes related to *Prevotella* species, corroborating the results found with T-RFLP.

In addition to the molecular analyses, NMR-based metabolic profiling was performed. The major acetate-isotopomer produced from the labeled starch was the M + 2 isotopomer (acetate with two ^{13}C-atoms), whereas there was only limited production of the M + 1-isotopomer (acetate with only one of the two C-atoms labeled). The major butyrate-isotopomer after 8 hours was the M + 2-isotopomer, indicating coupling of a fully ^{13}C-labeled acetate to a fully unlabeled acetate. This is most likely to occur when the labeled and unlabeled metabolite pool of acetate mix extracellularly, with subsequent coupling of these two molecules by a different microorganism, indicative of cross-feeding. Integrating molecular and metabolic data, it can be hypothesized that there is indeed metabolic cross-feeding in the studied system, where species closely related to *R. bromii* and the Bacteroidetes group ferment starch to acetate (and propionate), which is subsequently utilized by *E. rectale* for the formation of butyrate (Figure 6.3b). This hypothesis is currently being investigated using pure cultures of *R. bromii* and *E. rectale*.

Human intestinal samples seem particularly suited to an RNA-SIP approach because these samples contain large numbers of highly active cells, resulting in quick and sufficient labeling of RNA. However, in view of these same large numbers of cells in the human GI tract, together with the nutrient-rich environment and the broad range of potential substrates (e.g., carbohydrates, proteins, host-derived substrates) that can be fermented, human intestinal samples necessitate a sensitive RNA-analysis approach to cope with label dilution. The use of in vitro gut models which closely mimic the environmental conditions in the GI tract and which are easy to sample enables detailed analyses of successive label incorporation into the RNA of different community members over time. Such cross-feeding effects (i.e., the use by one member of the microbiota of (labeled) metabolites derived from the initially added substrate produced by a different member) will help to identify food chains in intestinal systems. This may lead to the generations of hypotheses that need to be tested in vivo. Yet, application of SIP in human trials is challenging. It remains to be shown (1) whether a labeled substrate can be effectively delivered through the intestinal tract into the target region and homogenously distributed there; and (2) whether the (singly or pulsed) applied substrate concentrations can be adjusted in a way that prevents dilution within the colon, while still enabling sufficient labeling of microbial (16S r)RNA for subsequent analysis.

Considerable efforts have been made to accurately quantify the microbial SCFA production. This is not an easy task due to the very active metabolism of the colonocytes and the liver which interferes as soon as the SCFA are released in the gut lumen (Figure 6.4). In an elegant protocol, Kien et al. (1996) used [d_3]-acetate and [$1\text{–}^{13}C$] sugars infused into the colonic lumen of pigs to determine the rate of microbial

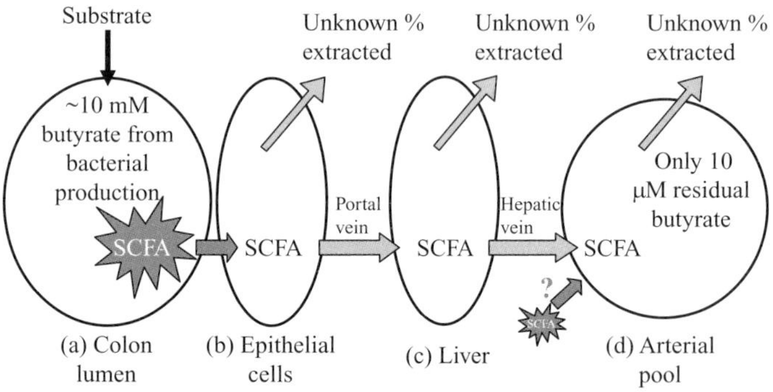

Figure 6.4. Schematic representation of microbial and host metabolism of SCFA. For example, luminal concentrations of butyrate are ~10 mM. An unknown amount of butyrate that is taken up by the epithelium is extracted by the coloncytes that use butyrate as an energy source. Subsequently, another unknown amount is taken up by the liver from the portal vein that drains the intestines. Similarly, an unknown amount is used by the systemic organs, leaving only μM quantities in venous blood. For some metabolites (e.g., acetate, lactate), there also is production by the host. These pools mix in venous blood.

acetate synthesis as well as the fraction of the sugars metabolized to acetate in a single experiment. Thereafter, Pouteau and colleagues (Pouteau et al., 2003) have developed and applied protocols to determine SCFA production in humans, including intragastric infusion to evaluate first-pass splanchnic retention of SCFA.

Isotopic tracers have been highly instrumental for the clarification of metabolic pathways involved in biosynthesis of compounds, as, e.g., described in an exquisite review by Bacher et al. (1999). By detecting doubly-^{13}C-labeled acetate produced from [3–^{13}C]glucose, Wolin et al. (1998) could establish that in a fecal suspension isolated from an infant, the *Bifidobacterium*-pathway was the major glucose fermentation pathway used. These authors also demonstrated the operation of the Embden-Meyerhof-Parnas as the major glycolytic pathway leading to SCFA in fecal suspensions of adults (Wolin et al., 1998). In these experiments they found that a considerable portion of microbially produced acetate was formed via the Wood–Ljungdahl pathway of CO_2 reduction.

Energy Balance

Large intestinal fermentation can account for approximately 10% of our daily energy supply (Bergman, 1990). Thus, the colonic microbiota plays a very significant role in whole body energy supply. Studies with germfree and conventional rodents have shown that conventional animals were able to extract significantly more energy from their diets than germfree counterparts, as judged from the fact that they had 40% more total body fat while consuming less chow per day (Bäckhed et al., 2004). This corroborates findings from other studies (Scheppach et al., 1991; Pouteau et al., 2005) that showed increased serum acetate concentrations and turnover correlating with colonic carbohydrate fermentation.

Microbially produced butyrate is the preferred and most important energy source for colonocytes (Csordas, 1996). This all more or less reflects a direct effect, i.e., additional energy produced by microbial fermentation of substrates entering the colon that would otherwise be useless to the host. There is however reason to believe that indirect effects may be at least as important. The colon functions within the whole of the intestine and associated visceral organs in controlling body energy balance (Badman and Flier, 2005). Gut and organs together play a key sensing and signaling role in the physiology of energy homeostasis. The gut, the pancreatic islets of Langerhans, elements in the portal vasculature, and even visceral adipose tissue communicate via neural and endocrine pathways with the controllers of energy balance in the brain. Signals reflecting energy stores, recent nutritional state, and other parameters are integrated in the central nervous system, particularly in the hypothalamus, to coordinate energy intake and expenditure (Badman and Flier, 2005). We are only beginning to uncover all the different features of this complex regulatory network, and expectations are high that understanding of the mechanisms that control energy balance will provide clues for therapies to fight obesity and metabolic syndrome.

Intestinal Microbiota: Is There a Link with Obesity?

As pointed out above, the colon (with its associated microbiota) functions within the whole of the intestine and associated visceral organs in maintaining the host energy balance. So, clearly a role for the microbiota in energy harvesting is there. It has been found that especially the SCFA acetate can interfere directly with host-lipid metabolism. On the one hand, acetate production (presumably microbial) after lactulose ingestion in overweight subjects was recently shown to result in short-term decrease in free fatty acid level and glycerol turnover related to a decrease of lipolysis (Ferchaud-Roucher et al., 2005), both factors believed to help in preventing insulin resistance. On the other hand, however, acetate may also stimulate lipid synthesis (Wolever et al., 1995), and it remains to be settled whether acetate has a long-term beneficial effect. Interestingly, the latter study also showed that another SCFA, propionate, inhibited lipid synthesis from acetate. Bäckhed et al. (Bäckhed et al., 2004) found that conventionalization of germfree mice with a microbiota harvested from the cecum of conventional animals produced a 60% increase in body fat content and insulin resistance despite reduced food intake. An increased absorption of monosaccharides from the intestine was detected in conventionalized mice, resulting in *de novo* lipogenesis in the liver. In addition, they showed that Fiaf, a lipase inhibitor, was suppressed in the intestinal epithelium of conventionalized mice and that this suppression was essential for the microbiota-induced deposition of fat in adipose tissue. These findings suggest that, at least in the mouse, the gut microbiota affects energy harvest from the diet and energy storage in adipose tissue. The same group has shown that the ratio between Bacteroidetes and Firmicutes might explain differences in energy harvesting (Ley et al., 2005). Whether this is also the case in humans remains to be determined. Conflicting results have been obtained. The Gordon-group corroborated their mouse-findings in human obese individuals (although with a limited number of volunteers). Subsequent results obtained in Europe showed different (Duncan et al., 2008) or opposite (Blaut et al., personal communication) effects. Considerable attention is being devoted to clarifying the picture.

Our intestine developed during evolution for optimal survival on natural diets. The recent rise to epidemic dimensions of obesity-linked diseases correlates with a shift in dietary habits toward a reduced intake of dietary fiber, an increased intake of simple sugars, a high intake of refined grain products, an altered fat composition of the diet, and a dietary pattern characterized by a high glycemic load (Suter, 2005). Recent epidemiological research (Maskarinec et al., 2006) of a large ethnically diverse population showed that on an individual level fiber intake predicted a lower body mass index (BMI). Similarly, a higher consumption of meat, poultry, and fish (fat and protein) was related to excess weight, whereas fruit and vegetable intake were inversely associated with excess weight. A recent interesting review (Hyman, 2006) pointed out that it is the glycemic load, rather than the glycemic index, that affects the neuroendocrine-immune signaling. Dietary fiber is one of the main factors lowering the glycemic load, whence it's beneficial effects in reducing overweight. This author also points out that bacterially produced fatty acids lower cholesterol production in the liver. In line with the observations above, it is tempting to hypothesize that obesity may—at least in part—be associated with deprival of proper fermentable substrate for colonic microbial fermentation. Therefore, it may be wise to distinguish between fermentable and nonfermentable (or slowly fermentable) fiber with regard to energy harvest. On the other hand, as discussed in the beginning of the chapter, it would be advisable for the microbiota to ferment carbohydrates instead of protein in the distal colon to reduce or prevent production of toxic metabolites. Clearly, more research is required in this area.

Thus, the effects of colonic SCFA production or microbial activity on host metabolism in general and lipid metabolism specifically is difficult to predict. The true role of the colon in regulation of lipid metabolism very likely is an even much more complex one, involving multiple neural and endocrine pathways. The recent finding that ingestion of dietary fat stimulates cholecystokinin (CCK) receptors, but at the same time leads to attenuation of the inflammatory response by way of the efferent vagus nerve and nicotinic receptors, may be an interesting foretaste of the type of regulatory hardwiring that can be expected. Also, the fact that microbially produced propionate influences adipokine production, such as leptin and resistin, in visceral adipose tissue, (Lahham et al., 2009) indicates that there is an intricate link between microbe activity and human physiology. These are just two recent examples of novel neuroimmunologic pathways, controlled by nutrition (whether or not through the colonic microbiota) that may help to unravel the complexity of microbiota–host interactions.

Role for Stable Isotopes

Can stable isotopes contribute to knowledge in the field of host–microbe interaction? Yes, they can. The key advantage of stable isotope methods is that they are very potent in tracing the fate of substrates entering the colon on the metabolic level, and therefore allow for a specific correlation of host responses to colon-derived metabolic events. This helps in discriminating colonic microbiota-related effects against those having an endogenous origin, and, therefore, allows to obtain a clearer picture of host–microbe interaction.

For instance, supplementation with resistant starch has been shown to improve colonic lesions in a dextran sulfate sodium (DSS)-induced colitis model in rats. To find out whether it is the increased colonic butyrate production that accelerates the healing process, Moreau et al., (2004) measured the ceco-colonic uptake of butyrate and its oxidation into CO_2 and ketone bodies in control, and DSS-treated rats fed a fiber-free basal diet or a resistant starch-supplemented diet. After cecal infusion of [1–^{13}C]butyrate, concentrations and ^{13}C-enrichment of butyrate, ketone bodies, and CO_2 were quantified in the abdominal aorta and portal vein, and portal blood flow was measured. These measurements allowed the authors to determine the utilization of butyrate specifically by the colonic mucosa, and to conclude that increased utilization of butyrate by the mucosa is subsequent to evidence of healing, and appears to be a consequence rather than a cause of the healing effect of resistant starch (Moreau et al., 2004).

In another study (Pouteau et al., 2005), it was tested whether acetate from colonic fermentation of inulin would stimulate peripheral acetate turnover in dogs. Dogs were simultaneously administered infusions of

[1–^{13}C] acetate intravenously (i.v.) and [1,2–$^{13}C_2$] acetate intrarectally. After switching from a control diet to a 3% inulin-enriched diet, initially no changes in whole body acetate concentration and turnover were seen after 4 days, but after 21 days the whole-body acetate turnover had increased significantly by 31%. While it was determined that a significant acetate production occurred in the colon, no [1,2–$^{13}C_2$] acetate tracer was recovered in the peripheral circulation. This led to the conclusion that the occurring colonic fermentation of inulin indirectly stimulated whole body acetate turnover (Pouteau et al., 2005).

Recently, in an elegant study Bloemen et al. (2009) determined the potential of the healthy liver to extract acetate and propionate from the portal vein. In this first in vivo study on the role of the gut and liver in SCFAs exchange in humans in vivo it was shown that intestinal SCFAs release by the gut is equaled by hepatic uptake. This is an important step to full understanding of the systemic effect of microbially produced SCFA.

Determining the Amount of kcal Produced Per Gram of Substrate After Fermentation

Experiments with stable isotopes not only allow the characterization of the metabolites that are produced from specific substrates and the metabolic routes employed, it is also possible to precisely determine how much energy the host can derive from such substrates. It is generally believed that approximately 95% of the SCFA produced by the microbiota are taken up by the epithelium, and that 5% are excreted in feces. However, it is likely that those SCFA excreted in feces reflected the SCFA produced in the distal colon, while those produced in the proximal colon will have been taken up during transit from proximal to distal colon (which takes >24 hours). In the calculations below we have assumed that all metabolites produced in the in vitro colon model (TIM-2) would have been taken up by the epithelium, as fermentation of the substrates was quick (<2 hours for lactose and inulin; ~8 hours for starch). Measuring the mass-isotopomers with liquid chromatography (LC), and the location of the label with NMR (exemplified for butyrate in Figure 6.5), allows one to calculate how much of the original ^{13}C-labeled substrate becomes available for uptake by the body. Since most of the label appeared in the SCFA acetate, propionate, and butyrate (>95%; not shown), the calculation was restricted to the amount of energy available from these SCFA. For the M + 1 isotopomer of

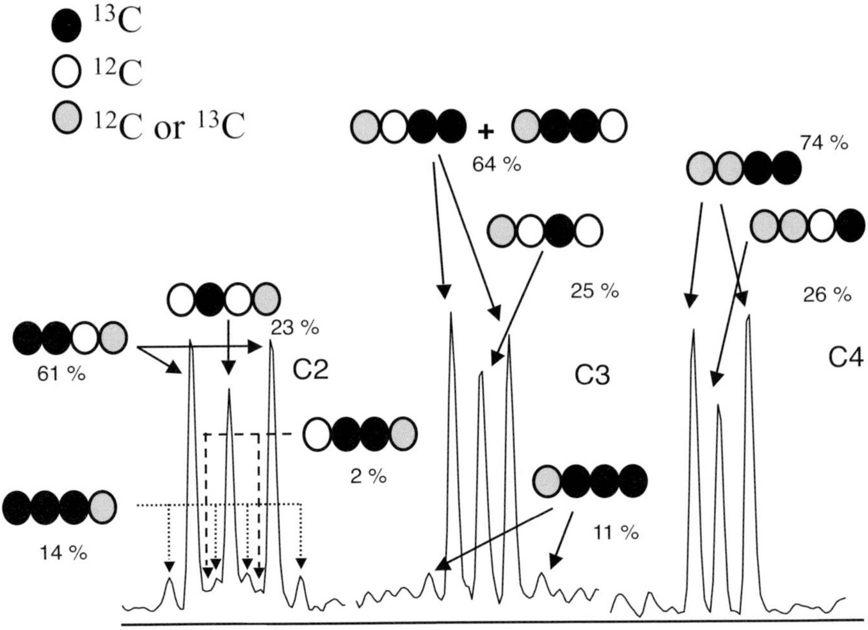

Figure 6.5. Illustrative example of the information provided by 2D 500 MHz proton-^{13}C NMR-coupling on the butyrate-isotopomers formed by the microbiota. Examples are shown for the carbon-atoms C2 to C4, as C1 does not have protons. Integration of the NMR-peaks allows quantification of the different isotopomers produced. Since only neighboring atoms can be characterized in this manner, only the black and white carbon-atoms can be identified with certainty as ^{13}C or ^{12}C, respectively, from these parts of the spectrum. Grey-colored atoms can be either ^{13}C or ^{12}C.

acetate (irrespectively of whether the C1 or C2-atom were labeled), half the amount of kcal was taken than for the fully labeled M + 2-isotopomer. Similarly, a third of the amount of kcal was taken for the M + 1 isotopomer of propionate, etc. Table 6.3 indicates the amount of kcal derived from the individual SCFA and the sum of these. The table shows that not only in terms of amount of kcal produced per gram of substrates the three different substrates differ, but also the contribution of the different SCFA to the sum differs significantly between the substrates. This might be relevant in terms of obesity and the amount of energy supplied to the host from fermentable substrates. Also, the contribution of the different SCFA to this energy might have importance, as we have, for instance, shown that propionate is able to decrease leptin secretion by adipocytes, while at the same time increasing resistin

Table 6.3. Amount of kcal obtained from fermentation of carbohydrates by the colonic microbiota through production of short-chain fatty acids (SCFA)*.

	M + 1	M + 2	M + 3	M + 4	Sum	kcal/mol	kcal
Acetate	0.362	4.317	–	–	4.679	209	0.978
Propionate	0.125	0.392	1.868	–	2.386	367	0.876
Butyrate	0.043	0.302	0.000	0.026	0.372	524	0.195
Energetic value of 1 g inulin (in kcal):							2.048
Energetic value of 1 g lactose (in kcal):							2.410
Energetic value of 1 g starch (in kcal):							1.692

*Metabolite concentration in mmol.

(Lahham et al., 2009), which both is beneficial for metabolic syndrome. On the other hand, butyrate is important for gut health.

Butyrate is Important

As already eluded to above, SCFA (primarily acetate, propionate, and butyrate) are the major end products of bacterial fermentation of fibers in the intestine and affect key functions of the colonic epithelium in vivo. These compounds are probably key participants in gut maintenance (Bergman, 1990; Kruh et al., 1991; Mariadason et al., 2000) and may also be beneficial contributors to the peripheral metabolism in humans (Macfarlane and Cummings, 1991). Therefore, SCFA have been the subject of numerous investigations. Butyrate is considered the most important of the SCFA (Hamer et al., 2008). In the large intestine, butyrate is present in mM concentrations. Butyrate (exclusively produced by the microbiota, not by the host) is metabolized by epithelial cells and is responsible for 70% of their energy need (Roediger, 1982; Scheppach, 1994). In addition, butyrate acts as a signaling metabolite, affecting epithelial cell proliferation and differentiation (Gamet et al., 1992; Gibson et al., 1992). In addition to its established role in regulating viability, differentiation and proliferation of the epithelium (Velazquez et al., 1997) butyrate was reported to be effective in cancer suppression (McIntyre et al., 1993; Singh et al., 1997). Furthermore, butyrate might be beneficial in preventing mucosal inflammation because decreased availability of butyrate has been associated with distinct forms of colitis (Harig et al., 1989; Chapman et al., 1994; Ahmad et al., 2000). Moreover, butyrate enemas have been shown to be an effective treatment for mucosal inflammation in both humans and animal models of colitis in some studies (e.g., Scheppach et al., 1992; Gibson and Rosella, 1995; Okamoto et al., 2000; Andoh et al., 2003; Cherbut et al., 2003), an increase in antioxidant status of the epithelium (Hamer et al., 2009), as well as in a reduction in pain-perception in human individuals (Vanhoutvin et al., 2009a). Transcriptional analyses in human colonic mucosa after a 2-week-daily-butyrate-administration by enema showed that 500 genes were differentially expressed (Vanhoutvin et al., 2009b). Pathway analysis after colonic administration of butyrate in concentrations that can be achieved by consumption of a high-fiber diet showed that the butyrate intervention mainly resulted in an increased transcriptional regulation of the pathways representing fatty acid oxidation, electron transport chain, and oxidative stress. In addition, several genes associated with epithelial integrity and apoptosis were found to be differentially expressed after the butyrate intervention. Essentially, the authors concluded that butyrate enhanced the maintenance of colonic homeostasis in these healthy subjects (Vanhoutvin et al., 2009b). Despite the fact that we know that fiber (especially, resistant starch) is important for butyrate production, the (regulation of the) flux of butyrate (and the other SCFA for that matter) by (specific) intestinal microorganisms in the colon and the use of the SCFA in the rest of the body however is still largely unknown. It is important therefore to know which bacterial metabolic routes exist in vivo for the synthesis of SCFA. Knowledge of the regulation of these metabolic routes then will allow the development of dietary strategies to influence SCFA metabolism, likely with fibers playing an important role, possibly even for therapeutic purposes.

Gut Environmental Parameters Matter

Fermentation of fermentable fiber results in lowered cecal pH concomitant with the production of SCFA (Cummings and Bingham, 1987). In sudden death individuals, a significant trend from high to low concentrations of SCFA has been found on passing distally from cecum to descending colon, while pH changed from 5.6 ± 0.2 in the cecum to 6.6 ± 0.1 in the descending colon (Cummings et al., 1987). Interestingly, there seems to be an inverse effect of pH on SCFA metabolism as well; the response of human fecal microbial communities in anaerobic continuous culture showed markedly higher final butyrate concentrations at pH 5.5 compared with pH 6.5, whereas acetate and propionate were higher at pH 6.5 (Walker et al., 2005). Changes in colon pH are also reported to alter the metabolism of protein, bile acids, nitrate, sulfate, and other substances (Cummings and Bingham, 1987).

In contrast to SCFA, products of protein fermentation, such as ammonia, branched chain fatty acids, and phenolic compounds, progressively increase from the proximal (right) to the distal (left) colon, as does the pH of gut contents (Macfarlane et al., 1992a). Epidemiological studies found a marked correlation between pH and the incidence of colon cancer (e.g., Levy et al., 1994), which in a subsequent study appeared to be associated with higher animal protein and fat consumption (O'Keefe et al., 1999). However, a direct linkage between colon cancer and alkaline colonic pH was questioned in other experimental studies (Hove et al., 1993). This demonstrates that it is difficult to establish unequivocally the cause–effect relationships between intestinal processes and human health due to the many interactions present, but proposes that fermentation of fiber reaching the distal colon might certainly be beneficial. Namely, this would locally decrease the pH, with a concomitant increase in butyrate production (see paragraph above) and a lowering of putrefactive metabolites from protein fermentation. Clearly, extensive additional systematic research is necessary to elucidate the regulation of intestinal bacterial metabolism, the reaction of the host, and the interaction between both.

Metabolic Routes for Propionate Production and Pathway Reconstruction

In an attempt to dig deeper into regulation of bacterial metabolism, propionate production was recently studied by us (unpublished results). Within the intestinal microbiota, propionate can be synthesized via two routes (Figure 6.6), (1) a short route with lactate as precursor—the

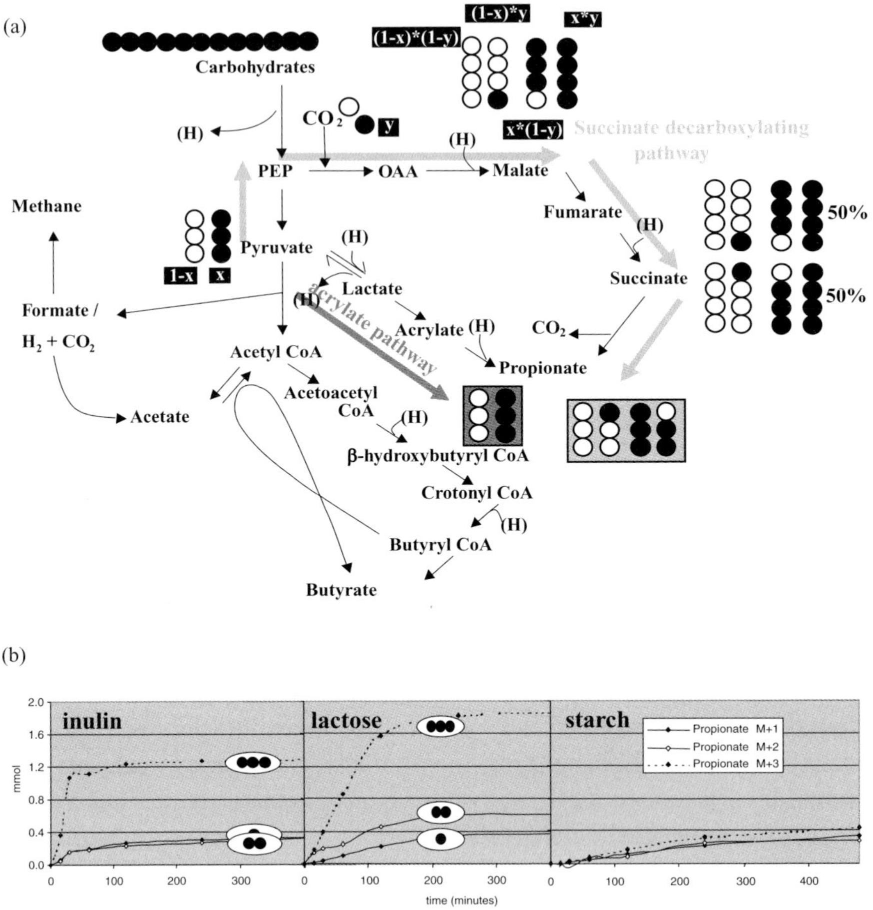

Figure 6.6. Detailed overview of anaerobic bacterial metabolic pathways involved in carbohydrate metabolism, specifically focused on the two routes for propionate synthesis. Based on different isotope mixing in the acrylate *versus* the succinate decarboxylating pathway (indicated in the figure), the isotopomers produced by the different pathways can be predicted. White circle: ^{12}C; black circle ^{13}C.

acrylate pathway; and (2) a longer route with succinate as precursor—the succinate decarboxylation pathway. The two routes can be distinguished from one another by the ^{13}C labeling patterns that they create as explained also in Figure 6.6a.

Recent experiments have been performed in TIM-2 with three different ^{13}C-labeled substrates, namely, inulin, lactose, and starch (unpublished data). Samples were taken for NMR and (liquid

Table 6.4. Contribution of the acrylate and succinate decarboxylating pathways involved in propionate production.

Substrate	Acrylate Pathway Contribution (%)	Succinate Decarboxylation Pathway Contribution (%)
Inulin	60	40
Lactose	50	50
Starch	0	100

chromatography-mass spectrometry) LC-MS analyses to determine isotopomer distribution for propionate. From the NMR and LC-MS data the relative abundancies of all eight possible carbon isotopomeric forms of the propionate molecule could be determined, making it possible to calculate the pathway distribution from the data. For illustrative purposes, Figure 6.6b shows the mass isotopomeric composition of propionate for the three investigated substrates (note that the mass isotopomers do not reveal the position of the label in the molecule, NMR does). The figure shows a dramatic difference in the abundance of especially the triply-labeled form of propionate, and to a lesser extent also the doubly labeled form. Comparing with Figure 6.6a, it can be observed that the triply labeled molecule reflects primarily the acrylate pathway activity. The formulas depicted in the figure (white text in the black boxes) were used together with the full isotopomeric composition data of propionate (LC + NMR data) to estimate the pathway activities on the three substrates. The results are shown in Table 6.4. These results seem to indicate that a faster fermentation (as was the case for inulin and lactose) goes along with an increased contribution of the acrylate pathway, i.e., the pathway that has lactate as a precursor. Of note, lactate accumulated during the early stages of both the inulin and lactose fermentation. The results as described above and similar data for other metabolites has resulted in the construction of a metabolic network for the colonic microbiota.

Future Perspectives

Despite the enormous progress in recent years in unraveling the role of the composition and activity of the microbiota on heath and disease, and

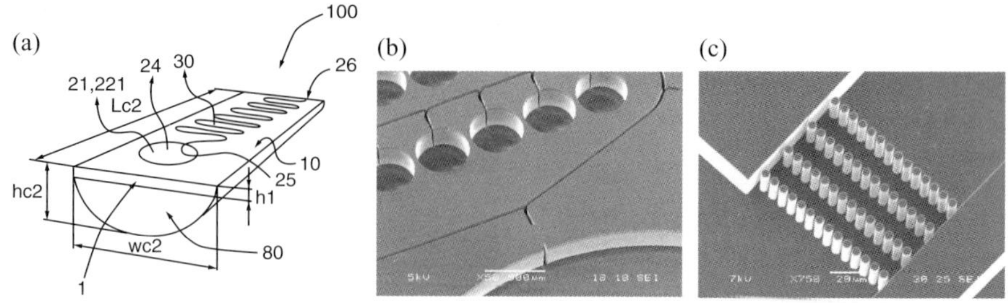

Figure 6.7. (a) Schematic overview of the lab-on-a-chip implanted on a capsule; (b) electron-microscopy photo of the channel and chambers on the chip; (c) electron-microscopy photo of the pillars in the channel that direct the flow of intestinal liquid.

the potential role of fiber to modulate this, the full picture is far from clear. A multidisciplinary approach is required to completely solve this black box. For instance, it might be necessary to incorporate the field of nanotechnology as well. We are currently developing a lab-on-a-chip (Sprenkels and Venema, unpublished results; Figure 6.7) which, upon implantation on a capsule, is swallowed and samples intestinal fluid in time, while traveling through the GI-tract. After recovery of the chip, these samples can be analyzed, e.g., for SCFA, and will give information on the SCFA concentrations in different parts of the colon. This allows us with minimally invasive techniques to study microbial metabolism in situ.

Also, progress in stable-isotope technology holds a lot of for the future. Determination of specific microbial pathways activated by selected members of the microbiota on different dietary fibers will soon be possible.

In conclusion, it may be possible in the future to direct the metabolic activity of the microbiota toward production of certain SCFA from certain fibers using the tools and approaches discussed here.

References

Ahmad, M.S., Krishnan, S., Ramakrishna, B.S., Mathan, M., Pulimood, A.B., Murthy, S.N. (2000) Butyrate and glucose metabolism by colonocytes in experimental colitis in mice. *Gut* 46(4), 493–499.

Andoh, A., Tsujikawa, T., Fujiyama, Y. (2003) Role of dietary fiber and short-chain fatty acids in the colon. *Curr. Pharm. Des.* 9(4), 347–358.

Annison, G., Topping, D.L. (1994) Nutritional role of resistant starch: chemical structure vs physiological function. *Annu. Rev. Nutr.* 14, 297–320.

Argenzio, R.A., Meuten, D.J. (1991) Short-chain fatty-acids induce reversible injury of porcine colon. *Dig. Dis. Sci.* 36(10), 1459–1468.

Aries, V., Hill, M.J. (1970) Degradation of steroids by intestinal bacteria. I. Deconjugation of bile salts. *Biochim. Biophys. Acta* 202(3), 526–534.

Babidge, W., Millard, S., Roediger, W. (1998) Sulfides impair short chain fatty acid beta-oxidation at acyl-CoA dehydrogenase level in colonocytes: implications for ulcerative colitis. *Mol. Cell. Biochem.* 181(1–2), 117–124.

Bacher, A., Rieder, C., Eichinger, D., Fuchs, G., Arigoni, D., Eisenreich, W. (1999) Elucidation of biosynthetic pathways and metabolic flux patterns via retrobiosynthetic NMR analysis. *FEMS Microbiol. Rev.* 22(5), 567–598.

Bäckhed, F., Ding, H., Wang, T., et al. (2004) The gut microbiota as an environmental factor that regulates fat storage. *Proc. Natl. Acad. Sci. USA* 101(44), 15718–15723.

Badman, M.K., Flier, J.S. (2005) The gut and energy balance: visceral allies in the obesity wars. *Science* 307(5717), 1909–1914.

Bergman, E.N. (1990) Energy contributions of volatile fatty acids from the gastrointestinal tract in various species. *Physiol. Rev.* 70(2), 567–590.

Bird, A.R., Brown, I.L., Topping, D.L. (2000) Starches, resistant starches, the gut microflora and human health. *Curr. Issues Intest. Microbiol.* 1(1), 25–37.

Birkett, A., Muir, J., Phillips, J., Jones, G., O'Dea, K. (1996) Resistant starch lowers fecal concentrations of ammonia and phenols in humans. *Am. J. Clin. Nutr.* 63(5), 766–772.

Bloemen, J.G., Venema, K., van de Poll, M.C., Olde Damink, S.W., Buurman, W.A., Dejong, C.H. (2009) Short chain fatty acids exchange across the gut and liver in humans measured at surgery. *Clin. Nutr.* 28(6), 657–661. doi:10.1016/j.clnu.2009.05.011.

Chadwick, V.S., Anderson, R.P. (1995) The role of intestinal bacteria in etiology and maintenance of inflammatory bowel disease. In: Gibson, G.R., Macfarlane, G.T. (editors), *Human Colonic Bacteria: Role in Nutrition, Physiology, and Pathology*, pp. 227–256. CRC Press, Boca Raton, FL.

Chapman, M.A., Grahn, M.F., Boyle, M.A., Hutton, M., Rogers, J., Williams, N.S. (1994) Butyrate oxidation is impaired in the colonic mucosa of sufferers of quiescent ulcerative colitis. *Gut* 35(1), 73–76.

Cherbut, C., Michel, C., Lecannu, G. (2003) The prebiotic characteristics of fructooligosaccharides are necessary for reduction of TNBS-induced colitis in rats. *J. Nutr.* 133(1), 21–27.

Csordas, A. (1996) Butyrate, aspirin and colorectal cancer. *Eur. J. Cancer Prev.* 5(4), 221–231.

Cummings, J.H. (1991) Production and metabolism of short-chain fatty acids in humans. In: Roche, A.F. (editor), *Short-Chain Fatty Acids: Metabolism and Clinical Importance. Report of the Tenth Ross Conference on Medical Research*, pp. 11–17. Ross Laboratories, Columbus, OH.

Cummings, J.H. (1995) Short chain fatty acids. In: Gibson, G.R., Macfarlane, G.T. (editors), *Human Colonic Bacteria: Role in Nutrition, Physiology, and Pathology*, pp. 101–130. CRC Press, Boca Raton, FL.

Cummings, J.H., Bingham, S.A. (1987) Dietary fibre, fermentation and large bowel cancer. *Cancer Surv.* 6(4), 601–621.

Cummings, J.H., Macfarlane, G.T. (1997) Role of intestinal bacteria in nutrient metabolism. *J. Parenter. Enteral Nutr.* 21(6), 357–365.

Cummings, J.H., Pomare, E.W., Branch, W.J., Naylor, C.P., Macfarlane, G.T. (1987) Short chain fatty acids in human large intestine, portal, hepatic and venous blood. *Gut* 28(10), 1221–1227.

De Boever, P., Deplancke, B., Verstraete, W. (2000) Fermentation by gut microbiota cultured in a simulator of the human intestinal microbial ecosystem is improved by supplementing a soygerm powder. *J. Nutr.* 130(10), 2599–2606.

de Graaf, A.A. (2000) Use of ^{13}C labeling and NMR spectroscopy in metabolic flux analysis. In: Barbotin, J.N., Portais, J.C. (editors), *NMR in Microbiology: Theory and Application*. Horizon Scientific Press, Norwich, UK.

de Graaf, A.A., Venema, K. (2008) Gaining insight in microbial physiology in the large intestine: a special role for stable isotopes. *Adv. Microb. Physiol.* 53, 73–168.

de Vos, W.M. (2001) Advances in genomics for microbial food fermentations and safety. *Curr. Opin. Biotechnol.* 12(5), 493–498.

Derrien, M., Vaughan, E.E., Plugge, C.M., de Vos, W.M. (2004) *Akkermansia muciniphila* gen. nov., sp nov., a human intestinal mucin-degrading bacterium. *Int. J. Syst. Evol. Microbiol.* 54(Pt5), 1469–1476.

Dolnikowski, G.G., Marsh, J.B., Das, S.K., Welty, F.K. (2005) Stable isotopes in obesity research. *Mass Spectrom. Rev.* 24(3), 311–327.

Dumont, M.G., Murrell, J.C. (2005) Stable isotope probing – linking microbial identity to function. *Nat. Rev. Microbiol.* 3(6), 499–504.

Duncan, S.H., Lobley, G.E., Holtrop, G., et al. (2008) Human colonic microbiota associated with diet, obesity and weight loss. *Int. J. Obes. (Lond).* 32(11), 1720–1724.

Egert, M., de Graaf, A.A., Maathuis, A., et al. (2007) Identification of glucose-fermenting bacteria present in an in vitro model of the human intestine by RNA-stable isotope probing. *FEMS Microbiol. Ecol.* 60(1), 126–135.

Egert, M., de Graaf, A.A., Smidt, H., de Vos, W.M., Venema, K. (2006) Beyond diversity: functional microbiomics of the human colon. *Trends Microbiol.* 14(2), 86–91.

Favier, C.F., Vaughan, E.E., De Vos, W.M., Akkermans, A.D. (2002) Molecular monitoring of succession of bacterial communities in human neonates. *Appl. Environ. Microbiol.* 68(1), 219–226.

Ferchaud-Roucher, V., Pouteau, E., Piloquet, H., Zair, Y., Krempf, M. (2005) Colonic fermentation from lactulose inhibits lipolysis in overweight subjects. *Am. J. Physiol. Endocrinol. Metab.* 289(4), E716–E720.

Gamet, L., Daviaud, D., Denis-Pouxviel, C., Remesy, C., Murat, J.C. (1992) Effects of short-chain fatty acids on growth and differentiation of the human colon-cancer cell line HT29. *Int. J. Cancer* 52(2), 286–289.

Gibson, G.R., Cummings, J.H., MacFarlane, G.T. (1988) Use of a three-stage continuous culture system to study the effect of mucin on dissimilatory sulphate reduction and methanogenesis by mixed populations of human gut bacteria. *Appl. Environ. Microbiol.* 54(11), 2750–2755.

Gibson, P., Rosella, O. (1995) Interleukin 8 secretion by colonic crypt cells in vitro: response to injury suppressed by butyrate and enhanced in inflammatory bowel disease. *Gut* 37(4), 536–543.

Gibson, P.R., Moeller, I., Kagelari, O., Folino, M., Young, G.P. (1992) Contrasting effects of butyrate on the expression of phenotypic markers of differentiation in neoplastic and non-neoplastic colonic epithelial cells in vitro. *J. Gastroenterol. Hepatol.* 7(2), 165–172.

Guarner, F., Malagelada, J.R. (2003) Gut flora in health and disease. *Lancet* 361(9356), 512–519.

Hamer, H.M., Jonkers, D., Venema, K., Vanhoutvin, S., Troost, F.J., Brummer, R.-J. (2008) Review article: the role of butyrate on colonic function. *Aliment. Pharmacol. Ther.* 27(2), 104–119.

Hamer, H.M., Jonkers, D.M., Bast, A., et al. (2009) Butyrate modulates oxidative stress in the colonic mucosa of healthy humans. *Clin. Nutr.* 28(1), 88–93.

Harig, J.M., Soergel, K.H., Komorowski, R.A., Wood, C.M. (1989) Treatment of diversion colitis with short-chain fatty acid irrigation. *N. Engl. J. Med.* 320(1), 23–28.

Heilig, H.G., Zoetendal, E.G., Vaughan, E.E., Marteau, P., Akkermans, A.D., de Vos, W.M. (2002) Molecular diversity of *Lactobacillus* spp. and other lactic acid bacteria in the human intestine as determined by specific amplification of 16S ribosomal DNA. *Appl. Environ. Microbiol.* 68(1), 114–123.

Hill, M.J. (1995) The metabolic activity of gut bacteria. In: Hill, M.J. (editor), *Role of Gut Bacteria in Human Toxicology and Pharmacology*, pp. 19–25. Taylor & Francis, London, UK.

Hove, H., Rye Clausen, M., Brøbech Mortensen, P. (1993) Lactate and pH in faeces from patients with colonic adenomas or cancer. *Gut* 34(5), 625–629.

Hyman, M.A. (2006) Systems biology: the gut-brain-fat cell connection and obesity. *Altern. Ther. Health Med.* 12 (1), 10–16.

Jacobasch, G., Schmiedl, D., Kruschewski, M., Schmehl, K. (1999) Dietary resistant starch and chronic inflammatory bowel diseases. *Int. J. Colorectal Dis.* 14(4–5), 201–211.

Kelleher, J.K. (2004) Probing metabolic pathways with isotopic tracers: insights from mammalian metabolic physiology. *Metab. Eng.* 6(1), 1–5.

Kien, C.L., Murray, R.D., Ailabouni, A., et al. (1996) Stable isotope model for assessing production of short chain fatty acids from colon-derived sugar: application in pigs. *J. Nutr.* 126(12), 3069–3076.

Kleessen, B., Stoof, G., Proll, J., Schmiedl, D., Noack, J., Blaut, M. (1997) Feeding resistant starch affects fecal and cecal microflora and short-chain fatty acids in rats. *J. Anim. Sci.* 75(9), 2453–2462.

Kovatcheva-Datchary, P., Egert, M., Maathuis, A., et al. (2009) Linking phylogenetic identities of bacteria to starch fermentation in an in vitro model of the large intestine by RNA-based stable isotope probing. *Environ. Microbiol.* 11(4), 914–926.

Kruh, J. Defer, N., Tichonicky, L. (1991) Molecular and cellular effects of sodium butyrate. In: Roche, A.F. (editor), *Short-Chain Fatty Acids: Metabolism and Clinical Importance. Report of the Tenth Ross Conference on Medical Research*, pp. 45–50. Ross Laboratories, Columbus, OH.

Lahham, S., Weening, D., Rezaee, F., Roelofsen, H., Venema, K., Vonk, R. (2009) Propionic acid stimulates leptin production in human adipose tissue through G-Protein coupled receptor(s). *Obes. Facts* 40(5), 401–407.

Le Blay, G.M., Michel, C.D., Blottiere, H.M., Cherbut, C.J. (2003) Raw potato starch and short-chain fructo-oligosaccharides affect the composition and metabolic activity of rat intestinal microbiota differently depending on the caecocolonic segment involved. *J. Appl. Microbiol.* 94(2), 312–320.

Levine, J., Ellis, C.J., Furne, J.K., Springfield, J., Levitt, M.D. (1998) Fecal hydrogen sulfide production in ulcerative colitis. *Am. J. Gastroenterol.* 93(1), 83–87.

Levy, R.D., Segal, I., Hassan, H., Saadia, R. (1994) Stool weight and faecal pH in two South African populations with a dissimilar colon cancer risk. *S. Afr. J. Surg.* 32(4), 127–128; discussion 128–129.

Ley, R.E., Bäckhed, F., Turnbaugh, P., Lozupone, C.A., Knight, R.D., Gordon, J.I. (2005) Obesity alters gut microbial ecology. *Proc. Natl. Acad. Sci. USA.* 102(31), 11070–11075.

Lim, C.C., Ferguson, L.R., Tannock, G.W. (2005) Dietary fibres as "prebiotics": Implications for colorectal cancer. *Mol. Nutr. Food Res.* 49(6), 609–619.

Maathuis, A., Hoffman, A., Evans, A., Sanders, L., Venema, K. (2009) The effect of the undigested fraction of maize products on the activity and composition of the microbiota determined in a dynamic in vitro model of the human proximal large intestine. *J. Am. Coll. Nutr.* 28(6), 657–666.

Macfarlane, G.T., Cummings, J.H. (1991) The colonic flora, fermentation and large bowel digestive function. In: Phillips, S.F., Pemberton, J.H., Shorter, R.G. (editors), *The Large Intestine: Physiology, Pathophysiology, and Disease*, pp. 51–92. Raven Press, New York.

Macfarlane, G.T., Gibson, G.R., Cummings, J.H. (1992a) Comparison of fermentation reactions in different regions of the human colon. *J. Appl. Bacteriol.* 72(1), 57–64.

Macfarlane, G.T., Gibson, G.R., Cummings, J.H. (1992b) In vivo and in vitro studies on formation of phenols and phenolic acids by human colonic bacteria. *Growth Dev. Nutr.* A565.

Macfarlane, G.T., Macfarlane, S., Gibson, G.R. (1998) Validation of a three-stage compound continuous culture system for investigating the effect of retention time on the ecology and metabolism of bacteria in the human colon. *Microb. Ecol.* 35(2), 180–187.

Macfarlane, S., Woodmansey, E.J., Macfarlane, G.T. (2005) Colonization of mucin by human intestinal bacteria and establishment of biofilm communities in a two-stage continuous culture system. *Appl. Environ. Microbiol.* 71(11), 7483–7492.

MacFarlane, S., MacFarlane, G.T. (1995) Proteolysis and amino acid fermentation. In: Gibson, G.R., MacFarlane, G.T. (editors), *Human Colonic Bacteria: Role in Nutrition, Physiology and Pathology*, pp. 75–100. CRC Press, Boca Raton, FL.

Manefield, M., Whiteley, A.S., Bailey, M.J. (2004) What can stable isotope probing do for bioremediation? *Int. Biodeterior. Biodegrad.* 54(2–3), 163–166.

Manefield, M., Whiteley, A.S., Griffiths, R.I., Bailey, M.J. (2002) RNA stable isotope probing, a novel means of linking microbial community function to phylogeny. *Appl. Environ. Microbiol.* 68(11), 5367–5373.

Manning, B.W., Federle, T.W., Cerniglia, C.E. (1987) Use of a semicontinuous culture system as a model for determining the role of human intestinal microflora in the metabolism of Xenobiotics. *J. Microbiol. Methods* 6(2), 81–94.

Mariadason, J.M., Corner, G.A., Augenlicht, L.H. (2000) Genetic reprogramming in pathways of colonic cell maturation induced by short chain fatty acids: comparison with trichostatin A, sulindac, and curcumin and implications for chemoprevention of colon cancer. *Cancer Res.* 60(16), 4561–4572.

Maskarinec, G., Takata, Y., Pagano, I., et al. (2006) Trends and dietary determinants of overweight and obesity in a multiethnic population. *Obesity (Silver Spring)* 14(4), 717–726.

McCabe, B.J., Previs, S.F. (2004) Using isotope tracers to study metabolism: application in mouse models. *Metab. Eng.* 6(1), 25–35.

McIntyre, A., Gibson, P.R., Young, G.P. (1993) Butyrate production from dietary fibre and protection against large bowel cancer in a rat model. *Gut* 34(3), 386–391.

Miller, T.L., Wolin, M.J. (1981) Fermentation by the human large-intestine microbial community in an in vitro semicontinuous culture system. *Appl. Environ. Microbiol.* 42(3), 400–407.

Minekus, M., Smeets-Peeters, M., Bernalier, A., et al. (1999) A computer-controlled system to simulate conditions of the large intestine with peristaltic mixing, water absorption and absorption of fermentation products. *Appl. Microbiol. Biotechnol.* 53(1), 108–114.

Moran, B.J., Jackson, A.A. (1990) Metabolism of 15N-labeled urea in the functioning and defunctioned human colon. *Clin. Sci. (Lond)* 79(3), 253–258.

Moreau, N.M., Champ, M.M., Goupry, S.M., et al. (2004) Resistant starch modulates in vivo colonic butyrate uptake and its oxidation in rats with dextran sulfate sodium-induced colitis. *J. Nutr.* 134(3), 493–500.

Noverr, M.C., Huffnagle, G.B. (2004) Does the microbiota regulate immune responses outside the gut? *Trends Microbiol.* 12(12), 562–568.

Okamoto, T., Sasaki, M., Tsujikawa, T., Fujiyama, Y., Bamba, T., Kusunoki, M. (2000) Preventive efficacy of butyrate enemas and oral administration of *Clostridium butyricum* M588 in dextran sodium sulfate-induced colitis in rats. *J. Gastroenterol.* 35(5), 341–346.

O'Keefe, S.J., Kidd, M., Espitalier-Noel, G., Owira, P. (1999) Rarity of colon cancer in Africans is associated with low animal product consumption, not fiber. *Am. J. Gastroenterol.* 94(5), 1373–1380.

Ouwehand, A.C., Derrien, M., de Vos, W., Tiihonen, K., Rautonen, N. (2005) Prebiotics and other microbial substrates for gut functionality. *Curr. Opin. Biotechnol.* 16(2), 212–217.

Palmer, C., Bik, E.M., DiGiulio, D.B., Relman, D.A., Brown, P.O. (2007) Development of the human infant intestinal microbiota. *PLoS Biol.* 5(7), e177.

Palmer, C., Bik, E.M., Eisen, M.B., et al. (2006) Rapid quantitative profiling of complex microbial populations. *Nucleic Acids Res.* 34(1), e5.

Pouteau, E., Frenais, R., Dumon, H., Noah, L., Martin, L., Nguyen, P. (2005) Colonic fermentation of inulin increases whole-body acetate turnover in dogs. *J. Nutr.* 135(12), 2845–2851.

Pouteau, E., Nguyen, P., Ballevre, O., Krempf, M. (2003) Production rates and metabolism of short-chain fatty acids in the colon and whole body using stable isotopes. *Proc. Nutr. Soc.* 62(1), 87–93.

Rajilić-Stojanović, M. (2007) Diversity of the human gastrointestinal microbiota: novel perspectives from high throughput analyses. Ph.D. dissertation, Wageningen University, the Netherlands.

Rajilić-Stojanović, M., Heilig, H.G.H.J., Molenaar, D., et al. (2009) Development and application of the human intestinal tract chip, a phylogenetic microarray: analysis of universally conserved phylotypes in the abundant microbiota of young and elderly adults. Published online, doi:10.1111/j.1462-2920.2009.01900.x.

Ratcliffe, R.G., Shachar-Hill, Y. (2006) Measuring multiple fluxes through plant metabolic networks. *Plant J.* 45(4), 490–511.

Roediger, W.E. (1980) Role of anaerobic bacteria in the metabolic welfare of the colonic mucosa in man. *Gut* 21(9), 793–798.

Roediger, W.E. (1982) Utilization of nutrients by isolated epithelial cells of the rat colon. *Gastroenterology* 83(2), 424–429.

Rose, D.J., Venema, K., Keshavarzian, A., Hamaker, B.R. (2010) Starch-entrapped microspheres show a beneficial fermentation profile and decrease in potentially harmful bacteria during in vitro fermentation in faecal microbiota obtained from patients with inflammatory bowel disease. *Br. J. Nutr.* 103(10), 1514–1524.

Rowland, I.R. (1995) Toxicology of the colon: role of the intestinal microflora. In: Gibson, G.R., Macfarlane, G.T. (editors), *Human Colonic Bacteria: Role in Nutrition, Physiology, and Pathology*, pp. 155–174. CRC Press, Boca Raton, FL.

Satokari, R.M., Vaughan, E.E., Akkermans, A.D., Saarela, M., de Vos, W.M. (2001) Bifidobacterial diversity in human feces detected by genus-specific PCR and denaturing gradient gel electrophoresis. *Appl. Environ. Microbiol.* 67(2), 504–513.

Scheppach, W. (1994) Effects of short-chain fatty acids on gut morphology and function. *Gut* 35(1 Suppl), S35–S38.

Scheppach, W., Pomare, E.W., Elia, M., Cummings, J.H. (1991) The contribution of the large intestine to blood acetate in man. *Clin. Sci. (Lond)* 80(2), 177–182.

Scheppach, W., Sommer, H., Kirchner, T., et al. (1992) Effect of butyrate enemas on the colonic mucosa in distal ulcerative colitis. *Gastroenterology* 103(1), 51–56.

Scheppach, W., Weiler, F. (2004) The butyrate story: old wine in new bottles? *Curr. Opin. Clin. Nutr. Metab. Care* 7(5), 563–567.

Shulman, R.G., Rothman, D.L. (2001) ^{13}C NMR of intermediary metabolism: implications for systemic physiology. *Annu. Rev. Physiol.* 63, 15–48.

Singh, B., Halestrap, A.P., Paraskeva, C. (1997) Butyrate can act as a stimulator of growth or inducer of apoptosis in human colonic epithelial cell lines depending on the presence of alternative energy sources. *Carcinogenesis* 18(6), 1265–1270.

Suter, P.M. (2005) Carbohydrates and dietary fiber. *Handb. Exp. Pharmacol.* 170, 231–261.

Van de Wiele, T., Boon, N., Possemiers, S., Jacobs, H., Verstraete, W. (2004) Prebiotic effects of chicory inulin in the simulator of the human intestinal microbial ecosystem. *Fems Microbiol. Ecol.* 51(1), 143–153.

van Nuenen, M.H.M.C., Meyer, P.D., Venema, K. (2003) The effect of various inulins and *Clostridium difficile* on the metabolic activity of the human colonic microbiota in vivo. *Microb. Ecol. Health Dis.* 15(2), 137–144.

Van Nuenen, M., Venema, K., van der Woude, J.C., Kuipers, E.J. (2004) The metabolic activity of fecal microbiota from healthy individuals and patients with inflammatory bowel disease. *Dig. Dis. Sci.* 49(3), 485–491.

Vanhoutvin, S.A., Troost, F.J., Hamer, H.M., et al. (2009b) Butyrate-induced transcriptional changes in human colonic mucosa. *PLoS One* 4(8), e6759.

Vanhoutvin, S.A., Troost, F.J., Kilkens, T.O., et al. (2009a) The effects of butyrate enemas on visceral perception in healthy volunteers. *Neurogastroenterol. Motil.* 21(9), 952–976.

Velazquez, O.C., Lederer, H.M., Rombeau, J.L. (1997) Butyrate and the colonocyte. Production, absorption, metabolism, and therapeutic implications. *Adv. Exp. Med. Biol.* 427, 123–134.

Venema, K., Minekus, M., Havenaar, R. (2004) Advanced in vitro models of the gastro-intestinal tract: novel tools to study functionality of dietary fibres. In: Asp, N.-G., Miller Jones, J., Schaafsma, G., van der Kamp, J.W. (editors), *Dietary Fibre. Bio-active Carbohydrates for Food and Feed*, pp. 99–112. Wageningen Academic Publishers, Wageningen, The Netherlands.

Venema, K., van Nuenen, M.H.M.C., Van Den Heuvel, E.G., Pool, W., Van Der Vossen, J.M.B.M. (2003) The effect of lactulose on the composition of the intestinal microbiota and short-chain fatty acid production in human volunteers and a computer-controlled model of the proximal large intestine. *Microb. Ecol. Health Dis.* 15(2–3), 94–105.

Venema, K., Vermunt, S.H.F., Brink, E.J. (2005) D-tagatose increases butyrate production by the colonic microbiota in healthy men and women. *Microb. Ecol. Health Dis.* 17(1), 47–57.

Walker, A.W., Duncan, S.H., McWilliam Leitch, E.C., Child, M.W., Flint, H.J. (2005) pH and peptide supply can radically alter bacterial populations and short-chain fatty acid ratios within microbial communities from the human colon. *Appl. Environ. Microbiol.* 71(7), 3692–3700.

Wolever, T.M., Spadafora, P.J., Cunnane, S.C., Pencharz, P.B. (1995) Propionate inhibits incorporation of colonic [1,2-^{13}C]acetate into plasma lipids in humans. *Am. J. Clin. Nutr.* 61(6), 1241–1247.

Wolfe, R.R. (1992) *Radioactive and Stable Isotope Tracers in Biomedicine: Principles and Practice of Kinetic Analysis*. Academic Press, New York.

Wolin, M.J., Zhang, Y., Bank, S., Yerry, S., Miller, T.L. (1998) NMR detection of $^{13}CH_3{}^{13}COOH$ from 3-^{13}C-glucose: a signature for *Bifidobacterium* fermentation in the intestinal tract. *J. Nutr.* 128(1), 91–96.

Wong, C.S., Sengupta, S., Tjandra, J.J., Gibson, P.R. (1995) The influence of specific luminal factors on the colonic epithelium: high-dose butyrate and physical changes suppress early carcinogenic events in rats. *Dis. Colon Rectum* 48(3), 549–559.

Xu, J., Gordon, J.I. (2003) Honor thy symbionts. *Proc. Natl. Acad. Sci. USA* 100, 10452–10459

Zoetendal, E.G., Akkermans, A.D., De Vos, W.M. (1998) Temperature gradient gel electrophoresis analysis of 16S rRNA from human fecal samples reveals stable and host-specific communities of active bacteria. *Appl. Environ. Microbiol.* 64(10), 3854–3859.

Chapter 7

Fermentable Carbohydrates and Digestive Health

Joanne Slavin

Introduction

The role of fiber in the human body continues to be debated. Historically, carbohydrates that were not digested and absorbed and used for energy were considered a negative. It is now accepted that dietary fiber is an essential nutrient, on the basis of its protection against cardiovascular disease (Slavin, 2008). Even though we accept that delivering carbohydrate to the large intestine confers health benefits, the important properties of this fiber are unknown. If fiber survives gut transit, it will increase stool weight and aid laxation. The fermentation of this fiber by the gut microbiota produces intestinal gas, short-chain fatty acids (SCFAs), and lowers the gut pH. The Institute of Medicine (IOM) fiber committee suggested that rather than focus on the solubility of fiber, the important physiological effects of fiber are linked to viscosity and fermentability. Most likely, more viscous fiber lower blood lipids and slow glucose absorption in the small intestine. And the fermentation of fiber produces important physiological changes in the gut, although the extent of fermentation and the timing of that fermentation can also produce undesirable side effects, intestinal gas (Grabitske and Slavin, 2009).

How to define and measure dietary fiber continues to be debated. But there has been a movement away from chemical definition toward nutrition or physiological definitions. We accept that fiber is the

Nondigestible Carbohydrates and Digestive Health. Edited by T. M. Paeschke and W. R. Aimutis

carbohydrate leftovers of digestion, but we struggle to measure these components.

IOM definitions of fiber describe dietary fiber as intact and intrinsic in foods. Functional fibers are isolated carbohydrates with proven physiological effects. One important physiological effect of fiber is its ability to be fermented in the gut. Fibers can be completely fermented in the gut or not fermented at all. Additionally, fiber can be quickly fermented causing undesirable intestinal gas or fermented more slowly. Most models for gut fermentation use fecal homogenates and collect fermentation products. This chapter highlights fermentable fibers. Fermentable fibers are not necessarily prebiotic, but still have strong beneficial digestive health aspects. The two concepts are often confused and definitions are changing as the science moves forward.

Digestive Health

The term laxation describes a wide range of gastrointestinal effects, including stool weight, transit time, bloating and distention, flatus, constipation, and diarrhea. Fiber increases stool weight and promotes normal laxation (Grabitske and Slavin, 2008). Unfortunately, there are no standardized, accepted definitions for either of these conditions. Constipation is a prevalent, chronic condition in Western society, a common clinical complaint, but a poorly studied condition. It has been defined as less than three bowel movements per week, although most people define constipation as less than one bowel movement per day. Frequency of defecation is only one aspect of constipation. Ease of passage of stools and lack of straining are other components of normal laxation.

Bowel habit is affected by other variables, including medications, stress, physical activity, and volume of food, the type of food, fluid intake, hormones, and other environmental factors. Although subjective measures of bowel function are important variables to collect, objective measures such as wet and dry stool weight, gastrointestinal transit time, and frequency of defecation are useful biomarkers to study. Increased volume of bulk, softness or pliability of colonic contents, and increased intestinal motility may protect against constipation. Stool weight increases as fiber intake increases, but the additional fiber tends to normalize defecation frequency to once daily and gastrointestinal transit time to between 2 and 4 days. The increase in stool weight and volume is caused by the presence of the fiber, by the water that the fiber holds, and by partial fermentation of the fiber, which increases the amount of bacteria in stool.

Diarrhea is an unpleasant digestive disorder that can affect anybody at any time. Normally, when food is consumed, it remains in a liquid form during most of the digestive process; when the unabsorbed food residue passes through the large intestine, most of the fluids are absorbed and what remains is a semisolid stool. However, in diarrhea, the food and fluid ingested pass too quickly or in too large an amount (or both) through the large intestine. The fluids are not sufficiently absorbed and the result is a watery bowel movement. Commonly accepted criteria for clinical diarrhea are: elevated stool output (>200 g/day), watery, difficult to control bowel movements, and frequency of bowel movements exceeding three per day (Bliss et al., 1992). The colonic fermentation of dietary fiber may help to improve gastrointestinal tolerance and decrease diarrhea. Dietary fiber seems to reduce diarrhea by protecting from harmful bacterial overgrowth in intestine. A meta-analysis of randomized, controlled trials found no evidence that dietary fiber is effective in treating diarrhea (Homann et al., 1994).

Irritable bowel syndrome (IBS) is a functional gastrointestinal disorder defined as "a group of functional bowel disorders" and is characterized by chronic or recurrent abdominal pain or discomfort, usually in the lower abdomen, which is associated with disturbed bowel function (i.e., diarrhea or constipation alone or alternating) and feeling of abdominal distention and bloating (Drossman et al., 2002). Due to its persisting symptoms, IBS has a significant negative impact on health-related quality of life. The prevalence of IBS is estimated to range between 10 and 20% among adults in the United States and Europe (Drossman et al., 1993); however, this is an underestimate of prevalence indicated by the fact that 70% of symptomatic adults do not seek medical evaluation. Women with IBS report more symptoms of constipation and abdominal discomfort while men with IBS report more diarrhea. Psychological disturbances, such as anxiety and depression, are more common in individuals with IBS who seek medical consultation for their symptoms than in those who do not seek care for them, which suggests that psychological disturbance may amplify IBS symptoms and affect health care-seeking behavior.

Austin et al. (2009) found that a very-low-carbohydrate diet improved symptoms and quality of life in patients with diarrhea-predominant

IBS. A systematic review of dietary interventions for children with IBS concluded that there is a lack of high-quality evidence on the effectiveness of dietary interventions and there is no evidence that fiber supplements, lactose-free diets, or *Lactobacillus* supplements are effective in children with recurrent abdominal pain (Huertas-Ceballo et al., 2009). Bijkerk et al. (2004) conducted a systematic review of fiber in the management of IBS. They determined the following outcome measures: the proportion of patients reporting clinical relief (global IBS symptom improvements), the proportion of patients reporting improved IBS-related abdominal pain, and the proportion of patients reporting an improvement in IBS-related constipation.

A meta-analysis showed general fiber supplementation alleviates IBS symptoms globally, but there is no benefit in the relief of abdominal pain, which is the most important feature capable of distinguishing IBS from functional constipation or functional diarrhea (Longstreth et al., 2006). Bijkerk et al. (2004) reported most psyllium studies showed improvement of global IBS symptoms compared to the placebo. Improvement in global symptoms was shown in only two of the six insoluble wheat fiber trials, while miller bran treatment did not improve symptoms. Overall, bran was no better than placebo in regards to improvement of global symptoms, but improvement was shown in IBS patients with constipation. Neither probiotics nor prebiotics are effective in the treatment of IBS (Spiller, 2008).

Thus, although it is generally accepted that dietary fiber plays a role in digestive health (Slavin, 2008), few studies have been conducted on selected dietary fiber sources and their effects on gut health. Most studies are in diseased populations and even in these studies the role of dietary fiber on gut health remains inconsistent.

Gut Fermentation

What actually gets to the human large intestine? The digestive tract is quite efficient in extracting carbohydrates for energy use. In general, about 95% of available carbohydrate is digested and absorbed in the upper gastrointestinal tract in healthy individuals. Low digestible carbohydrates include dietary fiber, resistant starch, oligosaccharides, and sugar alcohols (Grabitske and Slavin, 2009). The fate of these products once they enter the large intestine is largely unknown in human subjects.

It has been known for hundreds of years that fiber is apparently "digested" in man. Although the importance of fermentation in the ruminant has been long accepted, the microbial ecosystem of the human large intestine is equally important (Wolin, 1981). Williams and Olmsted (1936) completed a series of experiments with three medical students. Subjects consumed cell wall preparations from carrot, cabbage, peas, bran, and other fibers. Only about 20% of the hemicellulose was recovered in the feces while 60% of the bran could be recovered. Larger studies were completed in the 1980s to compare the "digestibility" of fibers. Digestion of fiber was highest for cabbage and lowest for Solka Floc, an isolated cellulose (Van Soest, 1984). We also fed defined diets for 1 month, with and without 16 per day Solka Floc (Slavin et al., 1981). Mean digestibility of fiber on the low-fiber diet was 65% while on the added Solka Floc diet, it was 21%. Essentially, none of the Solka Floc was fermented during gut residence.

Chen et al. (1998) reported that the apparent digestibility of fiber in wheat bran was 56% while it was 95% for oat bran. Auerbach et al. (2007) suggested that 50% of the ingested dose of polydextrose was excreted undigested. Most pectin is degraded during gut passage (Holloway et al., 1983). Nutriose, a resistant starch made from wheat, is partially hydrolyzed and absorbed in the small intestine (15%). The rest passes to the colon where 75% of the initial amount is fermented in the large intestine and 10% is excreted (Van den Heuvel et al., 2004). Results with a mixed, high-fiber diet from fruits and vegetables find that digestibilities of hemicellulose, cellulose, and lignin were 88, 42, and 18%, respectively (Kelsay et al., 1981). Based on results like these, it is typically thought that hemicellulose is the most fermentable of traditional fiber, with cellulose being less fermentable, and lignin mostly inert. Gender differences have also been reported in fiber breakdown, with women breaking down more fiber, probably as a result of slower gut transit (Lampe et al., 1993). It is difficult to compare these studies as different methods were used to measure fiber in food and feces. But it is clear that certain fibers are extensively fermented while other fibers pass through the digestive tract essentially untouched.

Although often insoluble fibers are said to be "nonfermentable" and soluble fibers "fermentable", this is incorrect. As discussed above, most fibers are partially fermented in the gut, although there is a large range among fibers. Determining the breakdown of carbohydrates in the large intestine in human subjects is problematic. Most measures available examine only the input and output—thus, there is no information how rapidly the fermentation took place and at what part of the colon the products were absorbed. Eastwood (1992) describes the colon as two organs: the right side is a fermenter, the left side affects continence. The right side of the colon is involved in nutrient salvage so that dietary fiber, resistant starch, fat, and protein are utilized by bacteria and the end products absorbed and used by the body. A better understanding of this system would also provide important data on the calorie capture from the wide range of fibers currently consumed by humans.

Nutritional Classifications for Carbohydrates

The traditional chemical classification or categorization of carbohydrates by degree of polymerization (DP) is not useful when discussing digestion and the gastrointestinal tract. A nutritional perspective considers the behavior of carbohydrates in the gastrointestinal tract, whether they are absorbable, digestible, or fermentable. This classification is most predictive of potential gastrointestinal effects associated with carbohydrate intake (Table 7.1).

Absorbable carbohydrates are absorbed in the small intestine without any digestion in the mouth or small intestine. Monosaccharides are absorbable carbohydrates. Although a small percentage of monosaccharides are absorbed passively by the enterocytes, most monosaccharides are transported across the cell membrane by a carrier—glucose and galactose require the sodium-glucose transporter-1 (SGLT1) for secondary active transport, and fructose requires the glucose transporter-5 (GLUT5) for facilitated diffusion. All other carbohydrates must be enzymatically digested into monosaccharides before they can be absorbed in the small intestine and sent into circulation.

Digestible carbohydrates include the disaccharides lactose and sucrose as well as starch, a polysaccharide. Salivary alpha-amylase hydrolyzes the alpha-1,4 linkages in starch, yielding the disaccharide maltose, the trisaccharide maltriose, and dextrins with 5–9 glucose molecules. Pancreatic alpha-amylase continues to digest starch into maltose in the small intestine. Disaccharidases in the brush border of the small intestine (maltase, sucrase, and lactase) complete carbohydrate digestion by hydrolyzing maltose, sucrose, lactose, into the absorbable monosaccharides glucose, fructose, and galactose.

Table 7.1. Classifications of carbohydrates.

Carbohydrate	Chemical (Structural)	Nutritional	Physiological
Fructose, galactose, glucose	Monosaccharide	Absorbable	Glycemic
Lactose, sucrose	Disaccharide	Digestible/ fermentable	Glycemic
Lactulose	Disaccharide	Fermentable	Nonglycemic
Inulin, fructooligosaccharides	Oligosaccharide	Fermentable	Nonglycemic
Starch	Polysaccharide	Digestible	Glycemic
Pectin, gums, polydextrose, psyllium	Polysaccharide	Fermentable/ nonfermentable	Nonglycemic
Resistant starch	Polysaccharide	Fermentable/ nonfermentable	Nonglycemic
Cellulose, hémicellulose, resistant maltodextrin, lignin	Polysaccharide	Fermentable/ nonfermentable	Nonglycemic

Fermentable carbohydrates are those that are not digested and absorbed in the small intestine but are metabolized by colonic bacteria. Carbohydrate fermentation may yield SCFAs (acetate, propionate, and butyrate) and gases (carbon dioxide, hydrogen, and methane) (Cummings et al., 1984; Cummings and Macfarlane, 2002). These products of carbohydrate fermentation are absorbed in the large intestine (providing energy), used by the bacteria for energy, released as flatulence, or expelled as biomass in the feces. The category of fermentable carbohydrates includes carbohydrates that cannot be digested because they have linkages for which humans do not have digestive enzymes: pectin, beta-glucans, psyllium, gums, oligosaccharides, polydextrose, and resistant starch (all of which are included in the definition of fiber proposed by the Institute of Medicine (2001)). Other carbohydrates, such as the disaccharide sugar alcohols and rare sugars, are fermentable; they are digestible by human enzymes, but their breakdown may occur more slowly, so these carbohydrates may be incompletely digested and absorbed in the small intestine. In addition, the monosaccharide sugar alcohols and rare sugar D-tagatose may be incompletely absorbed due to a slower rate of absorption. With the exception of erythritol, these carbohydrates may be fermented if they reach the large intestine. In addition, carbohydrates such as sucrose and lactose, though typically digested and absorbed in the small intestine, may be maldigested or malabsorbed; if this occurs, they are fermented in the large intestine.

Fiber is fermented by anaerobic intestinal bacteria that generate SCFA, which serve as energy sources for colonic mucosal cells (Slavin, 2008). Fermentable dietary fibers are thought to alter the gut environment, not only by inducing the production of SCFA, but also by altering the gut microflora. Indeed, fermentable dietary fibers have a significant prebiotic effect by altering the intestinal microflora composition toward a more beneficial distribution by leading to selective stimulation of microbial growth, which eventually helps to increase the water-holding capacity of the colonic content and fecal moisture.

Acetate, propionate, and butyrate are the SCFA produced in the highest concentrations (Topping and Clifton, 2001). Acetate is a fuel for skeletal and cardiac muscle, kidney, and the brain. Butyrate is the preferred fuel of the colonic epithelium, in particular, the distal colon and rectum. Propionate is metabolized by the liver and may play a role in cholesterol lowering. Fibers produce varying proportions of individual SCFA and thus differing concentrations of total SCFA. Physiological status may be improved by consuming fermentable fiber, so it is important to understand the fermentability of each type of fiber.

Fiber fermentability is difficult to study in vivo due to the invasiveness of colon studies and the dynamic nature of the colon. Fiber fermentation can be estimated by measuring fiber consumed in the diet and then collecting fecal samples and measuring fiber left in feces. This is tedious and difficult since feces contain bacterial cell walls that are also isolated in fiber methods. No easy biomarkers exist to measure fiber fermentation in vivo, so generally in vitro models are used.

Particle size, solubility, surface area, and other factors affect the extent of fermentation and the nature of the SCFAs. In living systems, SCFA are absorbed from the colonic lumen shortly after they are produced. No method has been developed to accurately measure SCFA absorption in vivo and measuring SCFAs excreted in the feces is the best estimate of SCFAs being produced in the colon. However, 95–99% of SCFA are absorbed from the lumen, so excreted SCFA concentration represent a very small portion of SCFA produced. Therefore, studying the amount of SCFA in the feces of human volunteers would provide only a partial picture.

In vitro models that study fiber fermentability are currently the best models to assess SCFA production in humans. A closed laboratory system can provide an estimate of fiber fermentability without losing SCFA to colonic absorption and, therefore, in vitro fermentation with representative human colonic microflora is a proven, noninvasive, time-efficient means to estimate fiber fermentability. Indeed, batch fermentation has been shown to degrade nonstarch polysaccharides (NSP) to a similar extent as the human colon, on the basis of residual NSP in fecal samples and fermentation flasks (Wisker et al., 1998).

Disruption of cell walls can increase fermentability of dietary fiber. Coarse wheat bran has a greater fecal bulking effect than finely ground wheat bran when fed at the same dosage (Heaton et al., 1988), suggesting that the particle size is an important factor in determining physiological effect. Coarse bran delayed gastric emptying and accelerated small bowel transit. The effect seen with coarse bran was similar to the effect of inert plastic particles, suggesting that the coarse nature of whole grains as compared to refined grains has a unique physiological effect beyond composition differences between whole and refined grains (McIntyre et al, 1997).

McIntosh et al. (2003) fed rye and wheat foods to overweight middle-aged men and measured markers of bowel health. The men were fed low-fiber cereal grains foods providing 5 g of dietary fiber for the refined grain diet and 18 g of dietary fiber for the whole grain diet, either high in rye or wheat. This was in addition to a baseline diet that contained 14 g of dietary fiber. Both the high-fiber rye and wheat foods increased fecal output by 33–36% and reduced fecal β-glucuronidase activity by 29%. Postprandial plasma insulin was decreased by 46–49% and postprandial plasma glucose by 16–19%. Rye foods were associated with significantly increased plasma enterolactone and fecal butyrate, relative to wheat and low-fiber diets. The authors conclude that rye appears more effective than wheat in overall improvement of biomarkers of bowel health.

Nonfermentable or poorly fermentable carbohydrates pass through the gastrointestinal tract largely unchanged and are excreted in the feces. These carbohydrates are neither digested in the small intestine nor are they fermented to a significant degree in the large intestine. This category includes components of plant cell walls such as celluloses, hemicelluloses, and lignin.

Fermentation

Carbohydrates that reach the large intestine are fermented to different degrees, depending on the DP, solubility, and structure of the carbohydrates. Fermentation of the carbohydrates in the large intestine produces gases, which may cause bloating, distention, borborygmi, and flatulence. If the carbohydrates are not fermented in the large intestine, either because the bacteria do not metabolize the carbohydrates or because intake exceeds the fermentation capacity of the bacteria, the water remains bound to the carbohydrates that are eliminated in the feces, which increases fecal bulk (Stephen, 1991) but also may produce a watery stool or diarrhea (Livesey, 2001).

In general, individuals would first notice symptoms related to carbohydrate fermentation such as borborygmi and excessive flatus, followed by abdominal discomfort such as bloating and cramps; when the colonic microflora's capacity to ferment the carbohydrates has been exceeded, diarrhea develops (Marteau and Flourié, 2001). Low-digestible carbohydrates with relatively low molecular weights have a greater potential to affect laxation compared to those with higher molecular weights because they have a greater osmotic force and draw more water into the bowels (Marteau and Flourié, 2001).

Factors that increase gastrointestinal transit time generally increase gastrointestinal acceptability of fiber because they increase the time available for digestion and absorption. Tolerance increases when low-digestible carbohydrates are consumed as part of a meal (Livesey, 2001). The presence of other foods and nutrients such as fat and protein increase transit time and decrease the potential for gastrointestinal effects. Personal characteristics also affect bowel function and may contribute to gastrointestinal effects. Transit time, intestinal enzymes, and colonic microflora are affected by gastrointestinal health and other physiological characteristics as well as lifestyle choices. The composition and health of colonic microflora affect the fermentation of carbohydrates. Antibiotic treatments may alter colonic bacteria, reducing fermentation and causing diarrhea. In addition, viral or bacterial infections, common in children, cause secretory diarrhea in which increased chloride ions and water are secreted into the small intestine but not reabsorbed (Sarbin, 2005).

Methods to Measure Carbohydrate Malabsorption and Fermentation

Several methods have been devised to estimate the proportion of ingested carbohydrate that is malabsorbed. Intubation and ileostomy studies measure carbohydrate malabsorption directly (Cummings and Englyst, 1991; Strocchi and Levitt, 1991), whereas hydrogen breath studies measure malabsorption indirectly by measuring a product of colonic fermentation (Rumessen, 1992). Strocchi and Levitt (1991) describe intubation studies as the "gold standard" for determining carbohydrate malabsorption because this method measures malabsorption directly. Perfusion studies are conducted on subjects with a normal gastrointestinal tract. Subjects consume the test meal, and then the contents of the ileum are aspirated and analyzed for carbohydrates (Strocchi and Levitt, 1991). A polyethylene glycol (PEG) marker is added to the meal in order to compare the amount of carbohydrate in question with the completely malabsorbed PEG marker (Strocchi and Levitt, 1991). This technique is invasive as they require intubation and the insertion of a tube into the terminal ileum. In addition, the presence of the tube may affect the function of the gastrointestinal tract, and it is impossible to aspirate and analyze a constant fraction of the material that enters the terminal ileum.

Ileostomy studies measure the amount of carbohydrate that escapes digestion in the ileum by analyzing the effluent from the terminal ileum of subjects with ileostomies (Cummings and Englyst, 1991). Investigators should take precautions to prevent breakdown of the carbohydrate in the effluent bag (Cummings and Englyst, 1991). This method has been criticized because people with ileostomies have increased bacteria in the terminal ileum and absorption of carbohydrates may be elevated; however, Cummings and Englyst (1991) concluded, on the basis of their previous studies of NSP, that any increased fermentation in this model is negligible and that this technique is a reliable method for obtaining quantitative data on starch absorption. Despite these potentially confounding variables, studies of carbohydrate malabsorption using the ileostomy model have yielded results that are similar to those from intubation studies (Cummings and Englyst, 1991).

The hydrogen breath test assesses carbohydrate malabsorption indirectly (Rumessen, 1992). The fermentation of carbohydrates by bacteria in the large intestine produces hydrogen, carbon dioxide, and (sometimes) methane, some of which is expired in the breath (Rumessen, 1992). Some studies also measure breath methane concentrations, but not all people expire methane and results are generally more variable than hydrogen production (Rumessen, 1992). A baseline measurement of the hydrogen concentration in expired breath is taken (often after an overnight fast) followed by additional measurements at given intervals such as 30 minutes anywhere from 2 to 24 hours. Changes from baseline are calculated and often compared to a control carbohydrate like lactulose that is completely malabsorbed in the small intestine (Strocchi and Levitt, 1991; Rumessen, 1992). In addition, the area under the curve (AUC) may be calculated from the breath hydrogen measurements interval data to compare fermentation of different carbohydrates (Rumessen, 1992).

Some studies use hydrogen breath tests to quantify malabsorption in the small intestine, but many factors affect the amount of hydrogen expired, including the degree of carbohydrate that is neither digested in the small intestine nor fermented in the large intestine but, rather, excreted in the feces (Rumessen, 1992). Several factors may affect breath hydrogen expiration, including initial peaks, bacterial overgrowth, delayed gastric emptying, prolonged small intestinal transit, absent hydrogen excretion, elevated baseline hydrogen levels, antibiotics, laxatives/enemas, active diarrheal disease, acidic colonic pH, defective sampling or storage, hyperventilation, pulmonary disease, exercise, smoking, sleep, and mood (Rumessen, 1992). Study exclusion/inclusion criteria and protocol should address these variables. The capacity of the bacteria to ferment undigested carbohydrate also affects breath hydrogen measurements. Undigested carbohydrate that is excreted in the feces must be accounted for by analyzing fecal composition.

When using breath hydrogen results to quantify carbohydrate malabsorption (such as starch), results do not always reflect the results from direct measurements with ileostomy and intubation studies (Cummings and Englyst, 1991). In addition, measurements of breath hydrogen often do not correlate with the occurrence or severity of gastrointestinal effects reported in study questionnaires or diaries. Therefore, studies that measure malabsorption based only on data from hydrogen breath measurements should be considered with caution and are only minimally useful for assessment of fermentation.

Although hydrogen breath tests provide some information about carbohydrate fermentation and potential symptoms, they are only of limited value. Generally, subjects are fasted before they consume the test product, which may be dissolved in liquid, so conditions are not necessarily representative of normal use and may even increase symptoms. The breath hydrogen test cannot be performed on infants and young children and others who are unable to fast overnight (Strocchi and Levitt, 1991). In addition, not all people produce or expire hydrogen (Strocchi and Levitt, 1991), so this test would not reflect actual carbohydrate fermentation in these individuals.

Cummings and Englyst (1991) conclude that breath hydrogen studies are better used as qualitative rather than quantitative measurements of (starch) absorption; ileostomy studies and intubation studies should be used for quantitative measurements of the amount of carbohydrate that escapes digestion in the small intestine. Despite the disadvantages of the hydrogen breath test, it is a rather simple and less invasive method to assess the occurrence of carbohydrate fermentation and, indirectly, malabsorption (Rumessen, 1992).

Measuring acetate in venous blood (Rumessen, 1992) or SCFAs and the pH of feces may also provide qualitative or semiquantitative information about carbohydrate malabsorption; however, it is difficult to account for the extent of the absorption of SCFAs in the colon, so such fecal measurements may have limited value.

Carbohydrates present in fecal samples provide quantitative information of the amount of carbohydrates that escape both digestion and absorption in the small intestine and fermentation in the large intestine; however, this method does not provide measurements for the amount of carbohydrate that is fermented in the large intestine, so it is less useful for carbohydrate than fat assessments since the latter is fully excreted in the feces (Strocchi and Levitt, 1991).

Oral tolerance tests that measure the rise in serum glucose levels should be interpreted with caution with regard to carbohydrate malabsorption because they are measurements of carbohydrate absorption and do not provide information about the amount of carbohydrate that is malabsorbed and enters the large intestine (Strocchi and Levitt, 1991).

Table 7.2. Estimated acceptable daily doses for carbohydrates.

Carbohydrate	Dose (g/day)
Nonstarch polysaccharides	10–50
Resistant starch	45
Sugar alcohols	20–40
Rare sugars	30–50

Gastrointestinal Tolerance for Poorly-Digested Carbohydrates

Despite great variety in study designs, protocol, and types of results, some recommendations and generalizations can be made for carbohydrate-consumption-based studies reporting on the gastrointestinal effects (Table 7.2) (Grabitske and Slavin, 2009).

The NSP guar gum and inulin fructans may cause mild symptoms such as flatulence with intakes below 10 g per day, but most individuals would be able to tolerate daily intakes of 10–15 g with an increase in the occurrence and severity of symptoms with 20 or more grams per day. Consuming at least 40 g FOS would likely increase fecal volume to meet the definition of clinical diarrhea; studies of inulin and guar gum at higher levels would be needed to estimate such a threshold for these nondigestible carbohydrates. Compared to other nonglycemic carbohydrates, resistant starch has a high laxation threshold since reports of diarrhea were rare, even at levels as high as 80 g/day. The main side effect, excessive flatulence, is related to colonic fermentation and is significantly greater at doses above 45 g/day.

In Vitro Studies

To compare the fermentation of popular fiber sources, we use an in vitro fermentation system. The details of the system were first presented in Pylkas et al., (2005). Briefly, fibers are fermented for 0, 4, 8, 12, and 24 hours. No added fiber served as a control for SCFAs produced by substrate present in the feces. Fibers are hydrated 12 hours prior to the start of the fermentation with 40 mL of sterile trypticase peptone fermentation media at 4°C. Two hours prior to inoculation, the bottles were heated to 37°C in a shaking water bath. Fibers inoculated with

Table 7.3. In vitro results on fermentation of fiber with an in vitro system.

Study	Question	Fibers Fed	Results
Stewart and Slavin (2006)	Effect of molecular weight	Guar gum	Molecular weight affects fermentation and SCFA production
Stewart et al. (2008)	Effect of chain length	FOS, inulin	Short chains are more rapidly fermented—differences in SCFA with chain length
Stewart et al. (2009)	Does donor affect results	Wheat dextrin, psyllium inulin	Absolute values different, but patterns similar
Stewart et al. (2009)	Does adding probiotic (*Lactobacillus reuteri*) affect results	Wheat dextrin, psyllium inulin	Some change with inulin—but no affect on wheat dextrin or psyllium
Timm et al. (2008)	Is gas production linked to fermentation	Wheat dextrin, psyllium inulin	Inulin produced more hydrogen gas; differences in rate of SCFA production
Stewart and Slavin (2009)	Does particle size affect fermentation	Wheat bran, wheat by products	Fine by-product produced the greatest SCFA concentrations

feces from three individuals consuming a nonspecific Western diet were pooled and homogenized and collected under anaerobic conditions. SCFA concentrations are determined by gas chromatography using a flame ionizing detector. The results of five studies are summarized in Table 7.3.

In the first study, we determined whether molecular weight of guar gum altered fermentation and SCFA production (Stewart and Slavin, 2006). A medium chain guar gum produced the highest amount of butyrate at 24 hours. The molecular weight was positively correlated with acetate production and negatively correlated with propionate production. So, molecular weight of guar gum does alter fermentation results.

In the second study we compared fermentation of fructooligosaccharides and inulin (Stewart et al., 2008). Chain lengths varied from an average DP of 3 to greater than 20. The rate of FOS fermentation was higher than inulin fermentation during 0–4 hours and the rate of inulin fermentation was higher than FOS fermentation during 12–24 hours for all SCFAs. Chain length affects in vitro fermentation, with short chain being rapidly fermented and long chains being steadily fermented in an in vitro system.

We conducted two experiments in the third paper (Stewart et al., 2009). The same fermentation was run at two separate occasions, to determine the effect of fecal donor on results. Although SCFA production was not the same in the different batch runs, the patterns were consistent between the runs. We also added a probiotic to the fermentation system, *Lactobacillus reuteri*. Wheat dextrin and psyllium were largely unaffected by the probiotic addition. *L. reuteri* improved inulin's fermentation profile by reducing the total SCFA peak at 4 hours and enhancing fermentation at 8 and 12 hours.

In the fourth study, three common dietary fiber supplements wheat dextrin (WD, Benefiber®), psyllium (PS, Metamucil®), and inulin (Fiber Sure®) were examined for pH, SCFAs, and gas production (Timm et al., 2008). An established in vitro fermentation model was used to simulate colonic fermentation at 0, 4, 8, 12, and 24 hours. WD and inulin produced significantly larger decreases in pH at 24 hours compared to PS. Inulin produced significantly more hydrogen and total gas. All treatments produced similar total SCFA concentrations at 24 hours; however, the rate of production was different. PS had a declining rate of SCFA production from 12 to 24 hours, while WD and inulin had a higher rate during that period. Fast fermenting substrates may not provide as much SCFAs to the distal colon as slow fermenting substrates. Differences in fermentation rate, gas production, and SCFA production observed for WD, PS, and inulin may affect their tolerability and may have therapeutic implications in colon health which requires further study.

The effect of particle size and fraction of wheat bran on fermentation were measured in the final study (Stewart and Slavin, 2009). Fine by-product produced the great SCFA concentrations. By-product (both coarse and fine) produced greater SCFA concentrations than bran (both large and small particle sizes). Small/fine particle size and by-product fraction of bran increase SCFA production compared with large/coarse particle size. Bran characteristics and composition should be considered when manufacturing foods due to the diversity of physiological effects.

Conclusion

Measurement of fiber breakdown in vivo is difficult for both participants and investigators. To determine fiber breakdown, a defined diet must be fed for more than a week so that the digestive tract can adapt to the additional fiber. Fecal samples must be collected for at least 5 days and fiber measured in both the food consumed and the fecal samples. The breakdown of different fibers varies greatly and any estimated of "fiber fermentation" must be measured in this fashion.

Thus, there is an interest and need for in vitro systems to screen resistant carbohydrates. In these systems, a fecal slurry is used and fibers are anaerobically fermented with measures of SCFA production and gas production. These studies can compare different fibers for their ability to be fermented to SCFAs and help design fiber supplements that are well tolerated and also produce important SCFAs.

References

Auerbach, M.H., Craig, S.A.S., Howlett, J.F., Hayes, K.C. (2007) Caloric availability of polydextrose. *Nutr. Rev.* 65, 544–549.

Austin, G.L., Dalton, C.B., Morris, C.B., et al. (2009) A very-low-carbohydrate diet improves symptoms and quality of life in diarrhea-predominant irritable bowel syndrome. *Clin. Gastroenterol. Hepatol.* 7(6), 706–708.

Bijkerk, C.J., Muris, J.W.M., Knottnerus, J.A., Hoes, A.W., NeWit, N.J. (2004) Systematic review: the role of different types of fibre in the treatment of irritable bowel syndrome. *Aliment. Pharmacol. Ther.* 19, 245–251.

Bliss, D.Z., Guenter, P.A., Settle, R.G. 1992. Defining and reporting diarrhea in tube fed patients: what a mess! *Am. J. Clin. Nutr.* 55, 753–759.

Chen, H., Haack, V.S., Janecky, C.W., Vollendorf, N.W., Marlett, J.A. (1998) Mechanisms by which wheat bran and oat bran increase stool weight in humans. *Am. J. Clin. Nutr.* 68, 711–719.

Cummings, J.D. (1984) Microbial digestion of complex carbohydrates in man. *Proc. Nutr. Soc.* 43, 35–44.

Cummings, J.H., Englyst, H.N. (1991) Measurement of starch fermentation in the human large intestine. *Can. J. Physiol. Pharmacol.* 69, 121–129.

Cummings, J.H., Macfarlane, G.T. (2002) Gastrointestinal effects of prebiotics. *Br. J. Nutr.* 87(Suppl 2), S145–S151.

Drossman, D.A., Camilleri, M., Mayer, E.A., Whitehead, W.E. (2002) AGA technical review on irritable bowel syndrome. *Gastroenterology* 123, 2108–2131.

Drossman, D.A., Li, Z., Andruzzie, E., et al. (1993) US householder survey of functional gastrointestinal disorders: prevalence, sociodemography, and health impact. *Dig. Dis. Sci.* 38, 1569–1580.

Eastwood, M.A. (1992) The physiological effect of dietary fiber: an update. *Ann. Rev. Nutr.* 12, 19–35.

Grabitske, H.A., Slavin, J.L. (2008) Laxation and like: assessing digestive health. *Nutr. Today* 43, 193–198.

Grabitske, H.A., Slavin, J.L. (2009) Gastrointestinal effects of low-digestible carbohydrates. *Crit. Rev. Food Sci. Nutr.* 49, 327–360.

Heaton, K.W., Marcus, S.N., Emmett, P.M., Bolton, C.H. (1988) Particle size of wheat, maize, and oat test meals: effects on plasma glucose and insulin responses and on the rate of starch digestion in vitro. *Am. J. Clin. Nutr.* 47, 675–682.

Holloway, W.D., Tasman-Jones, C., Maher, K. (1983) Pectin digestion in humans. *Am. J. Clin. Nutr.* 37, 253–255.

Homann, H.H., Kemen, M., Fuessenich, C., et al. (1994) Reduction in diarrhea incidence by soluble fiber in patients receiving total or supplemental enteral nutrition. *J. Parent. Enteral Nutr.* 18, 486.

Huertas-Ceballos, A.A., Logan, S., Bennett, C., Macarthur, C. (2009) Dietary interventions for recurrent abdominal pain (RAP) and irritable bowel syndrome (IBS) in children. *Cochrane Database Syst. Rev.* (1), CD003019.

Institute of Medicine. (2001) *Dietary Reference Intakes: Proposed Definition of Dietary Fiber*. National Academy of Sciences, Washington, DC.

Kelsay, J.L., Goering, H.K., Behall, K.M., Prather, E.S. (1981) Effect of fiber from fruits and vegetables on metabolic responses of human subjects: fiber intakes, fecal excretions, and apparent digestibilities. *Am. J. Clin. Nutr.* 34, 1849–1852.

Lampe, J.W., Fredstrom, S.B., Slavin, J.L., Potter, J.D. (1993) Sex differences in colonic function: a randomized trial. *Gut* 34, 531–536.

Livesey, G. (2001) Tolerance of low-digestible carbohydrates: a general view. *Br. J. Nutr.* 85(Suppl 1), S7–S16.

Longstreth, G.F., Thompson, W.G., Chey, W.D., Houghton, L.A., Mearin, F., Spiller, R.C. (2006) Functional bowel disorders. *Gastroenterology* 130, 1480–1491.

Marteau, P., Flourié, B. (2001) Tolerance to low-digestible carbohydrates: symptomatology and methods. *Br. J. Nutr*. 85(Suppl 1), S17–S21.

McIntosh, G.H., Noakes, M., Royle, P.J., Foster, P.R. (2003) Whole-grain rye and wheat foods and markers of bowel health in overweight middle-aged men. *Am. J. Clin. Nutr.* 77, 967–974.

McIntyre, A., Vincent, R.M., Perkins, A.C., Spiller, R.C. (1997) Effect of bran, ispaghula, and inert plastic particles on gastric emptying and small bowel transit in humans: the role of physical factors. *Gut* 40, 223–227.

Pylkas, A.M., Juneja, L.R., Slavin, J.L. (2005) Comparison of different fibers for in vitro production of short chain fatty acids by intestinal microflora. *J. Med. Foods* 8, 113–116.

Rumessen, J.J. (1992) Hydrogen and methane breath tests for evaluation of resistant carbohydrates. *Eur. J. Clin. Nutr.* 46(Suppl 2), S77–S90.

Sarbin, J.S. (2005) Gastrointestinal disorders. In: Copstead, L.C., Banasik, J.L. (editors), *Pathophysiology*, pp. 887–911. Elsevier, Inc., Philadelphia.

Slavin, J.L., Brauer, P.M., Marlett, J.A. (1981) Neutral detergent fiber, hemicellulose, and cellulose digestibility in human subjects. *J. Nutr.* 111, 287–297.

Slavin, J.L. (2008) Position of the American Dietetic Association: health implications of dietary fiber. *J. Am. Diet. Assoc.* 108, 1716–1731.

Spiller, R. (2008) Review article: probiotics and prebiotics in irritable bowel syndrome. *Aliment. Pharmacol. Ther.* 28, 385–397.

Stephen, A.M. (1991) Starch and dietary fibre: their physiological and epidemiological interrelationships. *Can. J. Physiol. Pharmacol.* 69, 116–120.

Stewart, M., Slavin, J.L. (2006) Molecular weight of guar gum affects short-chain fatty acid profile in model intestinal fermentation. *Mol. Nutr. Food Res.* 50, 971–976.

Stewart, M.L., Savarino, V., Slavin, J.L. (2009) Assessment of dietary fiber fermentation: effect of *Lactobacillus reuteri* and reproducibility of short-chain fatty acid concentrations. *Mol. Nutr. Food Res.* 53, S114–S120.

Stewart M.L., Slavin J.L. (2009) Particle size and fraction of wheat bran influence short-chain fatty acid production in vitro. *Br. J. Nutr.* 102, 1404–1407.

Stewart, M.L., Timm, D., Slavin, J.L. (2008) Fructooligosaccharides exhibit more rapid fermentation than long-chain inulin in an in vitro fermentation system. *Nutr. Res.* 28, 329–334.

Strocchi, A., Levitt, M.D. (1991) Measurement of starch absorption in humans. *Can. J. Physiol. Pharmacol.* 69, 108–110.

Timm, D., Stewart, M., Paredes-Diaz, A., Slavin, J.L. (2008) Wheat dextrin, psyllium, and inulin produce distinct short-chain fatty acid (SCFA) profiles, fermentation patterns, and gas volumes in vitro. *Am. J. Gastroenterol.* 103, S173.

Topping, D.L., Clifton, P.M. (2001) Short-chain fatty acids and human colonic function: roles of resistant starch and nonstarch polysaccharides. *Physiol. Rev.* 81, 1031–1064.

Van Den Heuvel, E.G.H.M., Wils, D., Pasman, W.J., Bakker, M., Saniez, M-H., Kardinaal, A.F.M. (2004) Short-term digestive tolerance of different doses of NUTRIOSE FB, a food dextrin, in adult men. *Eur. J. Clin. Nutr.* 58, 1046–1055.

Van Soest, P.J. (1984) Some physical characteristics of dietary fibres and their influence on the microbial ecology of the human colon. *Proc. Nutr. Soc.* 43, 25–33.

Williams, R.D., Olmsted, W.H. (1936) The effect of cellulose, hemicellulose, and lignin on the weight of the stool: a contribution to the study of laxation in man. *J. Nutr.* 11, 433–449.

Wisker, E., Daniel, M., Rave, G., Feldheim, W. (1998) Fermentation of non-starch polysaccharides in mixed diets and single fibre sources: comparative studies in human subjects and in vitro. *Br. J. Nutr.* 80, 253–261.

Wolin, M.J. (1981) Fermentation in the rumen and human large intestine. *Science* 213, 1463–1468.

Chapter 8

Overview of Dietary Fiber and Its Influence on Gastrointestinal Health

Devin J. Rose and Bruce R. Hamaker

Introduction

The term "dietary fiber" was coined by Hipsley (1953) to include indigestible plant cell wall constituents. Up to this time, most experts believed that the benefits of dietary fiber were limited to alleviating constipation (Kritchevsky, 1988). Then, in 1956, Cleave published a paper hypothesizing many diseases, including diabetes, cardiovascular disease, diverticulitis, and colon cancer, were due to high consumption of sugar and low intake of dietary fiber (Cleave, 1956). This paper was the basis for a study by Burkitt et al. (1974) that detailed eight diseases common to the United States, but rare in Africa. They suggested that lack of dietary fiber in the Western diet was the underlying problem, laying the groundwork for dietary fiber research today (Kritchevsky, 1988).

Recent research has revealed dietary fiber plays roles in both glucose (Butt et al., 2007; Sharma et al., 2008) and lipid metabolism (Brown et al., 1999; Hosobuchi et al., 1999; Butt et al., 2007); its consumption is correlated with decreased risk of heart disease (King, 2005; Theuwissen and Mensink, 2008), diabetes (Nuttall, 1993; Cameron-Smith & Collier, 2001; Brennan, 2005), obesity (Slavin, 2005; Butt et al., 2007), and some types of cancer (Lim et al., 2005; Young et al., 2005; Suzuki et al., 2008), and it may prevent certain types of colonic diseases, such as inflammatory bowel disease (Galvez et al., 2005), travelers' diarrhea

Nondigestible Carbohydrates and Digestive Health. Edited by T. M. Paeschke and W. R. Aimutis

(Cummings et al., 2001), and irritable bowel syndrome (Bijkerk et al., 2004; Giannini et al., 2006).

While the benefits of dietary fiber are numerous and diverse in nature, the purpose of this chapter is to discuss the beneficial effects that dietary fiber exhibits on digestive health. First, the current intake of dietary fiber in the United States will be compared to recommendations, and then common dietary fibers that contribute to colonic health will be defined, followed by the effect dietary fibers have on conditions in the upper and lower gastrointestinal tract. Finally, the relationship between dietary fiber intake and two pathologies of the digestive tract, inflammatory bowel disease and colon cancer, will be discussed.

Dietary Fiber Intake

Modern food processing techniques have decreased the amount of fiber consumed in developed countries (Burkitt and Spiller, 1992). Experts hypothesize that primitive diets contained between 80 and 250 g of dietary fiber/day (Sobolik, 1994; Brand-Miller & Holt, 1998; Dreher, 2001; Leach, 2007), whereas today mean dietary fiber intakes are only 13–18 g/day (Institute of Medicine, 2002a). The typical diet contains about one serving of whole grain per day (Liese et al., 2003), and a mere 7% of Americans consume three or more servings of whole grain per day (United States Department of Agriculture, 2002).

The current recommendations for dietary fiber intake from the Food and Nutrition Board (FNB) of the Institute of Medicine are 38 and 25 g/day (14 g/1,000 kcal) for adult men and women, respectively (Institute of Medicine, 2002b). To help meet these requirements, the Dietary Guidelines for Americans suggest consuming a diet high in whole cereal grains, fruits, and vegetables (United States Department of Agriculture, 2005). These products, however, are often not convenient or as desirable to consumers as more processed foods. Therefore, there is a pressing need for food scientists to improve the functionality of dietary fibers, thus broadening their applications for the development of commonly consumed foods with high dietary fiber content.

Types of Dietary Fiber

Dietary fiber consists of a complex class of compounds for which numerous definitions have been proposed (Institute of Medicine, 2001).

The most universal definition of dietary fiber includes indigestible polysaccharides and lignin; however, controversy arises when deciding whether to include indigestible oligosaccharides, nonstarch polysaccharides from animal sources, or indigestible proteins and waxes.

The formal definition of fiber is important because it influences analytical method development, dietary recommendations, food composition databases, developments in the food industry, dietary fiber research emphases, and nutrition labeling (Institute of Medicine, 2001). Most definitions today include a stipulation that dietary fibers must exhibit certain health benefits (DeVries, 2001; Institute of Medicine, 2001).

Resistant Starch

Structure and sources

Resistant starch refers to the total amount of starch and products of starch degradation that resist starch digestion in the small intestine of healthy people, thus acting as dietary fiber (Asp, 1992). In the United States, resistant starch intake has been estimated at 4.9 g/day for adults, although this figure can be highly variable among individuals (Murphy et al., 2008). Controversy regarding whether to include resistant starch as a dietary fiber has arisen in the past. One of the main arguments against calling resistant starch dietary fiber is that resistant starch content can vary widely depending on food preparation conditions. For example, RS1 refers to dietary starch that is indigestible due to physical inaccessibility of the starch to digestive enzymes (Table 8.1).

Coarsely ground grains fall into this category; however, upon fine grinding, the resistant starch dramatically decreases. Similar changes in resistant starch content can occur with the other three types of resistant starch, which are: RS2 (high amylose starch), RS3 (retrograded starch), and RS4 (chemically modified starch) (Englyst et al., 1992). Nevertheless, according to the latest definitions set forth by the FNB of the Institute of Medicine, AACC International, and the Codex Committee on Nutrition and Foods for Special Dietary Uses, resistant starch has been included as a dietary fiber (Cummings and Stephen, 2007).

Behavior in the gastrointestinal tract

The effects of resistant starch on digestive processes in the upper gastrointestinal tract have been inconclusive. In an analysis of 25 studies, Nugent (2005) found that 15 of these studies showed a decrease in

Table 8.1. Dietary fiber structures, classifications, and sources.

Dietary fiber	Structure	Classifications	Sources
Resistant starch	α-(1→4)-D-Glc*p* (backbone and branches); α(1→6)-D-Glc*p* (branch points)	Insoluble, highly fermentable	
	RS1: physically inaccessible starch		Coarsely ground flour
	RS2: high amylose starch		Specialty starches
	RS3: retrograded starch		Gelatinized and cooled starch
	RS4: chemically modified starch		Specialty starches
Cellulose	β-(1→4)-D-Glc*p*	Insoluble, poorly fermentable	Whole grains, fruits, vegetables
β-Glucan	β-(1→3 and 1→4)-D-Glc*p* (backbone); β-(1→6)-D-Glc*p* (branches; only on some types)	Soluble, viscous, highly fermentable	Oat bran, barley flour
Chitin	β-(1→4)-D-GlcNAc	Insoluble, poorly fermentable	Shellfish
Chitosan	β-(1→4)-D-GlcNAc and GlcN	Soluble and viscous/gel-forming in acidic solutions; insoluble at neutral pH	Alkali-treated chitin
(Glucurono) arabinoxylan	β-(1→3 or →4)-D-Xyl*p* (backbone); complex branches of Ara, Xyl, GlcA, and Gal; branches can be esterified with phenolic compounds; backbone can be acetylated; polymers often cross-linked through oxidative di- and trimerization of phenolic compounds	Insoluble or soluble and poorly to highly fermentable depending on fine structure and processing conditions; soluble forms can be low (cereal arabinoxylans) or high viscosity/gel-forming (psyllium)	Whole grains, cereal bran, psyllium seed and husk

Arabinogalactan	β-(1→3 or →4)-D-Gal*p* (backbone); side groups of Ara, Gal, GlcA, Rha	Soluble, fermentable	Fruits and vegetables, *Acacia senegal* (gum arabic)
Pectins	α-(1→4)-D-GalA with varying degrees of methyl esterification (backbone) and side groups containing Rha, Gal	Soluble, viscous, highly fermentable	Fruits and vegetables
Guar gum	β-(1→4)-D-Man*p* (backbone); side groups of Gal	Soluble, viscous, highly fermentable	*Cyamopsis tetragonolobus* seeds
Xanthan gum	β-(1→4)-D-Glc*p* (backbone); side chain of Man, GlcA, pyruvate	Soluble, viscous	*Xanthamonas campestris* cultures
Gellan	Tetrasaccahride repeat of 3 β-(1→4)-D-Glc*p* interrupted with 1 α-(1→3)-L-Rha*f*; some acetylation	Soluble, viscous, gel forming	*Pseudomonas elodea* cultures
Alginate	Backbone of β-D-ManA and α-L-GulA	Soluble at neutral pH; insoluble at acidic pH; viscous; gel forming	Algae

(Continued)

Table 8.1. (*Continued*)

Dietary fiber	Structure	Classifications	Sources
Carrageenan	α(1→3 and 1→4)-D-Gal*p* with varying degrees of sulfate esters and presence of 3,6-anhydro-Gal*p*	Soluble, viscous, gel forming	Seaweed
Polydextrose	Random polymer of Glc, sorbitol, and citric acid	Soluble, low viscosity, partially fermentable	Chemically synthesized
Inulin/Fructo-oligosaccharides	Fructose oligo- or polymer linked β(2→1) with Glc at the non-reducing end	Inulin insoluble; FOS soluble and low viscosity; highly fermentable	Onions, Jerusalem artichokes, enzymatically synthesized
Galacto-oligosaccharides	Oligomers of galactose	Soluble, low viscosity, highly fermentable	β-galactosidase treatment of lactose
Lactulose	D-Gal*p*-β-(1→4)-D-Fru	Soluble, low viscosity, highly fermentable	Glucose isomerase treatment of lactose

Glc, glucose; Gal, galactose; Xyl, xylose; Ara, arabinose; Rha, Rhamnose; GlcA, D-glucuronic acid; GalA, galacturonic acid; GlcN, D-glucosamine; GlcNAc, N-acetyl-D-glucosamine.

postprandial glucose and insulin responses with resistant starch, while 10 studies showed no effect. Animal studies have consistently shown resistant starch decreases serum lipid and cholesterol concentrations, but no effect has been consistently shown in humans (Nugent, 2005).

In the large bowel, resistant starch is a highly fermentable dietary fiber. In fact, resistant starch is believed to be the most significant contributor to colonic short-chain fatty acid (SCFA) production (Bird et al., 2000; Topping and Clifton, 2001). This is based on estimations that 60–80 g/day of fermentable substrate are needed to sustain the metabolism of the colonic microbiota; yet, dietary fiber consumption in Westernized countries is rarely above 20 g/day (Institute of Medicine, 2002a). Oligosaccharides, endogenous secretions, and indigestible protein may contribute some fermentable substrate, but resistant starch is believed to contribute to the majority of this "carbohydrate gap" (Topping et al., 2003; Nugent, 2005).

Cellulose

Structure and sources

Cellulose is a homopolysaccharide consisting of β-(1→4)-linked glucose residues (Table 8.1). The nature of the β-(1→4) linkage allows cellulose polymers to arrange into rigid, insoluble microfibrils through multiple hydrogen bonds. These microfibrils provide structural integrity and mechanical support for plants (Mutwil et al., 2008). In its native state, cellulose is found in whole grains, fruits, and vegetables. It can be obtained commercially from wood pulp and added to food as an anticaking, thickening, or texturizing agent; however, its uses in food are limited because cellulose often imparts a gritty texture and demonstrates little functionality in food. Two strategies have been employed to overcome these shortcomings: (1) chemical modification; and (2) high shear accompanied by alkaline and hydrogen peroxide treatments.

Cellulose contains multiple hydroxyl groups, which may be chemically modified to change its functionality. These modifications can result in water- or oil-soluble polymers that can be used in food or drug delivery applications. These modified celluloses are further categorized on the basis of their degree of substitution, which is defined as the average number of substituted hydroxyl groups on each glucose residue of the cellulose chain. Because each glucose monomer contains three available hydroxyl groups, the maximum degree of substitution that modified cellulose can have is three.

High shear and alkaline conditions have also been utilized to completely disintegrate the structure of the cell walls of agricultural products (Inglett, 1997, 1998). The cellulose that is recovered has a gel-like consistency with high water-holding capacity, which can be used to increase the moistness of meats and as a fat replacer in baked goods (Warner and Inglett, 1997).

Behavior in the Gastrointestinal Tract

Cellulose and modified cellulose pass through the digestive tract with little degradation. In an in vitro study analyzing the breakdown of different types of carbohydrates during fermentation, Barry et al. (1995) found that only 6–7% of cellulose was degraded after 24 hours, and methyl cellulose produced among the lowest levels of SCFA compared to a range of commercially available dietary fibers (Pylkas et al., 2005). The multiple intermolecular hydrogen bonds, in the case of cellulose, or the chemical substituents, in the case of modified cellulose, impede hydrolysis even by bacterial hydrolases.

β-*Glucan*

Structure and sources

Cellulose and β-glucan are similar in that they are both β-linked glucans; however, whereas cellulose forms rigid water-insoluble microfibrils, β-glucan is water soluble. The water solubility of β-glucans can be attributed to its glycosidic linkages, which, depending on source, can differ markedly. The most important and widely studied β-glucans, from the dietary perspective, are the cereal β-glucans that are found in the highest concentrations in oats and barley. This β-glucan is a linear polysaccharide consisting of β-(1→4) linkages, which are interrupted usually every 2–4 glucose units by a β(1→3) linkage. Bacteria can also produce β-glucans, the most prevalent of which is curdlan, an essentially linear homopolysaccharide of β-(1→3)-linked glucopyranosyl units. This type of β-glucan is also produced by some types of algae. Some yeast and mushrooms produce a β-glucan consisting of a β-(1→3) linked backbone, with β-(1→6)-branches on about 5% of the glucose units (Bell et al., 1999; Table 8.1). As expected, these differences in structure not only affect the functional properties in foods, such as gelling and viscosity (McIntosh et al., 2005; Lazaridou and Biliaderis, 2007), but also affect the physiological properties (Wood et al., 2002).

Behavior in the gastrointestinal tract
β-Glucan has a marked effect on viscosity of digesta in the upper gastrointestinal tract, affecting gastric emptying, absorption of nutrients, and small intestinal motility (Malkki and Virtanen, 2001). Viscosity can be affected by structure, in particular the ratio of (1→3) to (1→4) linkages, the presence of (1→6) linkages, and molecular weight (Wood, 2007). In the lower gastrointestinal tract, β-glucan is a completely fermentable dietary fiber.

Chitin and Chitosan

Structure and sources
Chitin is found in the exoskeletons of crustaceans and in the cell walls of some fungi and insects. It has a cellulosic backbone, except that the monomers that make up chitin are aminated (Table 8.1). In nature, the amino groups are completely acetylated, contributing to the rigid, insoluble nature of chitinous materials. Chitosan is the product of partial de-N-acetylation of chitin in hot alkali. This product is soluble in dilute acid solutions, and unique among polysaccharides because of its cationic nature. Chitin and its derivatives exhibit numerous industrial applications, including a moisture retainer in skin care products, a binding material for toxic metals, an immobilizing agent for enzymes, and an antimicrobial film-former (Tharanathan and Kittur, 2003; Krajewska, 2004; No et al., 2007; Prashanth and Tharanathan, 2007; Anal et al., 2008); chitosan has also received considerable attention as an encapsulant for drugs, phytochemicals, or probiotics (George and Abraham, 2006; Varshosaz, 2007). The use in chitosan in food, however, has been confined mostly to chitosan oligomers and has been extremely limited (Jeon et al., 2000).

Behavior in the gastrointestinal tract
Chitosan is unique among the soluble dietary fibers because of its cationic nature. Consequently, it is soluble in the acidic conditions in the stomach, but becomes insoluble in the small intestine. Under simulated upper gastrointestinal conditions, the viscosity of chitosan-

containing solutions is elevated in the acidic conditions of the stomach, but, after neutralization, the viscosity is significantly reduced compared to pectin, another soluble dietary fiber (Razdan et al., 1997). Chitin is an insoluble dietary fiber and has little effect on upper gastrointestinal conditions.

In the large intestine, chitosan acts as a fermentable dietary fiber, and chitin acts as a poorly fermentable dietary fiber. In dogs, $<10\%$ of chitosan was recovered in feces, while chitin was completely recovered (Okamoto et al., 2001).

Hemicellulose

Structure and sources

The term hemicellulose was coined by Schulze in 1891 to include polysaccharide substances extractable from agricultural biomass with aqueous alkali solutions (Ebringerova et al., 2005). Hemicelluloses include a number of complex polysaccharides made up of a variety of pentoses and hexoses, including arabinose, fucose, galactose, glucose, mannose, and uronic acids. There are two main types of hemicelluloses that occur in food: xylans and galactans. β-Glucans and mannans are also important food polysaccharides that have sometimes been grouped under hemicelluloses (Ebringerova et al., 2005). These carbohydrates are discussed elsewhere in this chapter.

Xylans contain a β-(1→4)-linked xylopyranosyl backbone with varying degrees of acetylation (Table 8.1). The xylan backbone can be substituted with arabinose, glucose, galactose, mannose, glucuronic acid, and other moieties. Xylans can be found in hardwoods, herbaceous plants, and grasses, but the most important food sources of xylans are cereals. Cereal xylans are referred to as arabinoxylans, (glucurono) arabinoxylans, or heteroxylans, depending on the nature of the side chains along the xylan backbone. One of the most unique aspects of these xylans is the presence of feruloyl crosslinks. In their native structure, these crosslinks serve to provide structure and rigidity to the cell wall. Unfortunately for food formulators, this key function makes the polymers insoluble and poorly functional in foods. Feruloyl crosslinks can be removed by alkali saponification (Doner et al., 1998), resulting in a soluble, more functional arabinoxylan (Yadav et al., 2007); however, commercial production of alkali extractable arabinoxylan has been limited by the costs associated with processing.

Galactans are divided into two groups: arabinogalactan type I and arabinogalactan type II (Table 8.1). Arabinogalactan type I is found intimately associated with pectin in fruits and vegetables, such as orange, apple, and potato. It is comprised of a linear β-(1→4)-linked galactopyranosyl backbone, with about 20–40% substitution of α(1→3)-linked arabinofuranosyl side chains. The main type of arabinogalactan type II is larch arabinogalactan, but can also be found in cell walls of dicots and cereals, usually associated with proteins (Serpe and Nothnagel, 1995). It has a β-(1→3)-linked galactopyranosyl backbone with short side chains of arabinose and galactose linked at position 6 (Ebringerova et al., 2005).

Behavior in the gastrointestinal tract

Because of their diverse structure, hemicelluloses can either be completely or poorly fermented. Most native hemicelluloses are insoluble, and contribute little to the digestive process in the small intestine (Dikeman et al., 2006). In the large intestine, these hemicelluloses are poorly fermented, and act mainly as bulking agents (Dintzis et al., 1979). Hemicelluloses that have been solubilized with alkali show low viscosity (Yadav et al., 2007), and are highly fermentable by colonic bacteria (Glitso et al., 2000).

Psyllium

Structure and sources

Psyllium is an important dietary fiber supplement; ground husks, seeds, or a mixture of the two are mixed with water and consumed as a beverage. Unlike the rest of the dietary fibers described in this chapter, psyllium refers to the crude source of dietary fiber, rather than the polysaccharide fraction itself.

The gel-forming component of the husk, representing 55–60% of the dry weight, is a neutral arabinoxylan consisting of a β-(1→4)-linked xylopyranosyl backbone, substituted at position 2 with single xylopyranosyl residues, and at position 3 with a trisaccharide moiety: α-L-arabinofuranosyl-(1→3)-β-D-xylopyranosyl-(1→3)-α-L-arabinofuranosyl-(1→ (Fischer et al., 2004; Table 8.1). In contrast, the major polysaccharide in psyllium seeds is an acidic glucuronoarabinoxylan, the backbone of which consists of β(1→3)-xylopyranosyl linkages interspersed among the normal β(1→4)-linkages. The backbone is heavily substituted at positions 2 and 3 with single xylose units and two major disaccharide side groups: α-L-arabinofuranosyl-(1→3)-β-D-xylopyranosyl-(1→, and α-D-glucuronopyranosyl-(1→3)-α-L-arabinofuranosyl-(1→ (Samuelsen et al., 1999).

Behavior in the gastrointestinal tract
As mentioned, the main reason for consuming psyllium is as a dietary fiber supplement. Psyllium is partially fermented by colonic bacteria, but the gel-forming polysaccharide component is not fermented. Consequently, psyllium forms thick gels, and increases the viscosity of both digesta and feces (Marlett et al., 2000).

Pectins

Structure and sources
Commercial pectin, which is the most abundant type of pectin, is harvested from lemon, lime and orange albedo, and apple pommace, but is contained in many other fruits and vegetables. It is a linear polymer of up to 100 degrees of polymerization, consisting of α(1→4)-linked galacturonic acid with varying amounts of methyl esterification on the acid moiety (Table 8.1). The degree of methylation determines its gel-forming properties. Other types of pectin include rhamnogalacturonan I and II, xylogalacturonan, and apiogalacturonan, based on the type of side groups they contain (Mohnen, 2008).

Traditionally, pectin has been used for thickening jams and jellies. High methoxy pectins form gels in the presence of acid and adequate amounts of sugar. Low methoxy pectins form gels with calcium and are used in low sugar jams and jellies. A more recent usage of pectin has been as a drug delivery system. Pectic acid, which is completely demethylated pectin, can be electrostatically crosslinked with a divalent cation. Its use, in conjunction with other polysaccharides or proteins, forms web-like structures capable of delivering drugs and other compounds to any region of the gastrointestinal tract for controlled drug delivery (Liu et al., 2006). Partially hydrolyzed pectins with low degrees of methylation, sometimes called modified citrus pectins, have also received considerable attention as substitutes for human milk oligosaccharides in baby formula (Fanaro et al., 2005), as anticancer agents due to their proposed influence on apoptosis (Nangia-Makker et al., 2002), and as heavy metal chelators (Eliaz et al., 2006).

Behavior in the gastrointestinal tract

In the upper gastrointestinal tract, pectins increase viscosity; high methoxyl pectin increases the viscosity of duodenal juice more than low methoxyl pectin (Isaksson et al., 1982). In the lower bowel, pectins act as a substrate for colonic bacteria. Low methoxyl pectin is more rapidly fermented than high methoxyl pectin (Dongowski et al., 2002).

Gums

Structure and sources

Polysaccharide gums are numerous and diverse in nature. They can be harvested from seeds, plant exudates, bacterial cultures, and algae. Although these polysaccharides are technically classified under dietary fibers, they are generally used in very small quantities in foods, and therefore contribute little to total dietary fiber intake.

The most popular seed gums include guar gum and locust bean gum. Guar gum is harvested from the seeds of *Cyamopsis tetragonolobus*, and locust bean gum comes from *Ceratonia siliqua*. While these gums are made up of the same basic chemical structure, a β(1→4)-linked D-mannopyranosyl backbone with α-D-galactose side chains linked (1→6) to the mannose backbone (Fox, 1992; Table 8.1), their physical properties differ markedly depending not only on the degree of galactosyl branching, but on the pattern of branching (Daas et al., 2002).

Plant exudate gums include gum arabic, gum tragacanth, gum karya, and gum ghatti. The most prevalent of these is gum arabic, which is harvested mainly from *Acacia senegal* (Imeson, 1992). The polysaccharide fraction of the crude gum consists of an arabinogalactan with a β-(1→3)-linked galactopyranosyl backbone, heavily substituted at C-6 with side chains consisting of arabinose, galactose, glucuronic acid, and rhamnose (Anderson et al., 1966; Table 8.1).

Bacterial exopolysaccharides represent an important source of food gums. The most common bacterial gum is xanthan. Xanthan gum is produced from fermentation of a select strain of *Xanthamonas campestris*. It consists of a cellulosic backbone that is substituted with a trisaccharide side chain: α-D-mannopyranosyl-(1→4)-α-D-glucuronopyranosyl-(1→2)-6-acetyl-β-D-mannopyranosyl-(1→3)-, with about half of the terminal mannose units containing pyruvate (Table 8.1). Gellan, another bacterial polysaccharide, is produced from *Pseudomonas elodea*, and is a linear anionic polysaccharide based on a tetrasaccharide repeat: →3)-β-D-glucopyranosyl-(1→4)-β-D-glucuropyransoyl-(1→4)-β-D-glucopyranosyl(1→4)-α-L-rhamnofuranosyl(1→. The backbone is acylated on one of the glucose moieties per tetrasaccharide repeat with either one or both of acetate and glycerate. The degree of acylation determines the strength of gellan gels; native gellan produces soft gels and deacylated gellan produces hard, brittle gels (Gibson, 1992; Fiaho et al., 2008).

Polysaccharide gums can also be harvested from algal sources. The two most prevalent food polysaccharides that fit into this category are alginates and carrageenans. Alginates are salts of β-D-mannuronic acid and α-L-guluronic acid, which are harvested from brown algae (Table 8.1). The monomers can be joined together in homopolymeric groups of mannuronic acid, termed M-blocks, and guluronic acid, termed G-blocks, or in heteropolymeric groups of alternating mannuromic and guluronic acid. The distribution of these uronic acids influences the functional properties: alginates with long G-blocks give stronger gels than those with long M-blocks. In the food industry, alginates are mainly used as a binding agent in restructured foods, and as a stabilizer in frozen foods (Onsoyen, 1992). Major research interest has also surrounded the use of alginates alone or in conjunction with other polymers, as controlled-release, drug-delivery agents (George and Abraham, 2006). Carrageenans are composed of a galactose backbone, linked together by alternating α(1→3) and α(1→4) bonds, with varying degrees of sulfate esterification and presence of 3,6-anhydro-D-galactose: κ-Carrageenan contains the lowest level of sulfate esters and the highest level of 3,6-anhydro-D-galactose, while λ-carrageenan contains the highest level of sulfate esters and the lowest level of 3,6-anhydro-D-galactose; ι-carrageenan falls in between κ- and λ-carrageenan. κ-Carrageenan forms strong, slightly opaque gels; ι-carrageenan forms elastic, clear gels; and λ-carrageenan forms viscous solutions.

Behavior in the gastrointestinal tract

As mentioned, polysaccharide gums are generally not added to foods in large amounts; therefore, research on the effect that these gums may have on the gastrointestinal tract are limited. The most significant work has been on the digestive properties of guar gum. In the upper gastrointestinal tract, guar gum acts as a viscous polysaccharide (Theuwissen and Mensink, 2008) and it is fermented in the large bowel (Stewart and Slavin, 2006). Because the high viscosity of guar gum makes it difficult to add to foods in high levels, some manufacturers have marketed partially hydrolyzed guar gum as a dietary fiber supplement (Yoon et al., 2008), although this is likely to affect its viscous influence on the gastrointestinal tract (Stewart and Slavin, 2006).

The algal gums have been studied to a limited extent with respect to their influence on the gastrointestinal tract. Alginates are partially fermented by colonic bacteria (Brownlee et al., 2005). Alginates that have been electrostatically crosslinked are not fermented by the colonic bacteria (Rose et al., 2009). Carrageenans may have a detrimental effect on the lower bowel in patients with inflammatory bowel disease due to the high sulfate content, which may increase hydrogen sulfide production in the lower bowel and interfere with butyrate metabolism (Roediger et al., 1997). Indeed, higher levels of sulfate-reducing bacteria have been found in patients with inflammatory bowel disease (Loubinoux et al., 2002).

Polydextrose

Sources and structure

Polydextrose is a synthetic oligosaccharide prepared from the condensation of sorbitol and glucose in the presence of citric or phosphoric acid (Table 8.1). Monomers of polydextrose are linked in a random fashion, making the polymer indigestible. The average molecular weight of polydextrose is 2,000, with an average degree of polymerization of 12. It is used in the food industry mostly as a bulking agent and sugar replacer (Flood et al., 2004), but more recently has been gaining interest as a possible prebiotic.

Behavior in the gastrointestinal tract

Polydextrose is a soluble dietary fiber; however, because of its low molecular weight, it imparts little viscosity to the upper gastrointestinal tract. In the lower gastrointestinal tract, approximately 50% of the ingested dose is excreted in the feces (Auerbach et al., 2007). This is slightly higher than poorly fermentable dietary fibers such as cellulose and hemicellulose.

Prebiotics

Structure and sources

A class of dietary fibers, which have received considerable attention in recent years, is prebiotics. The term "prebiotics" was coined by Gibson and Roberfroid (1995) to include specific types of indigestible carbohydrates (i.e., dietary fibers) that selectively stimulate the growth of beneficial bacteria in the colon, thus improving colonic health. Under the current definition, only a select few indigestible oligosaccharides may be classified as true prebiotics, fructooligosaccharides (FOS) (including inulin), galactooligosaccharides, and lactulose (Gibson et al., 2004; Table 8.1). Inulin and FOS have the same chemical structure, a $\beta(2\rightarrow 1)$-linked polymer of fructose with a single glucose unit at the reducing end, but differ in molecular weight. Inulin is commercially extracted from chicory root, and FOS are produced by hydrolysis of inulin (Singh and Gill, 2006) or enzymatically by the microbial enzyme fructosyl transferase with sucrose as substrate (Maiorano et al., 2008). Galactooligosaccharides are produced from the transglycosylation of lactose with β-galactosidase, wherein the galactose moiety of one lactose molecule is transferred to the reducing end of another lactose moiety with the subsequent release of glucose. Lactulose is also produced from lactose, except the reducing glucose unit is isomerized to a ketone under alkaline conditions (Yang and Silva, 1995). A number of other products are seen as potential prebiotics pending further research: isomaltooligosaccharides, lactosucrose, xylooligosaccharides, soybean oligosaccharides, glucooligosaccharides, and resistant starch (Topping et al., 2003; Gibson et al., 2004; Crittenden, 2006).

Behavior in the gastrointestinal tract

As oligosaccharides, most prebiotics exhibit little influence on small intestinal viscosity, and thus have little influence on small intestinal conditions. The major effects of prebiotics are, by definition, in the large intestine. These oligosaccharides stimulate the growth of beneficial bacteria in the colon; the beneficial bacteria implied in this definition have been understood to be of the genera *Bifidobacterium* and *Lactobacillus* (Gibson, 1999). This definition is somewhat restrictive due to our limited understanding of the intestinal microbiota and their influence on health and disease. As more is revealed about the microbiota through developing PCR-based methodologies, new beneficial types of bacteria that proliferate with specific dietary fibers or oligosaccharides are likely to emerge (Crittenden, 2006). For example, dietary fibers that selectively stimulate the growth of butyrogenic bacteria may prove quite beneficial to patients with inflammatory bowel disease, due to the anti-inflammatory properties of this SCFA.

Influence of Dietary Fiber on the Upper Gastrointestinal Tract

In the gastrointestinal tract, dietary fibers can have numerous effects, depending on type. By definition, dietary fiber is not metabolized in the upper gastrointestinal tract; however, dietary fiber can have a substantial influence on digestive processes in the stomach and small intestine. These influences are mainly due to the viscosity of the dietary fiber; therefore, soluble dietary fibers have more of an effect in the upper gastrointestinal tract than do insoluble types (Dikeman and Fahey, 2006).

Stomach

The acidic conditions in the stomach necessitate the formation of a mucous layer to protect the gastric lining from ulceration. Viscous dietary fibers, such as pectin and guar gum, may have a marked effect on the thickness of this mucous layer, effectively increasing its protective effects against certain types of ulcers (Rydning et al., 1982; Tasman-Jones, 1992).

Soluble dietary fibers also delay gastric emptying as a result of their water-holding capacity and ability to form gels (Dikeman and Fahey, 2006). The decrease in gastric emptying rate has two major effects: (1) it prolongs satiety, which may have implications for weight control and obesity (Howarth et al., 2001; Slavin, 2005); and (2) it may retard the rate of nutrient absorption in the small intestine, possibly influencing susceptibility to type 2 diabetes (Chaikomin et al., 2006; discussed below).

Small Intestine

Physiologically, soluble dietary fibers have been shown to stimulate mucosal growth and cell proliferation and exfoliation in the small intestine in rats (Tasman-Jones, 1992). Studies on the direct influence of dietary fiber on the intestinal epithelium in humans is lacking, although it is recognized that total parenteral nutrition in humans induces atrophy and a restructuring of the intestinal mucosa resulting in a decrease in absorptive surfaces (Groos et al., 1996).

Absorption of glucose

Soluble dietary fibers decrease the rate of intraluminal absorption of glucose; they do not inhibit transport across the epithelium (Eastwood, 1992), but delay gastric emptying and decrease mixing of digesta through increased viscosity (Dikeman and Fahey, 2006) resulting in a delay in glucose uptake.

Leclere et al. (1994) hypothesized that the delay in gastric emptying was the major mechanism whereby soluble dietary fibers decrease the rate of glucose absorption in the small intestine. They fed healthy volunteers diets containing either low or high viscosity guar gum orally, or had the treatment diets duodenally infused using an enteral feeding tube. Significant differences were found between the low and high viscosity treatments at all time points tested only when the guar gum was ingested orally. This suggested that gastric emptying was the major reason for the decrease in the rate of glucose absorption. Conversely, the experiments of Ehrlein and Stockmann (1998) lend support to the hypothesis that increased digesta viscosity reduces the rate of nutrient absorption, rather than gastric emptying. They measured the absorption rates of carbohydrate in pigs by perfusing a portion of the jejunum with simulated digesta containing different concentrations of guar gum. The absorption rate decreased linearly or logarithmically as a function of the guar gum concentration or luminal viscosity.

Most likely, the rate of glucose and other nutrient absorption is a function of both gastric emptying and digesta viscosity. Meyer et al. (1988) studied the effects of gastric emptying rate on small intestinal absorption of glucose. They instilled glucose solutions alone or containing pectin or guar gum into the stomachs of dogs via orogastric intubation. The solutions were allowed to leave the stomach, and immediately replaced with fresh glucose or a glucose-containing dietary fiber solution at the same rate. This allowed the authors to observe the effects of each dietary fiber supplement as a function of gastric emptying only or digesta viscosity in the small intestine only. They found that the pectin and guar gum effectively slowed gastric emptying, but that the decreased glucose response was significant only when the fiber-containing solutions were instilled in both organs, indicating that gastric emptying rate as well as digesta viscosity was important in the reduction in glucose response.

Absorption of lipids

The absorption of lipids is impeded by soluble dietary fibers (Cassidy and Calvert, 1992). There are two likely reasons for the decrease in the rate of lipid uptake. First, the viscosity of digesta containing soluble dietary fibers impedes emulsification of fat, which is essential for absorption. Second, dietary fibers bind bile acids, which are necessary for lipid micelle formation prior to absorption. The rapid increase in bile acid excretion with soluble dietary fibers may also decrease the risk of heart disease due to interruption of enterohepatic recirculation of cholesterol, causing an increase in the body's requirement for cholesterol.

Absorption of amino acids
In humans, dietary fiber has not shown consistent effects on protein digestibility and utilization (Gallaher and Schneeman, 1992). It is clear that dietary fiber causes an increase in fecal nitrogen output; however, it is not clear whether this represents a decrease in protein digestibility or an increase in bacterial mass and sloughed off cells as a result of increased fermentation in the large bowel (Dikeman and Fahey, 2006).

Absorption of vitamins and minerals
In addition to their effects on macronutrients, soluble dietary fibers can also affect the absorption of micronutrients. In particular, dietary fibers that contain ionic groups, such as alginates, pectins, and carrageenans may bind minerals and prevent their absorption (Eastwood, 1992), although other components of many sources of dietary fibers, such as phytate and oxalate, also bind minerals, masking the effect of the dietary fiber itself (Eastwood, 1992; Munoz and Harland, 1992). Additionally, owing to the decrease in lipid absorption, fat-soluble vitamin absorption may be impeded with soluble dietary fibers (Kasper, 1992). Although these are apparently detrimental effects of dietary fibers, it is unlikely that these changes would result in deficiencies in individuals consuming adequate and varied diets (Eastwood, 1992).

Influence of Dietary Fiber on the Lower Gastrointestinal Tract

By far, dietary fiber exhibits most of its benefits on digestive health in the large intestine, the site of dietary fiber metabolism. In this region of the gastrointestinal tract, a dietary fiber's propensity for fermentation is important and thus can be divided into fermentable and poorly fermentable, instead of the traditional divisions of soluble and insoluble. In general, soluble dietary fibers are fermentable dietary fibers, and insoluble dietary fibers are poorly fermentable, although there are exceptions. Fermentable dietary fibers provide the main source of energy for the microbiota. Voluminous literature has been devoted to the interactions between host and the microbiota (Dethlefsen et al., 2007; Fukata and Abreu, 2008; Stecher and Hardt, 2008), and it is clear that maintenance of a healthy gut microbiota is important to whole body and gut health. Poorly fermentable dietary fibers also exhibit an important role in gut health by increasing fecal bulk and decreasing whole gut transit time.

Dietary Fiber Fermentation and the Microbiota

The large bowel is a teeming ecology of numerous bacteria, termed the microbiota. Bacterial levels increase from about 10^8/mL in the proximal colon to about 10^{11}/mL in the distal colon (Marteau et al., 2001). Molecular phylogenetic sequencing of the bacterial 16S rRNA gene has revealed that the human microbiota is composed of about 800 different species and over 7,000 strains of bacteria (Backhed et al., 2005). Interestingly, however, these species belong to a relatively few major divisions; Firmicutes and Bacteroidetes are dominant in most individuals, with Actinobacteria and Proteobacteria as minor groups. In some individuals, minor levels of other phyla may be found, including chlamydiae, cyanobacteria, deferribacteres, deinococcus-thermus, fusobacteria, spirochaetes, verrucomicrobia, and the candidate phyla TM7 and SR1 (Dethlefsen et al., 2007). In contrast, soil samples routinely contain bacteria from 20 or more different major divisions (Dethlefsen et al., 2007).

Once dietary fibers reach the large intestine, they can either be completely or poorly fermented by the microbiota. Those carbohydrates that are fermentable mostly include the soluble fibers, such as pectin, gums, and FOS; however, there are a number of soluble fibers that are poorly fermented, such as methylcellulose (Pylkas et al., 2005), psyllium (Marlett et al., 2000), and alginate (Brownlee et al., 2005). Poorly fermentable fibers mostly include the insoluble fibers, such as cellulose and crosslinked hemicellulose, although there are also some insoluble dietary fibers that are highly fermentable, such as resistant starch. Indeed, numerous researchers have suggested that resistant starch actually contributes to the majority of fermentative processes that occur in the colon (Topping and Clifton, 2001).

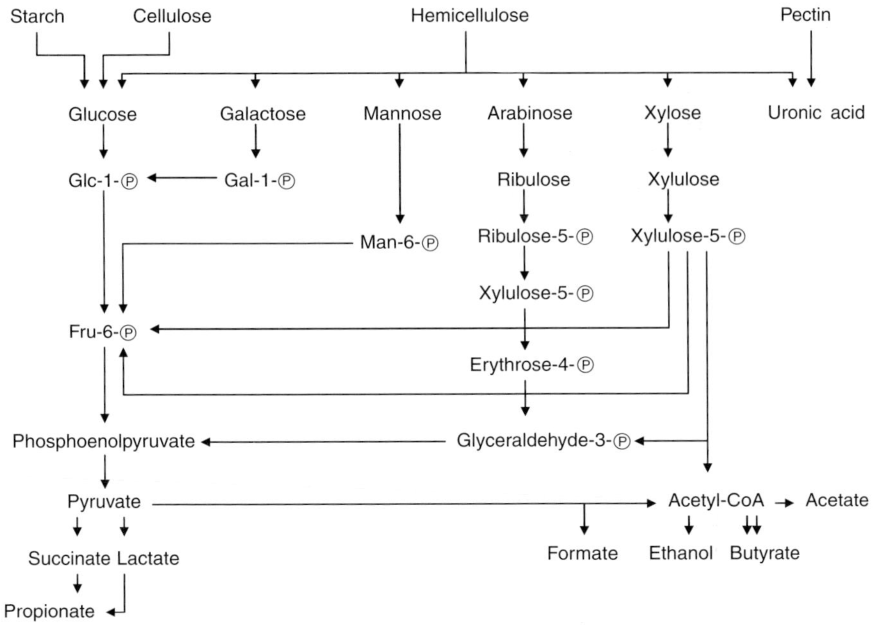

Figure 8.1. Simplified schematic of fermentative production of short chain fatty acids from dietary fiber Macfarlane and Gibson 1997. Ⓟ, phosphate; Glc, glucose; Gal, galactose; Man, mannose; Fru, fructose.

Short-chain fatty acids

The gut microbiota utilizes dietary fiber and other undigested material for growth through anaerobic fermentation, and produce SCFA and gases (Figure 8.1).

The amount and molar ratios of SCFAs produced depend on both the bacteria present and the dietary fiber source (McBurney and Thompson, 1987). For example, some individuals have an abnormal propensity for propionate production (Weaver et al., 1989), while others appear incapable of breaking down resistant starch (Cummings et al., 1996). Furthermore, disease states, such as inflammatory bowel disease can also affect fermentation variables. With respect to carbohydrate substrate, Kritchevsky (1995) reported the molar ratios of SCFAs produced during in vitro fermentation of a variety of carbohydrate substrates. From this report it is evident that resistant starch is highly butyrogenic, while guar gum and psyllium are the most propiogenic, and pectin the most acetogenic.

SCFAs exhibit a number of physiological effects. Less than 10% of SCFAs produced by colonic fermentation are recovered in the feces (Topping, 2007); therefore, they are largely utilized within the large bowel or absorbed. Within the large bowel, SCFAs reduce the luminal pH. This inhibits the growth of some opportunistic, pathogenic bacteria (Cummings et al., 2001), decreases the activity of co-carcinogenic enzymes such as glucuronidases, glycosidases, and 7α-hydroxylases (Goldin & Gorbach, 1976; Thornton, 1981), facilitates absorption of minerals (Lopez et al., 2000), and decreases the absorption of ammonia and other potentially toxic compounds by maintaining these compounds in their ionized form (Visek, 1978). Acetate and propionate are absorbed by diffusion, and anion exchange with sodium and potassium (Fleming et al., 1991). Once absorbed, they are taken up by the liver or peripheral tissues and metabolized in glucose or lipid catabolic pathways (Cook and Sellin, 1998; Wong et al., 2006). Unlike acetate and propionate, butyrate is not found in significant concentrations in circulation, but is utilized mainly by colonic epithelial cells for energy. The metabolism of butyrate by colonic epithelial cells is also hypothesized to facilitate in the mineral and water salvage functions of the large bowel (Canani et al., 2004). Butyrate may also play a role in prevention of colon cancer, through the repair of damaged DNA and induction of programmed cell death (apoptosis) in transformed epithelial cells, and prevention of colonic inflammation by inhibiting NF-κB activation (Inan et al., 2000; Luhrs et al., 2002).

As mentioned, acetate and propionate are metabolized by glycolytic or lipolytic pathways, and butyrate provides energy for colonic epithelial cells. This means that some energy is salvaged from dietary fiber despite the fact that it cannot be digested by human enzymes. Current data indicate that dietary fiber yields about 1.5–2.5 kcal/g (Livesey, 1990; Smith et al., 1998).

Importance of fermentation rate

Up to now, all fermentable dietary fibers have been lumped into one category; however, fermentable dietary fibers may be further divided on the basis of fermentation rate. Most fermentable dietary fibers such as pectin, FOS, and β-glucan, ferment rapidly in the cecum and proximal colon, leaving little carbohydrate substrate for bacteria in distal regions of the colon (Cummings and Macfarlane, 1991). This creates much different conditions in the proximal compared to the distal colon. In people consuming typical Western diets, fermentation in the proximal colon is saccharolytic (carbohydrate-based), producing SCFAs, whereas fermentation in the distal colon is largely putrefactive (protein-based), resulting in (relatively) high levels of toxic metabolites such as ammonia and phenol (Macfarlane et al., 1992). This is significant, as the most prevalent site of colonic disease is the distal colon (Topping & Clifton, 2001; Young et al., 2005).

A remedy to this problem may be to develop slowly fermentable dietary fibers with targeted health benefits. Based on the discussion above, an ideal dietary fiber is likely to: (1) have a low, steady rate of gas production to minimize bloating; (2) produce high levels of SCFAs to prevent the growth of pathogenic bacteria and reduce inflammation; (3) ferment slowly to deliver beneficial fermentation products to all regions of the colon and prevent the production of undesirable compounds (ammonia, phenols, etc.); and (4) promote the maintenance of a healthy colonic microbiota by selectively stimulating the growth of beneficial bacteria. Little research has been devoted to this area, although it is gaining momentum (Leach, 2007; Rose et al., 2007).

Fecal Bulk and Whole Gut Transit Time

Whole gut transit time refers to the elapsed time between ingestion of food and defecation. There are a number of factors that influence transit time that are not related to diet, e.g., anxiety and alcohol consumption tend to shorten transit time, while the use of oral contraceptives, depression, and consistently suppressing the urge to defecate prolong transit time (Gorard et al., 1996); men tend to have shorter transit times than women (Lampe et al., 1993), and particulate nonfood matter in the digestive tract, such as plastic beads, can speed up transit time (Lewis and Heaton, 1997). In the context of this review, however, dietary fiber is the greatest dietary component that affects transit time.

There are two main factors, in relation to dietary fiber, that influence transit time. First, poorly fermentable dietary fibers have a greater effect on transit time than fermentable dietary fibers. For example, in a randomized, crossover trial, cellulose intake of 15 g/day was shown to decrease stool transit time by 27%, compared to no change with pectin supplementation. A reason for this may be the greater effect that poorly fermentable dietary fibers have on fecal bulk than fermentable dietary fibers. Cummings (1992), in a summary of over 100 studies involving fecal bulk and dietary fiber, reported that cellulose increased fecal bulk by 3.5 g/g of cellulose. This was higher than other isolated, fermentable dietary fibers, such as pectin (1.2 g/g pectin), but lower than unmodified sources of poorly fermentable dietary fibers with a mixture of polysaccharides and other constituents, such as wheat bran (5.4 g/g wheat bran). Cellulose and other poorly fermentable dietary fibers influence stool weight because of their water-binding capacity, whereas fermentable dietary fibers influence stool weight by increasing bacterial mass due to increased fermentation.

The second factor to consider with regards to dietary fiber and transit time is that, when the same dietary fiber is consumed, the dietary fiber source with the larger particle size will decrease transit time to a greater extent than a small particle size (Cummings, 1992). The reason for this may be that larger particles are more slowly degraded, thus providing fecal bulk for a longer period of time during colonic transit (Cummings, 1992).

Slow intestinal transit can have a number of detrimental consequences (Lewis and Heaton, 1999). First, slow transit leads to greater SCFA absorption in the proximal colon, leading to even lower concentrations of SCFAs in the distal colon, a region chronically low in SCFAs. The consequences of low concentrations of SCFAs have been discussed previously; of particular mention here is that the lack of SCFAs and resultant high pH, as well as accompanying production of putrefactive metabolites, may explain why colon cancer and inflammatory bowel disease occur more prevalently in the distal colon. Additionally, the lower levels of SCFAs increase the deconjugation and dehydroxylation of bile acids due to the increased activity of cholylglycine hydrolase and 7-α-dehydroxylase as the pH nears 6.5. Not only are these secondary bile acids co-carcinogenic, but also increased levels of these compounds in the gall bladder upon resorption increase the risk of developing gall stones (Lewis and Heaton, 1999). Longer transit times also lead to constipation, due to the occurrence of small, firm stools that are difficult to pass. As mentioned, the decrease in stool weight can be explained by an increase in time for water absorption and a decrease in bacterial mass due to lack of carbohydrate substrate and bacterial cannibalism (Stephen et al., 1987). Conversely, rapid transit times are associated with high fecal weights, which are more easily passed with increased bacterial counts, water, and SCFAs (Lewis and Heaton, 1999).

Effect of Dietary Fiber on Colonic Diseases

Inflammatory Bowel Disease

Inflammatory bowel disease can be divided into two main diseases: Crohn's disease and ulcerative colitis, depending on site and severity of inflammation. The colonic microbiota can be simplistically divided into luminal and mucosal. Luminal bacteria are those bacteria present in the lumen of the bowel that move through with peristaltic movements and appear in the feces, while mucosal bacteria are present in the mucosal lining of the large intestine, and do not move to a great extent. Most likely, inflammatory bowel disease is caused by an improper immune response to a luminal agent (Galvez et al., 2005); either the mucosal bacteria fail to protect the epithelium from inflammatory agents, or there is some unique pro-inflammatory compound or bacteria in the lumen. Regardless, dietary fiber is likely to influence inflammatory bowel disease through fermentation, which can either selectively stimulate the growth of a more protective, healthy microbiota (Kanauchi et al., 2003) or lead to increased production of SCFAs, which have anti-inflammatory properties (Andoh et al., 2003).

A few studies have been published on the effects of crude dietary fiber sources on inflammatory bowel disease. Fernandez-Banares et al. (1999) recruited patients with quiescent ulcerative colitis to receive mesalazine, a drug for inflammatory bowel disease, psyllium (seeds), or a combination of mesalazine and psyllium. They found no significant differences between the three treatment groups, indicating that psyllium alone was as effective as mesalazine alone at maintaining remission of ulcerative colitis. In clinical trials with, germinated barley foodstuff, a source of hemicelluloses and cellulose, has been shown to improve clinical, endoscopic, and disease activity scores in patients with mild inflammation that continued to receive medication (Mitsuyama et al., 1998; Kanauchi et al., 2002). Interestingly, one study found that the severity of inflammation returned within 4 weeks of stopping germinated barley foodstuff supplementation (Mitsuyama et al., 1998).

FOS have shown positive effects against inflammatory bowel disease. Patients with active Crohn's disease receiving 15 g/day of FOS for 3 weeks have shown significant decreases in the Harvey Bradshaw index, a measure of disease severity (2006), and patients with active ulcerative colitis receiving 12 g/day of FOS plus mesalazine showed significant decreases in fecal calprotectin, a measurement of the presence of leukocytes, which were not observed in a control group receiving 12 g/day of maltodextrin plus mesalazine (Casellas et al., 2007). These small clinical trials suggest that certain types of dietary fiber help prevent or alleviate inflammatory bowel disease.

Colon Cancer

The effects of dietary fiber on colon cancer have been debated perhaps since Burkitt (1971) first proposed a relationship. Animal models have shown a beneficial effect of dietary fiber, but epidemiological studies have been inconsistent (Young et al., 2005). Human studies unavoidably have shortcomings, including the lack of differentiation between the different types of dietary fibers and consideration of other compounds that may be present in crude forms of dietary fiber (Baron, 2005; Eastwood and Kritchevsky, 2005).

In a recently published meta-analysis that included 725,628 subjects, Park et al. (2005) showed no significant relationship between colon cancer and dietary fiber intake after correction for confounding variables attempted to show the relationship between dietary fiber intake and colorectal cancer. The authors divided the cohort of participants into quintiles on the basis of dietary fiber intake. They found that there was no significant relationship between dietary fiber intake and colorectal cancer after adjusting for confounding variables.

A potential problem with these findings is that the quintile with the highest dietary fiber consumption had a mean of 31 g/day for men and 25 g/day for women. This amount is barely at, or even may be slightly below, the current recommendations (Institute of Medicine, 2002b). Paleobotanists have estimated that our ancestors' diets consisted of well over 100 g of fiber/day (Leach, 2007). Because we evolved from this lineage, it may be that we need much more fiber than even the current recommendations before a relationship between dietary fiber and colon cancer can be clearly shown.

Burkitt (1971) first proposed that dietary fiber protected against colon cancer by increasing fecal bulk and causing a dilution of toxins in the colon and more rapid transit time. This hypothesis has since been disregarded and replaced with hypotheses related to the fermentation of dietary fiber, namely, that dietary fiber leads to higher production of butyrate in the large intestine, which can induce apoptosis and influence gene expression (Young et al., 2005) or that dietary fiber can sustain a more healthy bacterial community that can provide protection against colorectal cancer (Lim et al., 2005). Paradoxically, however, in animal models (Jacobs, 1987; McIntyre et al., 1993) and in a case-controlled human study (Freudenheim et al., 1990), cereal dietary fibers that are low in fermentable dietary fiber and high in poorly fermentable dietary fiber have shown greater benefit against colon cancer than soluble dietary fibers. One explanation for this is that insoluble dietary fibers do ferment to a small degree, and are more associated with distal colonic fermentation compared to fermentable dietary fibers, which ferment proximally. This could lead to lower pH and higher butyrate concentrations in the distal colon, the area of the colon most prone to cancer. Alternatively, cereal dietary fibers have shown a greater reduction in the production or excretion of secondary bile acids, which are carcinogenic, than soluble dietary fibers (Reddy et al., 2000).

Conclusions

Dietary fiber includes a broad range of polysaccharides with differing physical and physiological properties. In foods, soluble dietary fibers, including β-glucan, pectin, and guar gum, can contribute desirable characteristics to foods such as viscosity or gelation, while insoluble dietary fibers, e.g., resistant starch, cellulose, and hemicellulose, generally impart undesirable sensory characteristics. Physiologically, in the upper gastrointestinal tract, soluble dietary fibers exhibit greater effects than insoluble dietary fibers. These effects include delayed gastric emptying and changes in small bowel morphology. Due to increased viscosity of digesta, soluble dietary fibers also delay absorption of both macro- and micronutrients. The influence that soluble dietary fibers have on glucose and lipid metabolism in particular, may decrease the risk of diabetes and heart disease. In the large bowel, soluble dietary fibers are generally highly fermentable and provide energy for the colonic microbiota. Adequate carbohydrate substrate for bacteria may improve colonic conditions by preventing the growth of undesirable microorganisms and promoting proper immune response to luminal bacteria. As the bacteria ferment dietary fiber, they produce SCFA, which have numerous trophic effects both systemically and on the gastrointestinal tract itself. Slowly fermentable dietary fibers may be particularly beneficial for their ability to sustain high SCFA production in distal regions of the colon. Poorly fermentable dietary fibers are also beneficial in the large intestine by providing fecal bulk and laxation. Greater fecal bulk results in stools that are more easily passed, and laxation, or a decreased transit time creates more uniform conditions in the proximal compared to the distal colon.

Two major diseases that plague the gastrointestinal tract, inflammatory bowel disease and colon cancer, may be prevented or alleviated by dietary fiber. Specific relationships are speculative, but current hypotheses are related to large bowel fermentation and the physiological effects of fermentation metabolites in the colon.

Despite the numerous beneficial effects that dietary fiber has on the gastrointestinal tract, dietary fiber consumption remains far below recommendations. Further research and education may reverse this trend.

References

Anal, A.K., Tobiassen, A., Flanagan, J., Singh, H. (2008) Preparation and characterization of nanoparticles formed by chitosan–caseinate interactions. *Colloid. Surf. B.* 64, 104–110.

Anderson, D.M., Hirst, E., Stoddart, J.F. (1966) Studies on uronic acid materials. 17. Some structural features of *Acacia senegal* gum (gum arabic). *J. Chem. Soc. C.* 21, 1959–1966.

Andoh, A., Tsujikawa, T., Fujiyama, Y. (2003) Role of dietary fiber and short-chain fatty acids in the colon. *Curr. Pharm. Des.* 9, 347–358.

Asp, N.G. (1992) Resistant starch – Proceedings from the 2nd plenary meeting of Euresta – European flair concerted action 11 on physiological implications of the consumption of resistant starch in man. *Eur. J. Clin. Nutr.* 46(Suppl 2), S1.

Auerbach, M.H., Craig, S.A.S., Howlett, J.F., Hayes, K.C. (2007) Caloric availability of polydextrose. *Nutr. Rev.* 65, 544–549.

Backhed, F., Ley, R.E., Sonnenburg, J.L., Peterson, D.A., Gordon, J.I. (2005) Host-bacterial mutualism in the human intestine. *Science* 307, 1915–1920.

Baron, J.A. (2005) Dietary fiber and colorectal cancer – an ongoing saga. *J. Am. Med. Assoc.* 294, 2904–2906.

Barry, J.L., Hoebler, C., Macfarlane, G.T., C.J. (1995) Estimation of the fermentability of dietary fiber *in vitro* – a European interlaboratory study. *Br. J. Nutr.* 74, 303–322.

Bell, S., Goldman, V.M., Bistrian, B.R., Arnold, A.H., Ostroff, G., Forse, R.A. (1999) Effect of beta-glucan from oats and yeast on serum lipids. *Crit. Rev. Food Sci.* 39, 189–202.

Bijkerk, C.J., Muris, J.W.M., Knottnerus, J.A., Hoes, A.W., De Wit, N.J. (2004) Systematic review: the role of different types of fibre in the treatment of irritable bowel syndrome. *Aliment. Pharmacol. Ther.* 19, 245–251.

Bird, A.R., Brown, I.L., Topping, D.L. (2000) Starches, resistant starches, the gut microflora and human health. *Curr. Issues Intest. Microbiol.* 1, 25–37.

Brand-Miller, J.C., Holt, S.H.A. (1998) Australian aboriginal plant foods: a consideration of their nutritional compositional and health implications. *Nutr. Res. Rev.* 11, 5–23.

Brennan, C.S. (2005) Dietary fiber, glycaemic response, and diabetes. *Mol. Nutr. Food Res.* 49, 560–570.

Brown, L., Rosner, B., Willett, W.W., Sacks, F.M. (1999) Cholesterol-lowering effects of dietary fiber: a meta-analysis. *Am. J. Clin. Nutr.* 69, 30–42.

Brownlee, I.A., Allen, A., Pearson, J.P., et al. (2005) Alginate as a source of dietary fiber. *Crit. Rev. Food Sci. Nutr.* 45, 497–510.

Burkitt, D.P. (1971) Epidemiology of cancer of the colon and rectum. *Cancer* 28, 3–13.

Burkitt, D.P., Spiller, G.A. (1992) Dietary fiber: From early hunter-gatherers to the 1990s. In: Spiller, G.A. (editor), *Dietary Fiber in Human Nutrition*, pp. 359–360. CRC Press, London.

Burkitt, D.P., Walker, A.R.P., Painter, N.S. (1974) Dietary fiber and disease. *J. Am. Med. Assoc.* 229, 1068–1074.

Butt, M.S., Shahzadi, N., Sharif, M.K., Nasir M. (2007) Guar gum: a miracle therapy for hypercholesterolemia, hyperglycemia and obesity. *Crit. Rev. Food Sci.* 47, 389–396.

Cameron-Smith, D., Collier, G.R. (2001) Dietary fiber and glucose metabolism and diabetes. In: Cho, S.S., Dreher, M.L. (editors), *Handbook of Dietary Fiber*, pp. 107–123. Marcel Dekker, New York.

Canani, R.B., Terrin, G., Cirillo, P., et al. (2004) Butyrate as an effective treatment of congenital chloride diarrhea. *Gastroenterology* 127, 630–634.

Casellas, F., Borruel, N., Torrejon, A., Varela, E., Antolin, M., Guarner, F., Malagelada, J.R. (2007) Oral oligofructose-enriched inulin supplementation in acute ulcerative colitis is well tolerated and associated with lowered faecal calprotectin. *Aliment. Pharm. Therap.* 25, 1061–1067.

Cassidy, M.M., Calvert, R.J. (1992) Effect of dietary fiber on intestinal absorption of lipids. In: Spiller, G.A. (editor), *Dietary Fiber in Human Nutrition,* pp. 153–162. CRC Press, London.

Chaikomin, R., Rayner, C.K., Jones, K.L., Horowitz, M. (2006) Upper gastrointestinal function and glycemic control in diabetes mellitus. *World J. Gastroenterol.* 12, 5611–5621.

Cleave, T.L. (1956) The neglect of natural principles in current medical practice. *J. R. Nav. Med. Serv.* 42, 55–82.

Cook, S.I., Sellin, J.H. (1998) Review article: short chain fatty acids in health and disease. *Aliment. Pharm. Therap.* 12, 499–507.

Crittenden, R. (2006) Emerging prebiotic carbohydrates. In: Gibson, G.R., Rastall, R.A. (editors), *Prebiotics: Development & Applications*, pp. 111–134. John Wiler & Sons, West Sussex, England.

Cummings, J.H. (1992) The effects of dietary fiber on fecal weight and constipation. In: Spiller, G.A. (editor), *Dietary Fiber in Human Nutrition,* pp. 263–350. CRC Press, London.

Cummings, J.H., Beatty, E.R., Kingman, S.M., Bingham, S.A., Englyst, H.N. (1996) Digestion and physiological properties of resistant starch in the human large bowel. *Br. J. Nutr.* 75, 733–747.

Cummings, J.H., Christie, S., Cole, T.J. (2001) A study of fructooligosaccharides in the prevention of travellers' diarrhea. *Aliment. Pharmacol. Ther.* 15, 1139–1145.

Cummings, J.H., Macfarlane, G.T. (1991) The control and consequences of bacterial fermentation in the human colon. *J. Appl. Bacteriol.* 70, 443–459.

Cummings, J.H., Stephen, A.M. (2007) Carbohydrate terminology and classification. *Eur. J. Clin. Nutr.* 61, S5–S18.

Daas, P.J.H., Grolle, K., van Vliet, T., Schols, H.A., de Jongh, H.H.J. (2002) Toward the recognition of structure-function relationships in galactomannans. *J. Agric. Food Chem.* 50, 4282–4289.

Dethlefsen, L., McFall-Ngai, M., Relman, D.A. (2007) An ecological and evolutionary perspective on human–microbe mutualism and disease. *Nature* 449, 811–818.

DeVries, J.W. (2001) The definition of dietary fiber. Cereal Food. *World* 46, 112–114, 116–118, 120–122, 124–129.

Dikeman, C.L., Fahey, G.C. (2006) Viscosity as related to dietary fiber: a review. *Crit. Rev. Food Sci. Nutr.* 46, 649–663.

Dikeman, C.L., Murphy, M.R., Fahey. (2006) Dietary fibers affect viscosity of solutions and simulated human gastric and small intestinal digesta. *J. Nutr.* 136, 913–919.

Dintzis, F.R., Legg, L.M., Deatherage, W.L., et al. (1979) Human gastro-intestinal action on wheat, maize, and soy hull bran – preliminary findings. *Cereal. Chem.* 56, 123–127.

Doner, L.W., Chau, H.K., Fishman, M.L., Hicks, K.B. (1998) An improved process for isolation of corn fiber gum. *Cereal. Chem.* 75, 408–411.

Dongowski, G., Lorenz, A., Proll, A. (2002) The degree of methylation influences the degradation of pectin in the intestinal tract of rats and *in vitro*. *J. Nutr.* 132, 1935–1944.

Dreher, M.L. (2001) Dietary fiber overview. In: Cho, S.S., Dreher, M.L. (editors), Handbook of Dietary Fiber, pp. 1–16. Marcel Dekker, New York.

Eastwood, M., Kritchevsky, D. (2005) Dietary fiber: how did we get where we are? *Annu. Rev. Nutr.* 25, 1–8.

Eastwood, M.A. (1992) The physiological effect of dietary fiber – an update. *Annu. Rev. Nutr.* 12, 19–35.

Ebringerova, A., Hromadkova, Z., Heinze, T. (2005) Hemicellulose. *Adv. Polym. Sci.* 186, 1–67.

Ehrlein, H., Stockmann, A. (1998) Absorption of nutrients is only slightly reduced by supplementing enteral formulas with viscous fiber in miniature pigs. *J. Nutr.* 128, 2446–2455.

Eliaz, I., Hotchkiss, A.T., Fishman, M.L., Rode, D. (2006) The effect of modified citrus pectin on urinary excretion of toxic elements. *Phytother. Res.* 20, 859–864.

Englyst, H.N., Kingsman, S.M., Cummings, J.H. (1992) Classification and measurement of nutritionally important starch fractions. *Eur. J. Clin. Nutr.* 46, 33–50.

Fanaro, S., Jelinek, T., Stahl, T., Boehm, T., Kock, R., Vigi, V. (2005) Acidic oligosaccharides from pectin hydrolysate as new component for infant formulae: effect on intestinal flora, stool characteristics, and pH. *J. Pediatr. Gastroenterol. Nutr.* 41, 186–190.

Fernandez-Banares, F., Hinojosa, J., Sanchez-Lombrana, J.L., et al. (1999) Randomized clinical trial of *Plantago ovata* seeds (dietary fiber) as compared with mesalamine in maintaining remission in ulcerative colitis. *Am. J. Gastroenterol.* 94, 427–433.

Fialho, A.M., Moreira, L.M., Granja, A.T., Popescu, A.O., Hoffmann, K., Sa-Correia, I. (2008) Occurrence, production, and applications of gellan: current state and perspectives. *Appl. Microbiol. Biotech.* 79, 889–900.

Fischer, M.H., Yu, N.X., Gray, G.R., Ralph, J., Anderson, L., Marlett, J.A. (2004) The gel-forming polysaccharide of psyllium husk (*Plantago ovata* Forsk). *Carbohydr. Res.* 339, 2009–2017.

Fleming, S.E., Choi, S.Y., Fitch, M.D. (1991) Absorption of short-chain fatty-acids from the rat cecum in vivo. *J. Nutr.* 121, 1787–1797.

Flood, M.T., Auerbach, M.H., Craig, S.A.S. (2004) A review of the clinical toleration studies of polydextrose in food. *Food Chem. Toxicol.* 42, 1531–1542.

Fox (1992) Seed gums. In: Imeson, A. (editor), *Thickening and Gelling Agents for Food,* pp. 262–283. Blackie Academic & Professional, New York.

Freudenheim, J.L., Graham, S., Horvath, P.J., Marshall, J.R., Haughey, B.P., Wilkinson, G. (1990) Risks associated with source of fiber and fiber components in cancer of the colon and rectum. *Cancer Res.* 50, 3295–3300.

Fukata, M., Abreu, M.T. (2008) Role of Toll-like receptors in gastrointestinal malignancies. *Oncogene* 27, 234–243.

Gallaher, D.D., Schneeman, B.O. (1992) Effect of dietary fiber on protein digestibility and utilization. In: Spiller, G.A. (editor), *Dietary Fiber in Human Nutrition,* pp. 179–208. CRC Press, London.

Galvez, J., Rodriguez-Cabezas, M.E., Zarzuelo, A. (2005) Effects of dietary fiber on inflammatory bowel disease. *Mol. Nutr. Food Res.* 49, 601–608.

George, M., Abraham, T.E. (2006) Polyionic hydrocolloids for the intestinal delivery of protein drugs: Alginate and chitosan – a review. *J. Control. Release.* 114, 1–14.

Giannini, E.G., Mansi, C., Dulbecco, P., Savarino, V. (2006) Role of partially hydrolyzed guar gum in the treatment of irritable bowel syndrome. *Nutrition* 22, 334–342.

Gibson, G.R. (1999) Dietary modulation of the human gut microflora using the prebiotics oligofructose and inulin. *J. Nutr.* 129, 1438S–1441S.

Gibson, G.R., Probert, H.M., Van Loo, J., Rastall, R.A., Roberfroid, M.B. (2004) Dietary modulation of the human colonic microbiota: updating the concept of prebiotics. *Nutr. Res. Rev.* 17, 259–275.

Gibson, G.R., Roberfroid, M.B. (1995) Dietary modulation of the human colonic microbiota – introducing the concept of prebiotics. *J. Nutr.* 125, 1401–1412.

Gibson, W. (1992) Gellan gum. In: Imeson, A. (editor), *Thickening and Gelling Agents for Food,* pp. 227–250. Blackie Academic & Professional, New York.

Glitso, L.V., Jensen, B.B., Knudsen, K.E.B. (2000) In vitro fermentation of rye carbohydrates including arabinoxylans of different structure. *J. Sci. Food Agric.* 80, 1211–1218.

Goldin, B.R., Gorbach, S.L. (1976) The relationship between diet and rat fecal bacterial enzymes implicated in colon cancer. *J. Natl. Cancer I.* 57, 371–375.

Gorard, D.A., Gomborone, J.E., Libby, G.W., Farthing, M.J.G. (1996) Intestinal transit in anxiety and depression. *Gut* 39, 551–555.

Groos, S., Hunefeld, G., Luciano, L. (1996) Parenteral versus enteral nutrition: morphological changes in human adult intestinal mucosa. *J. Submicrosc. Cytol. Pathol.* 28, 61–74.

Hipsley, E.H. (1953) Dietary "fibre" and pregnancy toxaemia. *Br. Med. J.* 2, 420–422.

Hosobuchi, C., Rutanassee, L., Bassin, S.L., Wong, N.D. (1999) Efficacy of acacia, pectin, and guar gum-based fiber supplementation in the control of hypercholesterolemia. *Nutr. Res.* 19, 643–649.

Howarth, N.C., Saltzman, E., Roberts, S.B. (2001) Dietary fiber and weight regulation. *Nutr. Rev.* 59, 129–139.

Imeson, A. (1992) Exudate gums. In: Imeson, A. (editor), *Thickening and Gelling Agents for Food,* pp. 66–97. Blackie Academic & Professional, New York.

Inan, M.S., Rasoulpour, R.J., Yin, L., Hubbard, A.K., Rosenberg, D.W., Giardina, C. (2000) The luminal short-chain fatty acid butyrate modulates NF-kappa B activity in a human colonic epithelial cell line. *Gastroenterology* 118, 724–734.

Inglett, G.E. (1997) Development of a dietary fiber gel for calorie-reduced foods. *Cereal. Food.* 42, 382–384.

Inglett, G.E. (1998) Dietary fiber gels for calorie reduced foods and method for preparing the same. United States patent 5,766,662.

Institute of Medicine. (2001) Definitions of dietary fiber. In: *Dietary Reference Intakes: Proposed Definition of Dietary Fiber,* pp. 3–11. National Academy Press, Washington, DC.

Institute of Medicine. (2002a) Appendix E: dietary intake data from the continuing survey of food intakes by individuals (1994–1996, 1998). In: *Dietary Reference Intakes for Energy, Carbohydrate, Fiber, Fat, Fatty Acids, Cholesterol, Protein, and Amino Acids,* pp. 1028–1065. National Academy Press, Washington, DC.

Institute of Medicine. (2002b) Dietary, functional, and total fiber. In: *Dietary Reference Intakes for Energy, Carbohydrate, Fiber, Fat, Fatty Acids, Cholesterol, Protein, and Amino Acids,* pp. 339–421. National Academy Press, Washington, DC.

Isaksson, G., Lundquist, I., Ihse, I. (1982) Effect of dietary fiber on pancreatic-enzyme activity in vitro – the importance of viscosity, pH, ionic-strength, adsorption, and time of incubation. *Gastroenterology* 82, 918–924.

Jacobs, L.R. (1987) Dietary fiber and cancer. *J. Nutr.* 117, 1319–1321.

Jeon, Y.J., Shahidi, F., Kim, S.K. (2000) Preparation of chitin and chitosan oligomers and their applications in physiological functional foods. *Food Rev. Int.* 16, 159–176.

Kanauchi, O., Mitsuyama, K., Araki, Y., Andoh, A. (2003) Modification of intestinal flora in the treatment of inflammatory bowel disease. *Curr. Pharm. Des.* 9, 333–346.

Kanauchi, O., Suga, T., Tochihara, M., et al. (2002) Treatment of ulcerative colitis by feeding with germinated barley foodstuff: first report of a multicenter open control trial. *J. Gastroenterol.* 37, 67–72.

Kasper, H. (1992) Effects of dietary fiber on vitamin metabolism. In: Spiller, G.A. (editor), *Dietary Fiber in Human Nutrition*, pp. 253–262. CRC Press, London.

King, D.E. (2005) Dietary fiber, inflammation, and cardiovascular disease. *Mol. Nutr. Food Res.* 49, 594–600.

Krajewska, B. (2004) Application of chitin- and chitosan-based materials for enzyme immobilizations: a review. *Enzyme Microb. Tech.* 35, 126–139.

Kritchevsky, D. (1988) Dietary fiber. *Ann. Rev. Nutr.* 8, 301–328.

Kritchevsky, D. (1995) Epidemiology of fibre, resistant starch and colorectal cancer. *Eur. J. Cancer Prev.* 4, 345–352.

Lampe, J.W., Fredstrom, S.B., Slavin, J.L., Potter, J.D. (1993) Sex-differences in colonic function – a randomized trial. *Gut* 34, 531–536.

Lazaridou, A., Biliaderis, C.G. (2007) Molecular aspects of cereal beta-glucan functionality: physical properties, technological applications and physiological effects. *J. Cereal. Sci.* 46, 101–118.

Leach, J.D. (2007) Evolutionary perspective on dietary intake of fibre and colorectal cancer. *Eur. J. Clin. Nutr.* 61, 140–142.

Leclere, C.J., Champ, M., Boillot, J., et al. (1994) Role of viscous guar gums in lowering the glycemic response after a solid meal. *Am. J. Clin. Nutr.* 59, 914–921.

Lewis, S.J., Heaton, K.W. (1997) The intestinal effects of bran-like plastic particles: is the concept of 'roughage' valid after all? *Eur. J. Gastroenterol. Hepatol.* 9, 553–557.

Lewis, S.J., Heaton, K.W. (1999) The metabolic consequences of slow colonic transit. *Am. J. Gastroenterol.* 94, 2010–2016.

Liese, A., Roach, A.K., Sparks, K.C., Marquart, L., D'Agostino, R.B. Jr., Mayer-Davis, E.J. (2003) Whole-grain intake and insulin sensitivity: the insulin resistance atherosclerosis study. *Am. J. Clin. Nutr.* 78, 965–971.

Lim, C.C., Ferguson, L.R., Tannock, G.W. (2005) Dietary fiber as "prebiotics": implications for colorectal cancer. *Mol. Nutr. Food Res.* 49, 609–619.

Lindsay, J.O., Whelan, K., Stagg, A.J., Gobin P, Al-Hassi, H.O., Rayment, N., Kamm, M.A., Knight, S.C., Forbes, A. (2006) Clinical, microbiological, and immunological effects of fructo-oligosaccharide in patients with Crohn's disease. *Gut* 55, 348–355.

Liu, L.S., Fishman, M.L., Hicks, K.B., Kende, M., Ruthel, G. (2006) Pectin/zein beads for potential colon-specific drug delivery: synthesis and in vitro evaluation. *Drug Deliv.* 13, 417–423.

Livesey, G. (1990) Energy values of unavailable carbohydrate and diets – an inquiry and analysis. *Am. J. Clin. Nutr.* 51, 617–637.

Lopez, H.W., Coudray, C., Bellanger, J., et al. (2000) Resistant starch improves mineral assimilation in rats adapted to a wheat bran diet. *Nutr. Res.* 20, 141–155.

Loubinoux, J., Bronowicki, J.P., Pereira, I.A.C., Mougenel, J.L., Le Faou, A.E. (2002) Sulfate-reducing bacteria in human feces and their association with inflammatory bowel diseases. *FEMS Microbiol. Ecol.* 40, 107–112.

Luhrs, H., Kudlich, T., Neumann, M., et al. (2002) Butyrate-enhanced TNF alpha-induced apoptosis is associated with inhibition of NF-kappa B. *Anticancer Res.* 22, 1561–1568.

Macfarlane, G.T., Gibson, G.R. (1997) Carbohydrate fermentation, energy transduction and gas metabolism in the human large intestine. In: Mackie, R.I., White, B.A. (editors), *Gastrointestinal Microbiology*, pp. 269–318. Chapman & Hall, New York.

Macfarlane, G.T., Gibson, G.R., Cummings, J.H. (1992) Comparison of fermentation reactions in different regions of the human colon. *J. Appl. Bacteriol.* 72, 57–64.

Maiorano, A.E., Piccoli, R.M., da Silva, E.S., Rodrigues, M.F.D. (2008) Microbial production of fructosyltransferases for synthesis of pre-biotics. *Biotechnol. Lett.* 30, 1867–1877.

Malkki, Y., Virtanen, E. (2001) Gastrointestinal effects of oat bran and oat gum—a review. *Lebensm. Wiss. Technol.* 34, 337–347.

Marlett, J.A., Kajs, T.M., Fischer, M.H. (2000) An unfermented gel component of psyllium seed husk promotes laxation as a lubricant in humans. *Am. J. Clin. Nutr.* 72, 784–789.

Marteau, P., Pochart, P., Dore, J., Bera-Maillet, C., Bernalier, A., Corthier, G. (2001) Comparative study of bacterial groups within the human cecal and fecal microbiota. *Appl. Environ. Microbiol.* 67, 4939–4942.

McBurney, M.I., Thompson, L.U. (1987) Effect of human fecal inoculums on in vitro fermentation variables. *Br. J. Nutr.* 58, 233–243.

McIntosh, M., Stone, B.A., Stanisich, V.A. (2005) Curdlan and other bacterial (1→3)-beta-D-glucans. *Appl. Microbiol. Biotech.* 68, 163–173.

McIntyre, A., Gibson, P.R., Young, G.P. (1993) Butyrate production from dietary fiber and protection against large-bowel cancer in a rat model. *Gut* 34, 386–391.

Meyer, J.H., Gu, Y.G., Jehn, D., Taylor, I.L. (1988) Intragastric vs. intraintestinal viscous polymers and tolerance after liquid meals of glucose. *Am. J. Clin. Nutr.* 48, 260–266.

Mitsuyama, K., Saiki, T., Kanauchi, O., et al (1998) Treatment of ulcerative colitis with germinated barley foodstuff feeding: a pilot study. *Aliment. Pharmacol. Ther.* 12, 1225–1230.

Mohnen, D. (2008) Pectin structure and biosynthesis. *Curr. Opin. Plant Biol.* 11, 266–277.

Munoz, J.M., Harland, B.F. (1992) Overview of the effects of dietary fiber on the utilization of minerals and trace elements. In: Spiller, G.A. (editor), *Dietary Fiber in Human Nutrition,* pp. 245–252. CRC Press, London.

Murphy, M.M., Douglass, J.S., Birkett, A. (2008) Resistant starch intakes in the United States. *J. Am. Diet. Assoc.* 108, 67–78.

Mutwil, M., Debolt, S., Persson, S. (2008) Cellulose synthesis: a complex complex. *Curr. Opin. Plant Biol.* 11, 252–257.

Nangia-Makker, P., Conklin, J., Hogan, V., Raz, A. (2002) Carbohydrate-binding proteins in cancer, and their ligands as therapeutic agents. *Trends Mol. Med.* 8, 187–192.

No, H.K., Meyers, S.P., Prinyawiwatkul, W., Xu, Z. (2007) Applications of chitosan for improvement of quality and shelf life of foods: a review. *J. Food Sci.* 72, R87–R100.

Nugent, A.P. (2005) Health properties of resistant starch. *Nutr. Bull.* 30, 27–54.

Nuttall, F.Q. (1993) Dietary fiber in the management of diabetes. *Diabetes* 42, 503–508.

Okamoto, Y., Nose, M., Miyatake, K., Sekine, J., Oura, R., Shigemasa, Y., Minami, S. (2001) Physical changes of chitin and chitosan in canine gastrointestinal tract. *Carbohydr. Polym.* 44, 211–215.

Onsoyen, E. (1992) Alginates. In: Imeson, A. (editor), *Thickening and Gelling Agents for Food,* pp. 1–23. Blackie Academic & Professional, New York.

Park, Y., Hunter, D.J., Spiegelman, D., et al. (2005) Dietary fiber intake and risk of colorectal cancer: a pooled analysis of prospective cohort studies. *J. Am. Med. Assoc.* 294, 2849–2857.

Prashanth, K.V.H., Tharanathan, R.N. (2007) Chitin/chitosan: modifications and their unlimited application potential – an overview. *Trends Food Sci. Tech.* 18, 117–131.

Pylkas, A.M., Juneja, L.R., Slavin, J.L. (2005) Comparison of different fibers for in vitro production of short chain fatty acids by intestinal microflora. *J. Med. Food.* 8, 113–116.

Razdan, A., Pettersson, D., Pettersson, J. (1997) Broiler chicken body weights, feed intakes, plasma lipid and small-intestinal bile acid concentrations in response to feeding of chitosan and pectin. *Brit. J. Nutr.* 78, 283–291.

Reddy, B.S., Hirose, Y., Cohen, L.A., Simi, B., Cooma, I., Rao, C.V. (2000) Preventive potential of wheat bran fractions against experimental colon carcinogenesis: implications for human colon cancer prevention. *Cancer Res.* 60, 4792–4797.

Roediger, W.E.W., Moore, J., Babidge, W. (1997) Colonic sulfide in pathogenesis and treatment of ulcerative colitis. *Dig. Dis. Sci.* 42, 1571–1579.

Rose, D.J., Demeo, M.T., Keshavarzian, A., Hamaker, B.R. (2007) Influence of dietary fiber on inflammatory bowel disease and colon cancer: importance of fermentation pattern. *Nutr. Rev.* 65, 51–62.

Rose, D.J., Keshavarzian, A., Patterson, J.A., Venkatachalam, M., Gillevet, P., Hamaker, B.R. (2009) Starch-entrapped microspheres extend in vitro fecal fermentation, increase butyrate production, and influence microbiota pattern. *Mol. Nutr. Food Res.* 53, S121–S130.

Rydning, A., Aadland, E., Berstad, A., Odegaard, B. (1982) Prophylactic effect of dietary fiber in duodenal-ulcer disease. *Lancet* 2, 736–739.

Samuelsen, A.B., Cohen, E.H., Paulsen, B.S., Brull, L.P., Thomas-Oates, J.E. (1999) Structural studies of a heteroxylan from plantago major L-seeds by partial hydrolysis, HPAEC-PAD, methylation and GC-MS, ESMS and ESMS MS. *Carbohydr. Res.* 315, 312–318.

Serpe, M.D., Nothnagel, E.A. (1995) Fractionation and structural characterization of arabinogalactan-proteinsfrom the cell-wall of rose cells. *Plant Physiol.* 109, 1007–1016.

Sharma, A., Yadav, B.S., Ritika. (2008) Resistant starch: physiological roles and food applications. *Food Rev. Int.* 24, 193–234.

Singh, P., Gill, P.K. (2006) Production of inulinases: recent advances. *Food Technol. Biotechnol.* 44, 151–162.

Slavin, J.L. (2005) Dietary fiber and body weight. *Nutrition* 21, 411–418.

Smith, T., Brown, J.C., Livesey, G. (1998) Energy balance and thermogenesis in rats consuming nonstarch polysaccharides of various fermentabilities. *Am. J. Clin. Nutr.* 68, 802–819.

Sobolik, K.D. (1994) Paleonutrition of the lower pecos region of the Chihuahuan Desert. In: Sobolik, K.D. (editor), *Paleonutrition: The Diet and Health of Prehistoric Americans,* pp. 247–264. Center for Archaeological Investigations: occasional paper no. 22. Center for Archaeological Investigations, University of Illinois, Urbana, IL.

Stecher, B., Hardt, W.D. (2008) The role of microbiota in infectious disease. *Trends Microbiol.* 16, 107–114.

Stephen, A.M., Wiggins, H.S., Cummings, J.H. (1987) Effect of changing transit time on colonic microbial metabolism in man. *Gut* 28, 601–609.

Stewart, M.L., Slavin, J.L. (2006) Molecular weight of guar gum affects short-chain fatty acid profile in model intestinal fermentation. *Mol. Nutr. Food Res.* 50, 971–976.

Suzuki, R., Rylander-Rudqvist, T., Ye, W., Saji, S., Adlercreutz, H., Wolk, A. (2008) Dietary fiber intake and risk of postmenopausal breast cancer defined by estrogen and progesterone receptor status – a prospective cohort study among Swedish women. *Int. J. Cancer* 122, 403–412.

Tasman-Jones, C. (1992) Effect of dietary fiber on the structure of the colon. In: Spiller, G.A. (editor), *Dietary Fiber in Human Nutrition,* pp. 359–360. CRC Press, London.

Tharanathan, R.N., Kittur, F.S. (2003) Chitin – the undisputed biomolecule of great potential. *Crit. Rev. Food Sci. Nutr.* 43, 61–87.

Theuwissen, E., Mensink, R.P. (2008) Water-soluble dietary fibers and cardiovascular disease. *Physiol. Behav.* 94, 285–292.

Thornton, J.R. (1981) High colonic pH promotes colorectal cancer. *Lancet* 1, 1081–1082.

Topping, D. (2007) Cereal complex carbohydrates and their contribution to human health. *J. Cereal Sci.* 46, 220–229.

Topping, D.L., Clifton, P.M. (2001) Short-chain fatty acids and human colonic function: roles of resistant starch and nonstarch polysaccharides. *Physiol. Rev.* 81, 1031–1064.

Topping, D.L., Fukushima, M., Bird, A.R. (2003) Resistant starch as a prebiotic and symbiotic: state of the art. *P. Nutr. Soc.* 62, 171–176.

United States Department of Agriculture. (2002) Profiling food consumption in America. In: *Agriculture Fact Book 2001–2002,* pp. 13–21. United States Department of Agriculture, Washington, DC. Available at: http://www.usda.gov/factbook/2002factbook.pdf.

United States Department of Agriculture. (2005) Food groups to encourage. In: *Dietary Guidelines for Americans,* pp. 23–28. United States Department of Agriculture, Washington, DC. Available at: http://www.health.gov/dietaryguidelines/dga2005/document/pdf/dga2005.pdf.

Varshosaz, J. (2007) The promise of chitosan microspheres in drug delivery systems. *Expert Opin. Drug Deliv.* 4, 263–273.

Visek, W.J. (1978) Diet and cell-growth modulation by ammonia. *Am. J. Clin. Nutr.* 31, S216–S220.

Warner, K., Inglett, G.E. (1997) Flavor and texture characteristics of foods containing Z-Trim corn and oat fibers as fat or flour replacers. *Cereal Food* 42, 821–825.

Weaver, G.A., Krause, J.A., Miller, T.L., Wolin, M.J. (1989) Constancy of glucose and starch fermentations by 2 different human fecal microbial communities. *Gut* 30, 19–25.

Wong, J.M.W., de Souza, R., Kendall, C.W.C., Emam, A., Jenkins, D.J.A. (2006)

Colonic health: fermentation and short chain fatty acids. *J. Clin. Gastroenterol.* 40, 235–243.

Wood, P.J. (2007) Cereal beta-glucans in diet and health. *J. Cereal Sci.* 46, 230–238.

Wood, P.J., Arrigoni, E., Miller, S.S., Amado, R. (2002) Fermentability of oat and wheat fractions enriched in beta-glucan using human fecal inoculation. *Cereal Chem.* 79, 445–454.

Yadav, M.P., Johnston, D.B., Hotchkiss, A.T., Hicks, K.B. (2007) Corn fiber gum: a potential gum arabic replacer for beverage flavor emulsification. *Food Hydrocolloid.* 21, 1022–1030.

Yang, S.T., Silva, E.M. (1995) Novel products and new technologies for use of a familiar carbohydrate, milk lactose. *J. Dairy Sci.* 78, 2541–2562.

Yoon, S.J., Chu, D.C., Juneja, L.R. (2008) Chemical and physical properties, safety and application of partially hydrolized guar gum as dietary fiber. *J. Clin. Biochem. Nutr.* 42, 1–7.

Young, G.P., Hu, Y., Le Leu, R.K., Nyskohus, L. (2005) Dietary fiber and colorectal cancer: a model for environment-gene interactions. *Mol. Nutr. Food Res.* 49, 571–584.

Chapter 9

Toward Second-Generation Carbohydrate Functional Food Ingredients

Robert A. Rastall

Prebiotic Oligosaccharides

There has been growing scientific and industrial interest in the concept of modulating the gut microbiota since the prebiotic concept was first outlined in 1995 (Gibson and Roberfroid, 1995). Prebiotics were defined as "a nondigestible food ingredient that beneficially affects the host by selectively stimulating the growth and/or activity of one or a limited number of bacteria in the colon, and thus improves host health." The colonic microbiota is an immensely complex ecosystem consisting of around 500–1,000 species (Ley, Peterson et al., 2006) with over 7,000 strains (Backhed, Ley et al., 2005). This total bacterial load of around 10^{12} bacteria per gram (Dethlefsen, Eckburg et al., 2006; Flint, Duncan et al., 2007) makes the colon one of the most metabolically active organs in the body. It is not surprising that this metabolically diverse ecosystem plays an important role in health (Flint, Duncan et al., 2007). Some members of the colonic microbiota, principally saccharolytic species such as *Bifidobacterium* and *Lactobacillus* are considered to be health positive in terms of their modulating effects on the immune system (Corthesy, Gaskins et al., 2007) and their production of desirable metabolites such as short-chain fatty acids (Cummings and MacFarlane, 1991). Other, mainly proteolytic, components of the colonic microbiota produce undesirable metabolites such as

Nondigestible Carbohydrates and Digestive Health. Edited by T. M. Paeschke and W. R. Aimutis

branched-chain fatty acids (Cummings and MacFarlane, 1991), toxic phenolic compounds (Heavey and Rowland, 2004) and genotoxic substances and enzymes (Heavey and Rowland, 2004). The human colon can be thought of in discrete regions, the proximal, transverse, and distal colon with steadily increasing pH and reducing water content. There is also a gradient of nutritional factors along the colon. These physiological conditions affect the microbiota that develops in these regions. The proximal colon tends to favor saccharolytic species with a lower pH and a ready supply of carbohydrates from the diet (Cummings and MacFarlane, 1991). The distal colon tends to be a much more proteolytic region with negative consequences for health. This region tends to be higher in products of proteolytic metabolism such a phenolics and branched-chain fatty acids (Cummings and MacFarlane, 1991).

To date, the main goal of prebiotic intervention has been to increase the relative populations of bifidobacteria and lactobacilli. This is likely, however, to change as we discover more about the functional ecology of the colon and become more adept at measuring the metabolic consequences of prebiotic consumption.

All prebiotics are currently carbohydrates. This is always likely to be the case as a requirement for prebiotic activity is to reach the colon largely intact and this is more easily achieved with oligosaccharides than with proteins or peptides, which tend to be readily digested in the small intestine. Further, as discussed above, the metabolic consequences of protein metabolism are not generally desirable. There are several carbohydrates currently marketed as prebiotics (Table 9.1). In Europe and the United States, there are only three. These are the fructans, inulin, and fructooligosaccharides (FOS) (derived from inulin by hydrolysis or from sucrose by synthesis) and galactooligosaccharides (GOS) derived from lactose by enzymatic synthesis. In Japan, there is a wider range of oligosaccharides on the market for their prebiotic effects although many of them do not have the support of good quality human trials with robust microbiological techniques.

Health Benefits of Prebiotics

Prebiotics have been investigated for their potential benefit in a variety of conditions. Table 9.2 summarizes some significant trials on health benefits of prebiotics. The evidence behind health effects is rather varied.

Table 9.1. Prebiotic carbohydrates.

*Established Prebiotics**
Inulin
Fructooligosaccharides
Galactooligosaccharides
Lactulose
Emerging Prebiotics†
Isomaltooligosaccharides
Soybean oligosaccharides
Resistant starch
Xylooligosaccharides
Polydextrose
Experimental Prebiotics‡
Pectic oligosaccharides
Glucooligosacharides
Gentiooligosacharides

*Prebiotics established in more than one well-designed human trial with modern microbiological methods.

†Carbohydrates marketed as prebiotics but with limited human data.

‡Carbohydrates under investigation for their prebiotic potential but with no human data.

Inflammatory Bowel Disease

Inflammatory bowel disease (IBD) is an uncontrolled inflammation of the intestinal tract. It generally takes two forms, Crohn's Disease (CD), which can affect any part of the gastrointestinal (GI) tract or ulcerative colitis (UC) which affects the large intestine (Hoentjen, Welling et al., 2005). Both conditions are an inappropriate immune response to the normal microbiota (Hoentjen, Welling et al., 2005; Geier, Butler et al., 2007). There are several animal studies on IBDs such as CD and UC, which show potential benefits via modulation of the immune system (Guarner, 2007). There are relatively few studies in humans, however. Studies have shown significant decreases in pro-inflammatory cytokines and in calprotectin, a marker of intestinal inflammation (Furrie, Macfarlane et al., 2005; Casellas, Borruel et al., 2007).

Table 9.2. Potential functional enhancements in second-generation prebiotic carbohydrates.

Desirable Attribute in Prebiotic	Properties of Oligosaccharides
Microbiological enhancements	
Active at low dosage	Highly selectively and efficiently metabolized by "beneficial" bacteria, e.g., bifidobacteria and lactobacilli
Lack of side effects	Highly selectively metabolized by lactic acid bacteria but not gas producers, putrefactive organisms, pathogens, etc.
Fine control of microflora modulation	Selectively metabolized by certain *species* of bifidobacteria and lactobacilli
Persistence throughout the colon, i.e., toward distal areas	High molecular weight, slow fermentation
Antipathogen effects	
Induce antipathogen activity in probiotic microorganisms	Not yet defined
Inhibit adhesion of pathogens	Possess receptor sequences
Technological enhancements	
Varying viscosity	Available in different molecular weights and linkages
Good storage and processing stability	Possess $1 \rightarrow 6$ linkages and pyranosyl sugar rings
Differing sweetness	Varied monosaccharide composition

Colon Cancer

There is clear evidence of the involvement of the gut microbiota in the development of colon cancer (Heavey and Rowland, 2004). Gut microbes can produce a range of genotoxic and tumor-promoting compounds and enzymes capable of liberating these from harmless precursors. For this reason, there is much interest in the possibility of using probiotics and prebiotics to reduce cancer risk. There are many data from animal studies showing potential in reduction of colon cancer risk on feeding prebiotics (Macfarlane, Macfarlane et al., 2006). This data tends to focus on effects biomarkers of cancer risk. There are not many studies in humans, however. One of the most comprehensive involved feeding a prebiotic, Synergy 1, together with two probiotics, *Lactobacillus rhamnosus* GG and *Bifidobacterium bifidum* Bb-12 (Rafter, Bennett et al., 2007) in a synbiotic form. This study found that after 12 weeks of feeding, cancer patients and polypectomized volunteers had a significantly reduced colorectal cell proliferation and genotoxicity and an increased intestinal barrier function. As the prebiotic was administered in a synbiotic form, however, it is not clear whether it was the prebiotic or the probiotics that were responsible.

Calcium Absorption

Perhaps the most well-supported health benefit of prebiotic consumption is increased calcium absorption (Coxam, 2007), (Scholz-Ahrens and Schrezenmeir, 2007). Most calcium absorption (95%) takes place in the small intestine but some reaches the colon. Prebiotics can increase the scavenging of this calcium and have an overall impact on mineral balance. In a recent study, FOS and inulin fed to adolescents, significantly increased calcium absorption during feeding and increased calcium bone mineral content and bone mineral density after a year (Abrams, Griffin et al., 2005).

Acute Infections

It is well established in vitro that bifidobacteria and lactobacilli, the normal targets for prebiotic intervention, produce a range of antimicrobial compounds including short-chain fatty acids and bacteriocin-type molecules, that can effectively inhibit pathogens in coculture systems (Ouwehand, Isolauri et al., 1999; Collado, Hernandez et al., 2005). There has consequently been a lot of interest in the use of prebiotics to increase the GI barrier effect against invading pathogens. Such a prebiotic-mediated increase in resistance to pathogens can be shown clearly using in vitro models. GOS, FOS, and inulin have been shown to induce resistance to *Clostridium difficile* in fecal cultures (Hopkins and Macfarlane, 2002). It is, however, rather difficult to establish such effects in humans. Direct challenges are obviously not possible and studies have been confined to traveler's diarrhea. A study on 244 volunteers fed 10 g/day of FOS or placebo whilst on holiday (Cummings, Christie et al., 2001). Volunteers reported a significantly increased perception of "well-being" but there was no impact on incidence of diarrhea. Traveler's diarrhea is, however, a difficult condition to target with prebiotics. Generally, the causative agent is not identified and in many cases, this will be a virus, or a bacterial pathogen that acts on the small intestine. Prebiotics act in the colon and so could not impact on these causative agents.

Structure–Function Relationships in Prebiotic Oligosaccharides

Despite the range of oligosaccharides with claimed prebiotic properties, we have very little understanding of the mechanisms by which prebiotics bring about selective microbiota changes or of the structure–function relationships in these materials. There have been several in vitro studies comparing prebiotics. The commercial prebiotics inulin, FOS, GOS, lactulose, xylooligosaccharides (XOS), soy oligosaccharides (SOS) and isomaltooligosaccharides (IMO) were studied in non-pH controlled fecal batch cultures (Rycroft, Jones et al., 2001). They found that XOS were the most selective for bifidobacteria followed by inulin and FOS and IMO. The fructans had the highest gas evolution and GOS represented the best compromise between gas generation and selectivity. Further comparative testing of inulin, FOS, GOS, lactulose, and IMO took place in a pH-controlled system at pH 6.0 and 6.8 and at doses of 1 or 2% (w/w). The fructan-based prebiotics had the highest selectivity at 1% and pH 6.8 whereas the oligosaccharides were most selective at pH 6.8 and at 2% substrate (Palframan, Gibson et al., 2002).

Whilst these studies give comparative information about prebiotics with different structures, they do not give an insight into the relationship between glycosyl linkage, anomeric configuration, and molecular weight with selectivity of fermentation. One approach to obtaining such information is to evaluate the selectivity of a comprehensive set of disaccharides (Sanz, Gibson et al., 2005). These carbohydrates are only available in small quantities and so this study necessitated the development of a microscale fermentation system. This was established in microcentrifuge tubes but pH-control was not possible. The study established the selectivity indices of the disaccharides (Figure 9.1). This

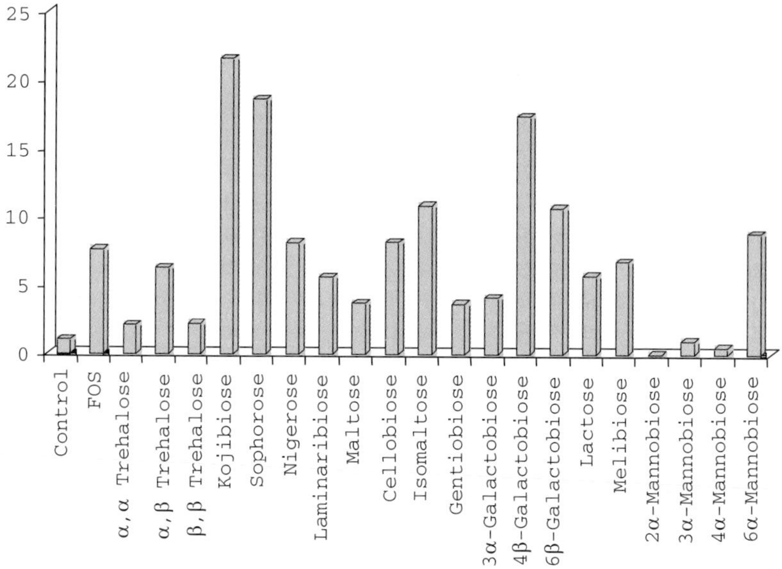

Figure 9.1. Fermentation selectivity of disaccharides as a function of structure. Disaccharides were used as sole carbon source in microscale non-pH-controlled fecal fermentations. Bacterial populations were monitored by fluorescent in situ hybridization. Selectivity index was defined by Palframan, Gibson, Rastall (2003a). Control fermentations had no added carbohydrate.

is a measure of the selectivity of the population changes, in this instance for bifidobacteria and lactobacilli (Palframan, Gibson, Rastall, 2003a). From this data, some trends emerge. For instance, in the case of the glucose–disaccharides, 1→2 linkages displayed the highest selectivity and the 1↔1 linked trehaloses the lowest. It is interesting to note the high selectivity displayed by the 1→2 linkages from both α- and β-linked disaccharides, despite the differences in structure and confirmation between these disaccharides. The mannose disaccharides displayed very low selectivity for bifidobacteria and lactobacilli with the exception of the α(1→6) linked molecule which was selective for lactobacilli.

There is also a deficit of good information on the influence of molecular weight on selectivity. Studies on fractionated FOS (Kaplan and Hutkins, 2000; Perrin, Warchol et al., 2001; Kaplan and Hutkins, 2003) have shown that some lactobacilli have specificity for trisaccharides and tetrasaccharides. Similar studies on GOS have shown similar results

(Gopal, Sullivan et al., 2001). Studies on synthetic candidate prebiotics made from sucrose with glycosyltranferases have also displayed marked selectivity for certain fractions, frequently tetrasaccharides (Sanz, Cote et al., 2005; Sanz, Cote et al., 2006a; 2006b).

It has been reported several times that hydrolysis of polysaccharides with low fermentation selectivity tends to increase the selectivity toward bifidobacteria and lactobacilli. This has been seen in the case of dextrans (Olano-Martin, Mountzouris et al., 2000), xylans (Palframan, Gibson et al., 2003b; Hughes, Shewry et al., 2007), and pectins (Olano-Martin, Gibson et al., 2002). This opens up the possibility of manufacturing novel prebiotics from biomass sources.

Potential for Enhanced Prebiotics

Increased understanding of the structure–function relationships in prebiotic oligosaccharides, of the functional composition of the gut microbiota and of the role of the gut microbiota in health, coupled with advances in carbohydrate biotechnology allow consideration of the development of "designer" prebiotics with specific functionality. Specific enhancements can be identified (Table 9.2).

Increased Persistence Through the Colon

Most of the current generation of prebiotics is low molecular weight oligosaccharides. The exception is inulin which has components with degree of polymerization (DP) of over 70 (van Loo, Coussement et al., 1995). Most of the mass of inulin, however, is considerably shorter. This means that most of them are fermented mainly in the proximal colon, as judged by the use of gut model systems. Some serious chronic gut disorders such as colon cancer and UC tend to originate in the distal colon, possibly due to the higher toxic load of bacterial metabolites. Consequently, there is interest in designing prebiotics which will give a selectively saccharolytic microbiota in this region. Beneo–Orafti have designed fructan-based products which attempt to have greater persistence. They have blended high molecular weight inulin with shorter FOS chains to produce a product called Synergy 1. This product displays better persistence in gut model systems (Roberfroid, Van Loo et al., 1998).

Control over persistence may be achieved by controlling the molecular weight of the prebiotic components. For this to be practical, the starting material must be a high molecular weight polysaccharide. An attractive place to look for suitable candidates is the plant cell wall. Plant cell wall polysaccharides represent a rich resource of complex carbohydrate chemistry for exploitation in the manufacture of prebiotics (Mohnen, 2008).

Cereal-derived materials can be used as polysaccharide sources for enzymatic hydrolysis to increase selectivity. There is clear potential for controlling the molecular weight of arabinoxylans to optimize selectivity of fermentation toward bifidobacteria (Hughes, Shewry et al., 2007). With commercially available wheat arabinoxylan fractions, the selectivity is strongly influenced by molecular weight. Fractions of 354, 278, and 66 kDa all displayed selectivity in vitro but the 66-kDa fraction had markedly greater selectivity (Figure 9.2). These studies in batch cultures were followed up with studies in a three-phase model of the human colon. This colonic model is formed from three continuous culture vessels connected in series and is inoculated with feces (Gibson, Cummings et al., 1988). It has been developed to mimic the lumenal microbiology of the ascending, transverse, and descending colon and has been validated against the colonic contents of sudden death victims (Macfarlane, Macfarlane et al., 1998). The 354- and 66-kDa fractions were tested in this model and the 66-kDa fraction displayed good persistence through to vessel 3, representing the distal colon, with highest selectivity in vessel two, representing the transverse colon (Figure 9.2).

Recently, there has been new commercial interest in making prebiotics from wheat arabinoxylans. A range of structures has been evaluated in studies in rats for their prebiotic potential (Van Craeyveld, Swennen et al., 2008). The oligosaccharides tested displayed good selectivity toward bifidobacteria with low DP oligosaccharides producing higher selectivity. Higher DP fractions, however, produced a shift away from protein metabolism as evidenced by lower production of branched-chain fatty acids. A fraction of DP 5 with an arabinose substitution ratio of 0.27 proved to give optimal properties. There is clearly much potential to develop economically viable processes to manufacture novel oligosaccharides from waste materials. Brewery spent grains, e.g., have been evaluated as a source of oligosaccharides using autohydrolysis procedures (Carvalheiro, Esteves et al., 2004; Carvalheiro, Garrote et al.,

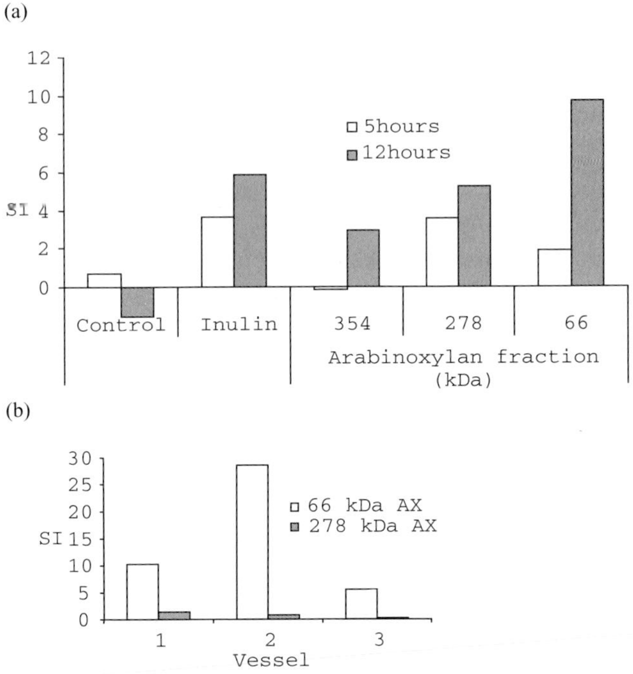

Figure 9.2. In vitro fermentation selectivity of arabinoxylans. Commercial arabinoxylan fractions were tested for fermentation selectivity using fecal inocula. Bacterial populations were monitored by fluorescent in situ hybridization. Selectivity index was defined by Palframan, Gibson, Rastall (2003a). Control fermentations had no added carbohydrate.(a) pH-controlled batch culture, pH 6.8. (b) Three-stage gut model system. Vessels 1–3 model the ascending, transverse, and descending colon, respectively.

2005). There is no data at present on the prebiotic potential of these oligosaccharides, however.

β-Glucans from cereals have also long been associated with health properties. One of the most well supported is their ability to reduce cholesterol levels (Braaten, Wood et al., 1994; Naumann, van Rees et al., 2006). The selectivity of a range of β-glucans from barley and oats has been investigated in vitro (Hughes, Shewry et al., 2008). These were commercial fractions having molecular weights of 137, 172, and 243 kDa for barley, and 150 and 230 for oats. These materials had no selectivity at all for any microbial groups recognized as having health-promoting properties. They did, however, produce relatively high ratios of propionate to acetate as fermentation end products (Hughes, Shewry et al., 2008). This may be relevant as the acetate to propionate ratio has been implicated in regulation of cholesterol metabolism in the liver (Hughes, Shewry et al., 2008). Attempts to increase the selectivity by hydrolysis of the glucans to short (DP 2–3) oligosaccharides increased selectivity slightly, but not to a level approaching that of FOS (Hughes, Shewry et al., 2008).

Pectins are commonly found in cell walls and seed mucilage of plants. An important aspect of pectins in the context of potential functional food ingredients is that they are very complex in terms of chemical structure and composition. They frequently include uronic acids and neutral sugars, mainly rhamnose, arabinose, galactose, and xylose. These monomers are linked by many different glycosidic linkages to form linear and branched polysaccharides.

Pectin is defined as a dietary fiber from fruits and vegetable or present in food as gelling agent, stabilizer, and bulking agent. It is highly fermented by bacteria in the colon and fermentation end products are short-chain fatty acids. Consumption has been reported to increase stool bulk and to reduce cholesterol and glucose levels in serum (Matheson and Story, 1994; Matheson, Colon et al., 1995; Marounek, Volek et al., 2007). Pectins have been developed commercially as dietary fiber products.

Pectins have also attracted attention as candidate prebiotics. Hydrolysis of orange albedo in water acidified with nitric acid at high temperature provided low molecular weight pectins (Manderson, Pinart et al., 2005). Pectic oligosaccharides (POS) of 3.5 kDa and 3.8 kDa were produced by hydrolysis of high methoxy citrus and low methoxy apple pectin with *Aspergillus pulverulentus* endo-polygalacturonase in an enzyme-membrane reactor (Olano-Martin, Mountzouris et al., 2001).

Oligosaccharides from pectin (POS) showed greater prebiotic potential (as judged by their selectivity for bifidobacteria) than the parent pectins (Olano-Martin, Gibson et al., 2002). The rate of substrate fermentation of POS was also significantly faster than pectins. Similar results were seen with POS (Manderson, Pinart et al., 2005). Oligosaccharides produced as a waste stream from a citrus pectin production process (Fishman, Chau et al., 2000) were evaluated for their prebiotic potential. POS had similar bifidogenic activity and gave the same amount and molar ratio of short-chain fatty acids as FOS. The persistence of these materials is currently being investigated.

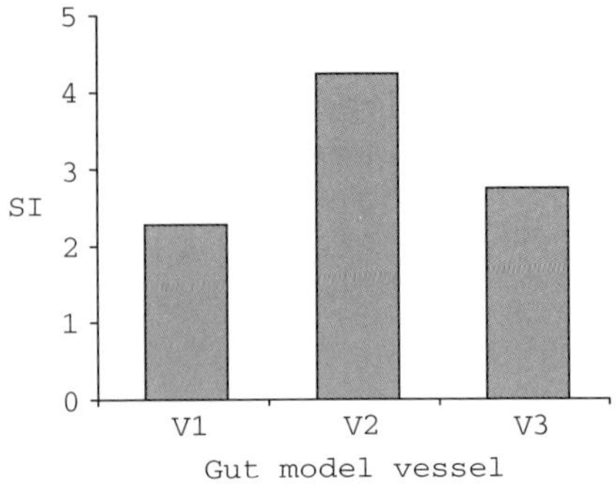

Figure 9.3. Fermentation persistence of *Gluconobacter oxydans* glucooligosaccharides. Mixed linkage ($\alpha(1\rightarrow6)$ and $\alpha(1\rightarrow4)$) glucooligosaccharides were synthesized with *G. oxydans* dextrin dextranase (Mountzouris, Gilmour et al., 1999) and evaluated in the three-stage gut model. Bacterial populations were monitored by fluorescent in situ hybridization. Selectivity index was defined by Palframan, Gibson, Rastall (2003a).

Microbial polysaccharides may also be sources of persistent prebiotics. The bacterium *Gluconobacter oxydans* expresses a dextrin dextranase enzyme which transfers glucosyl residues from $\alpha(1\rightarrow4)$-linked maltodextrins to maltodextrin acceptors to build up $\alpha(1\rightarrow6)$-linked dextran chains (Mountzouris, Gilmour et al., 1999). IMO are $\alpha(1\rightarrow6)$-linked glucooligosaccharides and these are marketed as prebiotics in Japan (Crittenden and Playne, 1996). Olano-Martin et al. (2000) also found that *Leuconostoc mesenteroides*-derived dextrans and oligodextrans were selective for bifidobacteria in vitro. For this reason the in vitro selectivity of *G. oxydans*-synthesized glucans have been investigated (Wichienchot, Prasertsan et al., 2006a). These materials had a disperse molecular weight from 1 to 42 kDa and a content of $\alpha(1\rightarrow6)$-linkages of approximately 60%. In the three-stage gut model, the glucans resulted in higher selectivity for bifidobacteria in vessel 2 (Figure 9.3, (Wichienchot, Prasertsan et al., 2006b).

Species-Targeted Prebiotics

The majority of studies on the fermentative properties of prebiotics have used either selective media or molecular techniques such as fluorescent in vitro hybridization (FISH). Both of these approaches give quantitative population information at the genus or group level. There are consequently very little data at the species level. This is an unfortunate situation as the contribution of bacteria to health and disease depends on species and strain. It is only in recent years, however, that suitable techniques such as denaturing gradient gel electrophoresis (DGGE) and quantitative PCR have been available for addressing species level changes in the gut microbiota (McCartney, 2002).

Despite this situation, it may be possible to exploit the specificity of bacterial enzyme and transport systems to design more selectivity into prebiotic oligosaccharides. Probiotic bacteria have been used to synthesize a range of GOS (Rabiu, Jay et al., 2001). These were manufactured from lactose by crude extracts of β-galactosidases. The original producing strains were then tested for their growth rates on the GOS produced. Most of the bacteria had a higher growth rate on their corresponding GOS. Although this work has not been repeated in mixed culture systems, the concept has been applied commercially in the development of a GOS product known as Bimuno® (Tzortzis, Goulas et al., 2005a). This is manufactured by whole cell catalysis with a *B. bifidum* strain. The GOS performs very well in gut models (Tzortzis, Goulas et al., 2005b) and in humans (Depeint, Tzortzis et al., 2008; Vulevic, Drakoularakou et al., 2008) displaying comparable selectivity with the leading commercial product but at lower dosage. In a human trial (Depeint, Tzortzis et al., 2008), when fecal isolates were identified to species level, 7 out of 15 were *B. bifidum* strains. Further, the *B. bifidum* strains isolated displayed higher growth rates on the Bimuno® than the comparative GOS. The other isolates had similar growth rates on both GOS products.

Antiadhesive Activity

Prebiotics are generally considered to exert their effects primarily by modulating the metabolism and populations of the gut microbiota. It may be possible, however, to design oligosaccharides that can modulate the microbiota via other, more direct mechanisms. Many intestinal pathogens utilize monosaccharides or short oligosaccharide sequences as receptors and knowledge of these receptor sites has relevance for biologically enhanced prebiotics (Table 9.3).

Table 9.3. Oligosaccharide pathogen receptor sequences.

Binding Epitope	Pathogen or Toxin
Galα→4Gal	*Escherichia coli* (P-piliated), Vero cytotoxin
Galβ→4GlcNAcβ→3Gal	*Streptococcus pneumoniae*
GalNAcβ→4Gal	*Pseudomonas aeruginosa, Haemophlius influenzae, Staphylococcus aureus, Klebsiella pneumoniae*
Sialic acids	*E. coli* (S-fimbriated)
Galα→3Galβ→4GlcNAc	*Clostridium difficile* toxin A
Fucose	*Vibrio cholerae*
GlcNAc	*E. coli, V. cholerae*
Mannose	*E. coli, K. aerogenes, Salmonella* spp., (Type 1-fimbriated)

Binding of pathogens to these receptors is the first step in the colonization process (Finlay and Falkow, 1989; Karlsson, 1989). There are currently several pharmaceutical preparations based upon such oligosaccharides in clinical trials. These agents are multivalent derivatives of the sugars and act as "blocking factors," dislodging the adherent pathogen (Heerze, Kelm et al., 1994). The idea of combining prebiotic properties with antiadhesive activities is currently under investigation. This would add major functionality to the approach of altering gut pathogenesis. This might be achieved by developing prebiotics which incorporate such a receptor oligosaccharide sequence. These molecules may have sufficient antiadhesive activity to inhibit binding of low levels of pathogens. While not a therapeutic option, such multifunctional prebiotics may increase host resistance to infection, reducing the likelihood of pathogen establishment and subsequent elaboration of virulence factors. They can thus be thought of as "decoy oligosaccharides."

Recently, GOS have been evaluated for their potential to inhibit the interaction between cholera toxin and host cells. A commercial GOS preparation (Vivinal GOS) was fractionated and tested for antiadhesive activity (Sinclair, de Slegte et al., 2009). The lactose component of the GOS was first removed by β-galactosidase digestion and then the mixture applied to an ion exchange resin. Although this does not produce clean fractionation of every component in these very complex mixtures, it does resolve components on the basis of size (Figure 9.4). The ability

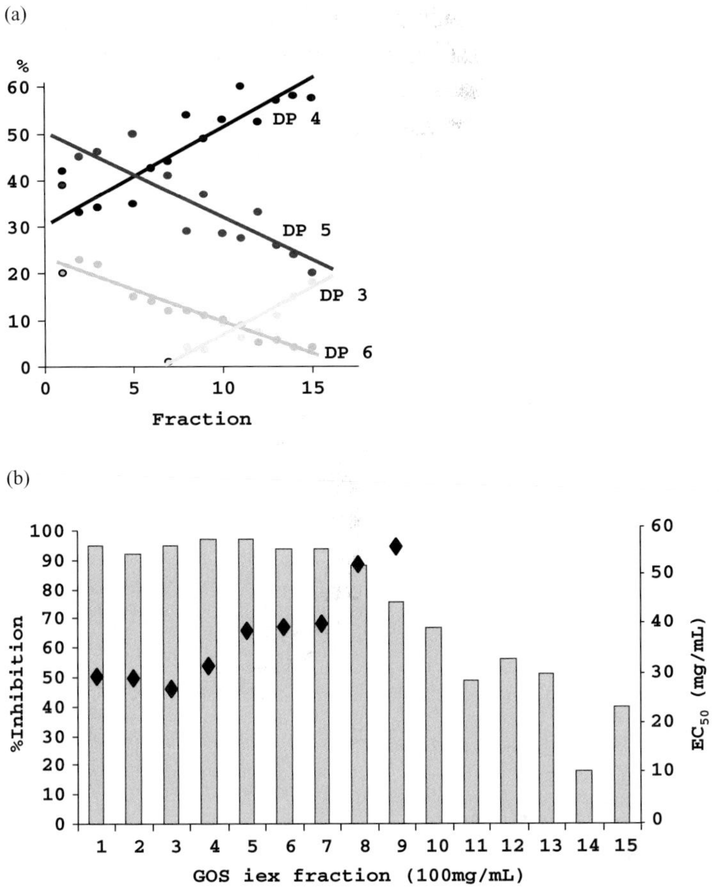

Figure 9.4. Antiadhesive activity of GOS fractions. (a) Commercial GOS was fractionated on an ion exchange column and the fractions analyzed for oligosaccharide molecular weight (expressed as DP) by mass spectrometry. (b) Antiadhesive activity of GOS fractions toward cholera toxin was assayed using a competitive ELISA assay against ganglioside GM_1, the cholera toxin cellular receptor.

of the various GOS fractions to inhibit the adhesion of cholera toxin to its cellular receptor, GM_1, was then tested using an indirect enzyme linked immunosorbent assay (ELISA) method. Over 90% inhibition was achieved with fractions 1–7 (Figure 9.4) which correlated to some degree to the content of oligosaccharides with DP 5 and 6 (Figure 9.4).

While this seems to offer a potential added functionality to GOS, the effective concentration, as measured as the EC_{50} (concentration to produce 50% inhibition) was around 30 mg/mL.

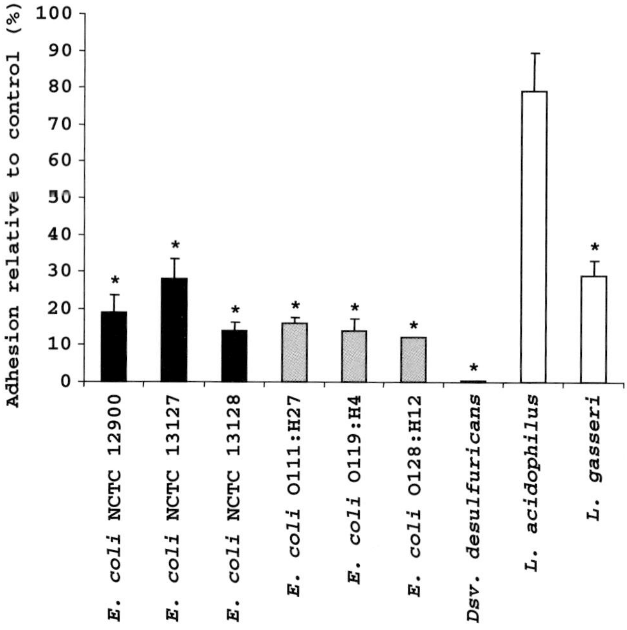

Figure 9.5. Antiadhesive activity of POS. Bacterial pathogens were adhered to HT29 cells in culture. POS were evaluated at 2.5 mg/mL and percentage inhibition calculated relative to phosphate buffered saline set at 100%.

Another candidate prebiotic that has been evaluated for antiadhesive potential are oligosaccharides derived from pectins. POS (Manderson, Pinart et al., 2005) were evaluated for their ability to inhibit adhesion of a range of *Escherichia coli* strains, *Desulfovibrio desulfuricans*, and two probiotic lactobacilli to human colonocytes in vitro (Rhoades, Manderson et al., 2008; Figure 9.5). Significant reductions in adhesion were seen with three verotoxigenic strains and three enteropathogenic strains of *E. coli*. The strain of *D. desulfuricans* tested was extremely inhibited. Of some concern, however, is that one of the two probiotic strains tested was also inhibited. These data show clear potential of POS to inhibit adhesion of pathogens; however, it is clear that the selectivity of the effect needs much further investigation.

POS preparations have also been shown to have protective effects toward tissue culture cells against E. coli verocytotoxins (Olano-Martin, Williams et al., 2003; Manderson, Pinart et al., 2005). Protection was achieved against both toxins, VT1 and VT2, at a concentration of 1 μg/mL.

It is clear that the concept of decoy oligosaccharides has some validity with at least some prebiotics; however, much more research is needed to clarify how useful this effect will be in real world situations.

Conclusions and Future Perspectives

There is a very active interest in prebiotic oligosaccharides at the present time in academic research circles and in the food and healthcare industries. An increasing number of studies are contributing to our understanding of the health effects and mechanisms of action of prebiotics. In parallel, increasingly sophisticated molecular microbiological studies are enhancing our understanding of the functional ecology of the human GI microbiota. Carbohydrate biotechnology meanwhile offers us the potential to tailor carbohydrates for specific health and functional attributes. Clearly, these three areas will produce novel prebiotics in the future.

To date, most prebiotics have had as their immediate end point an increase in the population of bifidobacteria and lactobacilli. This is understandable as these organisms are known to have health benefits in their own right and they represent a relatively easily enumerated target. Increasingly in the future, however, it is microbial metabolism that will be targeted.

It is, perhaps, timely to revisit the definition of a prebiotic. The term was first defined in 1995 as "a nondigestible food ingredient that beneficially affects the host by selectively stimulating the growth and/or activity of one or a limited number of bacteria in the colon, and thus improves host health" (Gibson and Roberfroid, 1995). This was subsequently amended slightly to be "a selectively fermented ingredient that allows specific changes both in the composition and/or activity in the GI microflora that confers benefits upon host well-being and health" (Gibson, Probert et al., 2004). This version included well-being, referred to the entire GI tract and removed the requirement for nondigestibility. In 2007, a technical meeting convened by the FAO produced a definition of "a nonviable food component that confers a health benefit on the host associated with modulation of the microbiota." This definition is very broad and is rather unhelpful in the number of compounds that would be included. A dietary fiber would, for instance, classify as a prebiotic under this definition. In order to address the situation the International Scientific Association for Probiotics and Prebiotics (ISAPP) met in London, Ontario, in 2008 to produce a more robust definition. This group defined a prebiotic as follows: "a *dietary* prebiotic is a selectively fermented ingredient that results in specific changes, in the composition and/or activity of the *GI* microbiota, thus conferring benefit(s) upon host health." This definition emphasized that a prebiotic brings about specific changes by being selectively metabolized, as distinct from dietary fibers. It also deliberately stressed the words "dietary" and "GI" to allow modified wording for extraintestinal applications such as the skin.

Whatever technological developments may be around the corner in the field of prebiotics, it is of prime importance that health claims are based upon sound rigorous science. For that reason any carbohydrate being touted as a prebiotic must have been evaluated in good quality human studies with robust microbiological methods.

References

Abrams, S.A., Griffin, I.J., Hawthorne, K.M., et al. (2005) A combination of prebiotic short- and long-chain inulin-type fructans enhances calcium absorption and bone mineralization in young adolescents. *Am. J. Clin. Nutr.* 82(2), 471–476.

Backhed, F., Ley, R.E., Sonnenburg, J.L., Peterson, D.A., Gordon, J.I. (2005) Host–bacterial mutualism in the human intestine. *Science* 307(5717), 1915–1920.

Braaten, J.T., Wood, P.J., Scott, F.W., et al. (1994) Oat beta-glucan reduces blood cholesterol concentration in hypercholesterolemic subjects. *Eur. J. Clin. Nutr.* 48(7), 465–474.

Carvalheiro, F., Esteves, M.P., Parajo, J.C., Pereira, H., Girio, F.M. (2004) Production of oligosaccharides by autohydrolysis of brewery's spent grain. *Bioresoure Technol.* 91(1), 93–100.

Carvalheiro, F., Garrote, G., Parajo, J.C., Pereira, H., Girio, F.M. (2005) Kinetic modeling of brewery's spent grain autohydrolysis. *Biotechnol. Prog.* 21(1), 233–243.

Casellas, F., Borruel, N., Torrejon, A., et al. (2007) Oral oligofructose-enriched inulin supplementation in acute ulcerative colitis is well tolerated and associated with lowered faecal calprotectin. *Aliment. Pharmacol. Ther.* 25(9), 1061–1067.

Collado, M.C., Hernandez, M., Sanz, Y. (2005) Production of bacteriocin-like inhibitory compounds by human fecal *Bifidobacterium* strains. *J. Food Prot.* 68(5), 1034–1040.

Corthesy, B., Gaskins, H.R., Mercenier, A. (2007) Cross-talk between probiotic bacteria and the host immune system. *J. Nutr.* 137(3), 781S–790S.

Coxam, V. (2007) Current data with inulin-type fructans and calcium, targeting bone health in adults. *J. Nutr.* 137(11Suppl), 2527S–2533S.

Crittenden, R.G., Playne, M.J. (1996) Production, properties and applications of food-grade oligosaccharides. *Trends Food Sci. Technol.* 7(11), 353–361.

Cummings, J.H., Christie, S., Cole, J.H. (2001) A study of fructo oligosaccharides in the prevention of travellers' diarrhoea. *Aliment. Pharmacol. Ther.* 15(8), 1139–1145.

Cummings, J.H., Macfarlane, G.T. (1991) The control and consequences of bacterial fermentation in the human colon. *J. Appl. Microbiol.* 70(6), 443–459.

Depeint, F., Tzortzis, G., Vulelvic, J., l'Anson, K., Gibson, G.R. (2008) Prebiotic evaluation of a novel galactooligosaccharide mixture produced by the enzymatic activity of *Bifidobacterium bifidum* NCIMB 41171, in healthy humans: a randomized, double-blind, crossover, placebo-controlled intervention study. *Am. J. Clin. Nutr.* 87(3), 785–791.

Dethlefsen, L., Eckburg, P.B., Bik, E.M., Relman, D.A. (2006) Assembly of the human intestinal microbiota. *Trends Ecol. Evol.* 21(9), 517–523.

Finlay, B.B., Falkow, S. (1989) Common themes in microbial pathogenicity. *Microbiol. Rev.* 53(2), 210–230.

Fishman, M.L., Chau, H.K., Hoagland, P., Ayyad, K. (2000) Characterization of pectin, flash-extracted from orange albedo by microwave heating, under pressure. *Carbohydr. Res.* 323(1–4), 126–138.

Flint, H.J., Duncan, S.H., Scott, P., Louis, P. (2007) Interactions and competition within the microbial community of the human colon: links between diet and health. *Environ. Microbiol.* 9(5), 1101–1111.

Furrie, E., Macfarlane, S., Kennedy, A., et al. (2005) Synbiotic therapy (*Bifidobacterium longum*/Synergy 1) initiates resolution of inflammation in patients with active ulcerative colitis: a randomised controlled pilot trial. *Gut* 54(2), 242–249.

Geier, M.S., Butler, R.N., Howarth, G.S. (2007) Inflammatory bowel disease: current insights into pathogenesis and new therapeutic options; probiotics, prebiotics and synbiotics. *Int. J. Food Microbiol.* 115(1), 1–11.

Gibson, G.R., Cummings, J.H., MacFarlane, G.T. (1988) Use of a three-stage continuous culture system to study the effect of mucin on dissimilatory sulfate reduction and methanogenesis by mixed populations of human gut bacteria. *Appl. Environ. Microbiol.* 54(11), 2750–2755.

Gibson, G.R., Probert, H.M., Rastall, R.A., Roberfroid, M.B. (2004) Dietary modulation of the human colonic microbiota: updating the concept of prebiotics. *Nutr. Res. Rev.* 17(2), 259–275.

Gibson, G.R., Roberfroid, M.B. (1995) Dietary modulation of the human colonic microbiota: introducing the concept of prebiotics. *J. Nutr.* 125(6), 1401–1412.

Gopal, P.K., Sullivan, P.A., Smart, J.B. (2001) Utilisation of galacto-oligosaccharides as selective substrates for growth by lactic acid bacteria including *Bifidobacterium lactis* DR10 and *Lactobacillus rhamnosus* DR20. *Int. Dairy J.* 11(1–2), 19–25.

Guarner, F. (2007) Prebiotics in inflammatory bowel diseases. *Br. J. Nutr.* 98(Suppl 1), S85–S89.

Heavey, P.M., Rowland, I.R. (2004) Gastrointestinal cancer. *Best Pract. Res. Clin. Gastroenterol.* 18(2), 323–336.

Heerze, L.D., Kelm, M.A., Talbot, J.A., Armstrong, G.D. (1994) Oligosaccharide sequences attached to an inert support (SYNSORB) as potential therapy for antibiotic-associated diarrhea and pseudomembranous colitis. *J. Infect. Dis.* 169(6), 1291–1296.

Hoentjen, F., Welling, G.W., Harmsen, H.J.M., et al. (2005) Reduction of colitis by prebiotics in HLA-B27 transgenic rats is associated with microflora changes and immunomodulation. *Inflamm. Bowel Dis.* 11(11), 977–985.

Hopkins, M.J., Macfarlane, G.T. (2002) Changes in predominant bacterial populations in human faeces with age and with *Clostridium difficile* infection. *J. Med. Microbiol.* 51(5), 448–454.

Hughes, S.A., Shewry, P.R., Li, L., Gibson, G.R., Sanz, M.L., Rastall, R.A. (2007) In vitro fermentation by human fecal microflora of wheat arabinoxylans. *J. Agric. Food Chem.* 55(11), 4589–4595.

Hughes, S.A., Shewry, P.R., Gibson, G.R., McCleary, B.V., Rastall, R. (2008) *In vitro* fermentation of oat and barley derived beta-glucans by human faecal microbiota. *FEMS Microbiol. Ecol.* 64(3), 482–493.

Kaplan, H., Hutkins, R.W. (2000) Fermentation of fructooligosaccharides by lactic acid bacteria and bifidobacteria. *Appl. Environ. Microbiol.* 66(6), 2682–2684.

Kaplan, H., Hutkins, R.W. (2003) Metabolism of fructooligosaccharides by *Lactobacillus paracasei* 1195. *Appl. Environ. Microbiol.* 69(4), 2217–2222.

Karlsson, K.A. (1989) Animal glycosphingolipids as membrane attachment sites for bacteria. *Annu. Rev. Biochem.* 58, 309–350.

Ley, R.E., Peterson, D.A., Gordon, J.I. (2006) Ecological and evolutionary forces shaping microbial diversity in the human intestine. *Cell* 124(4), 837–848.

Macfarlane, G.T., Macfarlane, S., Gibson, J.I. (1998) Validation of a three-stage compound continuous culture system for investigating the effect of retention time on the ecology and metabolism of bacteria in the human colon. *Microb. Ecol.* 35(2), 180–187.

Macfarlane, S., Macfarlane, G.T., Cummings, J.H. (2006) Review article: prebiotics in the gastrointestinal tract. *Aliment. Pharmacol. Ther.* 24(5), 701–714.

Manderson, K., Pinart, M., Touhy, K.M., et al. (2005) In vitro determination of prebiotic properties of oligosaccharides derived from an orange juice manufacturing by-product stream. *Appl. Environ. Microbiol.* 71(12), 8383–8389.

Marounek, M., Volek, Z., Skrivanova, E., Tuma, J., Duskova, D. (2007) Effect of pectin and amidated pectin on cholesterol homeostasis and cecal metabolism in rats fed a high-cholesterol diet. *Physiol. Res.* 56(4), 433–442.

Matheson, H.B., Colon, I.S., Story, J.A. (1995) Cholesterol 7 alpha-hydroxylase activity is increased by dietary modification with psyllium hydrocolloid, pectin, cholesterol and cholestyramine in rats. *J. Nutr.* 125(3), 454–458.

Matheson, H.B., Story, J.A. (1994) Dietary psyllium hydrocolloid and pectin increase bile acid pool size and change bile acid composition in rats. *J. Nutr.* 124(8), 1161–1165.

McCartney, A.L. (2002) Application of molecular biological methods for studying probiotics and the gut flora. *Br. J. Nutr.* 88(Suppl 1), S29–S37.

Mohnen, D. (2008) Pectin structure and biosynthesis. *Curr. Opin. Plant. Biol.* 11(3), 266–277.

Mountzouris, K.C., Gilmour, S.G., Jay, A.J., Rastall, R.A. (1999) A study of dextran production from maltodextrin by cell suspensions of *Gluconobacter oxydans* NCIB 4943. *J. Appl. Microbiol.* 87(4), 546–556.

Naumann, E., van Rees, A.B., Onning, G., Oste, R., Wydra, M., Mensink, R.P. (2006) Beta-glucan incorporated into a fruit drink effectively lowers serum LDL-cholesterol concentrations. *Am. J. Clin. Nutr.* 83(3), 601–605.

Olano-Martin, E., Gibson, G.R., Rastall, R.A. (2002) Comparison of the in vitro bifidogenic properties of pectins and pectic-oligosaccharides. *J. Appl. Microbiol.* 93(3), 505–511.

Olano-Martin, E., Mountzouris, K.C., Gibson, G.R., Rastall, R.A. (2000) In vitro fermentability of dextran, oligodextran and maltodextrin by human gut bacteria. *Br. J. Nutr.* 83(3), 247–255.

Olano-Martin, E., Mountzouris, K.C., Gibson, G.R., Rastall, R.A., (2001) Continuous production of pectic oligosaccharides in an enzyme membrane reactor. *J. Food Sci.* 66(7), 966–971.

Olano-Martin, E., Williams, M.R., Gibson, G.R., Rastall, R.A. (2003) Pectins and pectic-oligosaccharides inhibit *Escherichia coli* O157:H7 Shiga toxin as directed towards the human colonic cell line HT29. *FEMS Microbiol. Lett.* 218(1), 101–105.

Ouwehand, A.C., Isolauri, E., Kirjavainen, P.V., Selminen, S.J. (1999) Adhesion of four *Bifidobacterium* strains to human intestinal mucus from subjects in different age groups. *FEMS Microbiol. Lett.* 172(1), 61–64.

Palframan, R., Gibson, G.R., Rastall, R.A. (2003a) Development of a quantitative tool for the comparison of the prebiotic effect of dietary oligosaccharides. *Lett. Appl. Microbiol.* 37(4), 281–284.

Palframan, R.J., Gibson, G.R., Rastall, R.A. (2002) Effect of pH and dose on the growth of gut bacteria on prebiotic carbohydrates in vitro. *Anaerobe* 8(5), 287–292.

Palframan, R.J., Gibson, G.R., Rastall, R.A. (2003b) Carbohydrate preferences of *Bifidobacterium* species isolated from the human gut. *Curr. Issues Intest. Microbiol.* 4(2), 71–75.

Perrin, S., Warchol, M., Grill, J.P., Schneider, F., (2001) Fermentations of fructo-oligosaccharides and their components by *Bifidobacterium infantis* ATCC 15697 on batch culture in semi-synthetic medium. *J. Appl. Microbiol.* 90(6), 859–865.

Rabiu, B.A., Jay, A.J., Gibson, G.R., Rastall, R.A. (2001) Synthesis and fermentation properties of novel galacto-oligosaccharides by beta-galactosidases from *Bifidobacterium* species. *Appl. Environ. Microbiol.* 67(6), 2526–2530.

Rafter, J., Bennett, M., Caderni, G., et al. (2007) Dietary synbiotics reduce cancer risk factors in polypectomized and colon cancer patients. *Am. J. Clin. Nutr.* 85(2), 488–496.

Rhoades, J., Manderson, K., Wells, A., et al. (2008) Oligosaccharide-mediated inhibition of the adhesion of pathogenic *Escherichia coli* strains to human gut epithelial cells in vitro. *J. Food Prot.* 71(11), 2272–2277.

Roberfroid, M.B., Van Loo, J.A., Gibson, G.R. (1998) The bifidogenic nature of chicory inulin and its hydrolysis products. *J. Nutr.* 128(1), 11–19.

Rycroft, C.E., Jones, M.R., Gibson, G.R., Rastall, R.A. (2001) A comparative in vitro evaluation of the fermentation properties of prebiotic oligosaccharides. *J. Appl. Microbiol.* 91(5), 878–887.

Sanz, M.L., Cote, G.L., Gibson, G.R., Rastall, R.A. (2005) Prebiotic properties of alternansucrase maltose-acceptor oligosaccharides. *J. Agric. Food Chem.* 53(15), 5911–5916.

Sanz, M.L., Cote, G.L., Gibson, G.R., Rastall, R.A. (2006a) Influence of glycosidic linkages and molecular weight on the fermentation of maltose-based oligosaccharides by human gut bacteria. *J. Agric. Food Chem.* 54(26), 9779–9784.

Sanz, M.L., Cote, G.L., Gibson, G.R., Rastall, R.A. (2006b) Selective fermentation of gentiobiose-derived oligosaccharides by human gut bacteria and influence of molecular weight. *FEMS Microbiol. Ecol.* 56(3), 383–388.

Sanz, M.L., Gibson, G.R., Rastall, R.A. (2005) Influence of disaccharide structure on prebiotic selectivity in vitro. *J. Agric. Food Chem.* 53(13), 5192–5199.

Scholz-Ahrens, K.E., Schrezenmeir, J. (2007) Inulin and oligofructose and mineral metabolism: the evidence from animal trials. *J. Nutr.* 137(11Suppl), 2513S–2523S.

Sinclair, H.R., de Slegte, J., Gibson, G.R., Rastall, R.A. (2009) Galactooligosaccharides (GOS) inhibit *Vibrio cholerae* toxin binding to its GM1 receptor. *J. Agric. Food Chem.* 57(8), 3113–3119.

Tzortzis, G., Goulas, A.K., Gee, J.M., Gibson, G.R. (2005) A novel galactooligosaccharide mixture increases the bifidobacterial population numbers in a continuous in vitro fermentation system and in the proximal colonic contents of pigs in vivo. *J. Nutr.* 135(7), 1726–1731.

Tzortzis, G., Goulas, A.K., Gibson, G.R. (2005) Synthesis of prebiotic galactooligosaccharides using whole cells of a novel strain, *Bifidobacterium bifidum* NCIMB 41171. *Appl. Microbiol. Biotechnol.* 68(3), 412–416.

Van Craeyveld, V., Swennen, K., Durnez, E., et al. (2008) Structurally different wheat-derived arabinoxylooligosaccharides have different prebiotic and fermentation properties in rats. *J. Nutr.* 138(12), 2348–2355.

van Loo, J., Coussement, P., de Leenheer, L., Hoebregs, H., Smits, G. (1995) On the presence of inulin and oligofructose as natural ingredients in the western diet. *Crit. Rev. Food Sci. Nutr.* 35(6), 525–552.

Vulevic, J., Drakoularakou, A., Yaqoob, P., Tzortzis, G., Gibson, G.R. (2008) Modulation of the fecal microflora profile and immune function by a novel trans-galactooligosaccharide mixture (B-GOS) in healthy elderly volunteers. *Am. J. Clin. Nutr.* 88(5), 1438–1446.

Wichienchot, S., Prasertsan, P., Hongpattarakere, T., Gibson, G.R., Rastall, R.A. (2006a) In vitro fermentation of mixed linkage glucooligosaccharides produced by *Gluconobacter oxydans* NCIMB 4943 by the human colonic microflora. *Curr. Issues Intest. Microbiol.* 7(1), 7–12.

Wichienchot, S., Prasertsan, P., Hongpattarakere, T., Gibson, G.R., Rastall, R.A. (2006b) In vitro three-stage continuous fermentation of gluco-oligosaccharides produced by *Gluconobacter oxydans* NCIMB 4943 by the human colonic microflora. *Curr. Issues Intest. Microbiol.* 7(1), 13–18.

Chapter 10

Whole Grains and Digestive Health

Isabel Bondia-Pons, Jenni Lappi, Emilia Selinheimo, Marjukka Kolehmainen, Hannu Mykkänen, and Kaisa Poutanen

Whole Grain: Definition, Structure and Health Benefits

The current interest in whole grain foods stems from epidemiological evidence suggesting their beneficial role in health maintenance in comparison with refined cereal foods. This has led to a definition of whole grain and whole grain foods, and to health claims of such foods in United States, United Kingdom, and Sweden (Marquardt et al., 2004). According to the definition by the American Association of Cereal Chemists approved in 2000, whole grains consist of the intact, ground, cracked, or flaked caryopsis, whose principal anatomical components—the starchy endosperm, germ, and bran—are in the same relative proportion as they exist in the intact caryopsis (Richardson, 2003). Whole grains and foods made thereof contain thus considerably higher amounts of dietary fiber (DF) and many other nutrients and phytochemicals than refined cereals produced by milling processing discarding the outer layers.

Structure

The principal anatomic components of the grain are shown in Figure 10.1. The endosperm is the largest part (about 82%) of the grain, and consists predominantly of starch and proteins, but also some DF. The

Nondigestible Carbohydrates and Digestive Health. Edited by T. M. Paeschke and W. R. Aimutis

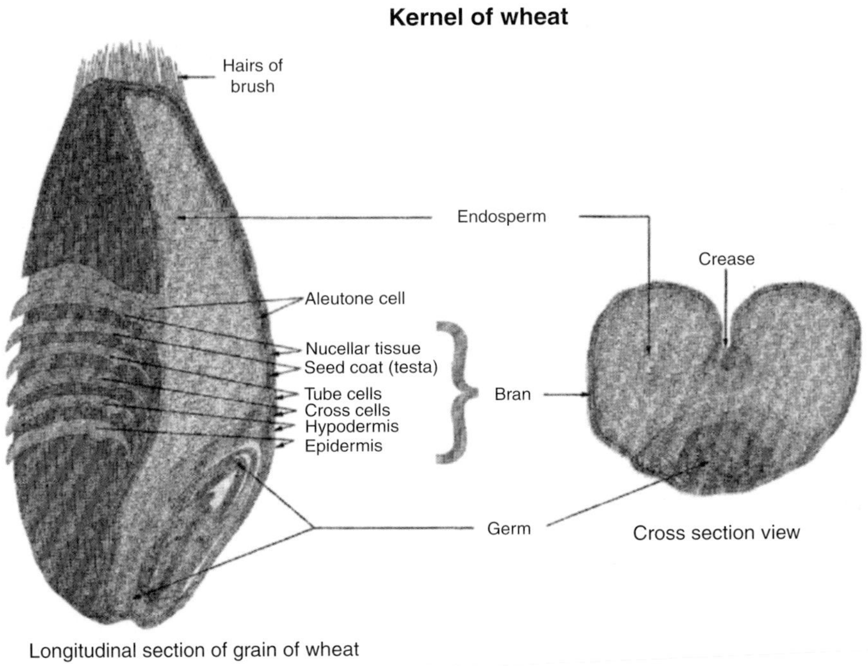

Figure 10.1. Principal anatomic components of a whole grain (Izydorczyk et al., 2002).

aleurone layer, which is a 1–4 cell layer surrounding the endosperm (Figure 10.2) has a high concentration of some amino acids, minerals, vitamins, and phenolic compounds (PC). The content of DF is also higher in aleurone layer than in endosperm. The outermost part of the grain can be divided into hyaline layer, testa, inner, and outer pericarp layers. The pericarp fraction is mostly made up of highly lignified cell walls with a high content of insoluble DF, such as cellulose and highly branched arabinoxylan (AX). Bran contains all the outmost layers of grain and, depending on the separation process, also part of the endosperm. The germ contains the plant embryo and has the majority of the grain lipids and sterols. The germ is also rich in vitamins and sugars.

Processing influences the tissue integrity and also chemical composition of whole grain foods, influencing, e.g., carbohydrate bioavailability and gastrointestinal conversions. Depolymerization of DF polymers may occur due to activation of endogenous enzymes or due to

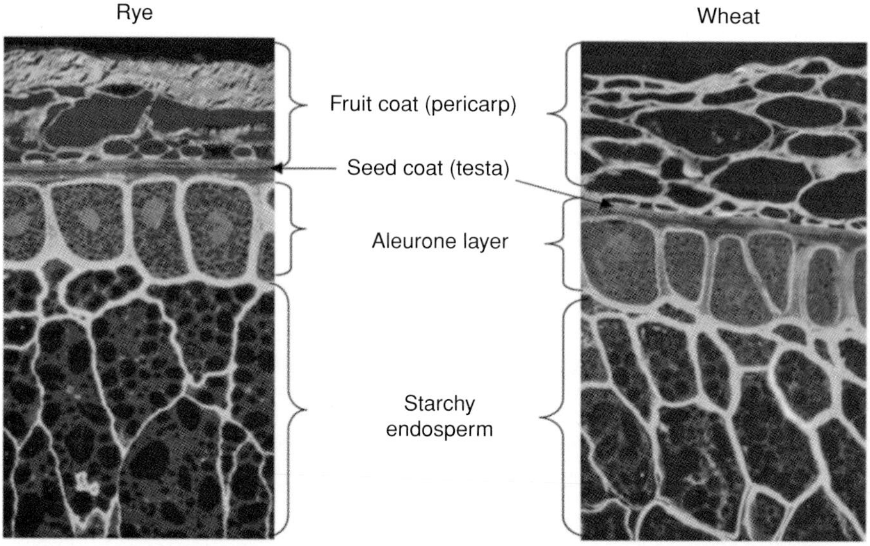

Figure 10.2. Microstructure of parts of intact grains of (a) rye and (b) wheat (Kamal-Eldin et al., 2009).

thermal degradation, and content of micronutrients may change. Starch bioavailability also is greatly influenced by processing.

Health Effects

Epidemiological studies show that intake of whole grain cereal foods reduces the risk of type 2 diabetes (de Munter et al., 2007) and cardiovascular disease (Jensen et al., 2004; Mellen et al., 2007). Bran seems to play an important role in this association (Jensen et al., 2006), and it has been suggested that grains should be consumed in a minimally refined form to achieve the benefit (Jones et al., 2004; Venn and Mann, 2004). The metabolic link between the consumption of cereal foods and disease risk is unclear. The observed health benefits of whole grain intake might be attributable to their ability to modulate hunger and satiety (Koh-Banerjee and Rimm, 2003), to the glycemic, lipidic, and inflammatory status of individuals (Hallfrisch et al., 2000; Jang et al., 2001; Riccardi et al., 2008), to directly or indirectly oxidative stress inhibition (Jacobs et al., 2007), and to the prebiotic activity of whole grain DF (Slavin, 2003; Costabile et al., 2008).

Gastrointestinal Effects

As whole grains are rich in different types of DF and associated compounds entering the large intestinal fermentation, it is also natural that they influence the gastrointestinal function, fermentation, and metabolism. Whole grain diets containing wheat or rye products have been reported to improve bowel function in intervention studies (Gråsten et al., 2000; Costabile et al., 2008). Consumption of whole grain wheat and rye leads to a greater fecal frequency and fecal bulking than consumption of refined grains (Gråsten et al., 2000; Pereira et al., 2002; McIntosh et al., 2003).

Wholegrain carbohydrates also play a role in treatment of gastrointestinal tract diseases, which include coeliac disease, Crohn's disease, ulcerative colitis, and irritable bowel syndrome (IBS) (Chudleigh and Hunter, 2005). In celiac disease, avoidance of rye, wheat, and barley is essential. However, oats has been shown to be tolerated by celiac sufferers (Janatuinen et al., 1995, 2000; Storsrud et al., 2003; Peräaho et al., 2004; Haboubi et al., 2006). In addition to increasing diet diversity diet, and thus quality of life, use of oats can serve as a rich source of grain fiber in a gluten-free diet, often devoid of DF (Poutanen, 2009).

The role of cereal fiber in the treatment of inflammatory bowel disease (IBD) is controversial, although increasing fiber in diet is often recommended in general practice. A systematic review and meta-analysis on fiber use in the treatment of IBS showed that the effects of wheat bran were not different from placebo or low-fiber diet (Ford et al., 2008). As reviewed by Bijkerk et al. (2004), insoluble fiber (corn fiber, wheat bran) may in some IBS patients be more harmful than a normal diet. In fact, a recent randomized, controlled intervention study on primary care patients with IBS points out bran is not beneficial in IBS treatment and may even worsen the symptoms (Bijkerk et al., 2009). Instead, soluble fiber seems to relieve symptoms of IBS in primary, secondary, and tertiary care patients (Bijkerk et al., 2004; Ford et al., 2008; Bijkerk et al., 2009).

Whole grains contain both insoluble and soluble fiber, but the effect of whole grain foods on the treatment of IBS symptoms has not been studied so far. Furthermore, there are no intervention studies on the effect of whole grain foods in treating the symptoms of other chronic diseases.

DF Constituents in Whole Grain

Whole grains are a rich and versatile source of DF. The main fiber constituents are the nonstarch polysaccharides (NSP), AX, β-glucans, fructans, and cellulose. AX and β-glucans are generally the most abundant DFs in cereal products. Table 10.1 shows the amount of the main DF components for the four main Western diet whole grains. The absolute amount of DF in whole grains depends on seed morphology and ranges from around 11% in Triticum species up to 37% in barley and rye (Vitaglioni et al., 2008; Ward et al., 2008). In regards to the individual fibers, e.g., rye grain contains 8.0–10.0% AX, 1.7–2.0% β-glucans, 1.0–1.7% cellulose, and 4.6–6.6% fructans whereas in wheat these polysaccharides make up, on average, 6.0, 0.6, 1.8, and 1.0%, respectively (Nilsson et al., 1997; Karppinen et al., 2003; Vitaglioni et al., 2008; Stone and Morell, 2009).

In addition to high fiber content, the wide and unique diversity of fiber constituents is one of the main differences between whole grains and refined grains, contributing to gastrointestinal events and health consequences of whole grain foods. The presence of different types of carbohydrates also influences the intestinal microflora metabolism and signaling, contributing to digestive health outcomes.

Arabinoxylans

AX constitute a major fraction of cereal cell wall polysaccharides with molecular weights from hundreds to thousands kDa. As reviewed by Kamal-Eldin and colleagues (2008), AX consists of a linear chain backbone of β-D-xylopyranose residues (X-residues) linked through (1→ 4) glycosidic linkages, substituted to varying degrees with single α-L-arabinofuranose residues (A-residues) at either the O-2 position or the O-3 position (Figure 10.3c). The majority of A-residues in AX are present as monomeric substituents. However, a small proportion of oligomeric side chains, consisting of two or more A-residues linked via 1→2, 1→3, and 1→5 linkages can be present (Izydorczyk and Dexter, 2008). The substitution of A-residues varies depending on the source of AX. The arabinose/xylose (A/X) ratio in cereal grains may vary from 0.3 to 1.1 (Glitsø et al., 1999). In wheat, the degree of branching

Table 10.1. Dietary fiber content of wholegrain wheat, rye, oat, and barley.

Grain	TDF (%)	IDF (%)	SDF (%)	AX (%)	β-glucan (%)	Cellulose (%)	Fructan (%)
Wheat	11.6–17.0[*]	10.2–14.7[*]	1.4–2.3[*]	6[†]	0.6–1[‡]	2[‡]	0.5–1.5[‡]
Rye	15.2–20.9[*]	11.1–16.0[*]	3.7–4.5[*]	8–10[‡]	1.7–2.0[**]	1.5[§§]	4.6–6.6[¶¶]
Oat	11.5–37.7[*]	8.6–33.9[*]	2.9–3.8[*]	7.6[§]	4.5–5.6[††]	Unknown	1.0[§]
Barley	14.6–27.1[*]	12.0–22.1[*]	2.6–5.0[*]	3.5–6.1[¶]	3.6–8.9[‡‡]	1.4[§§]	2.[illegible][§]

AX, arabinoxylan; TDF, total dietary fiber; IDF, insoluble dietary fiber; SDF, soluble dietary fiber.
[*] Vitaglioni et al., 2008.
[†] Stone and Morell, 2009.
[‡] Nilsson et al., 1997.
[§] Henry, 1985.
[¶] Izydorczyk and Dexter, 2008.
[**] Nyström et al., 2008.
[††] Shewry et al., 2008.
[‡‡] Andersson et al., 2008.
[§§] Englyst et al., 1983.
[¶¶] Karppinen et al., 2003.

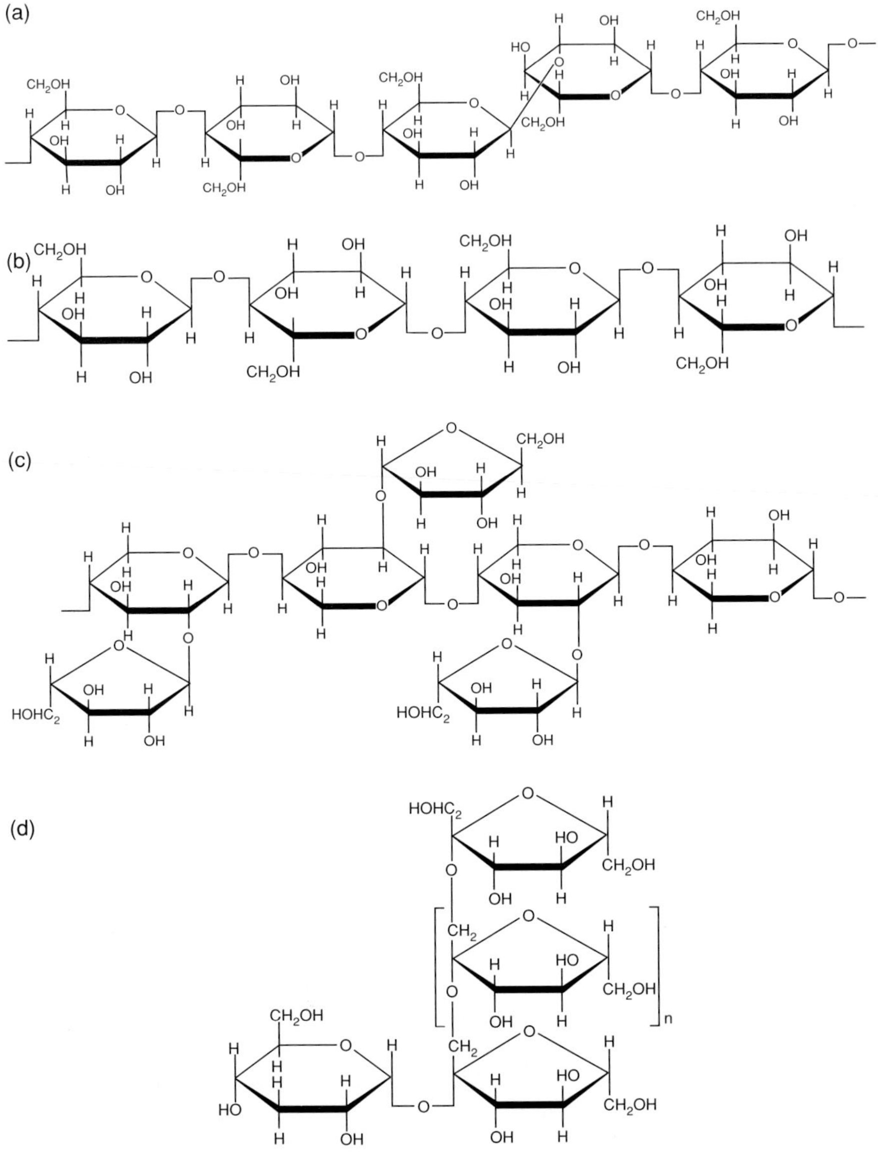

Figure 10.3. Dietary fiber constituents in whole grain: (a) *β*-glucan, (b) cellulose, (c) arabinoxylan, (d) fructan.

depends on the cell wall type. AX in outer pericarp layers (Figure 10.2) are relatively highly substituted with arabinose (A/X > 1), whereas those of aleurone (A/X < 0.5) and hyaline (A/X < 0.3) layer are poorly substituted. The A/X ratio in wheat endosperm is around 0.50–0.75 (Antoine et al., 2003; Barron et al., 2007). Small amounts of A-residues are substituted with hydroxycinnamic acids, mainly ferulic acid (FA), that are esterified to O-5 of A-residues linked to O-3 of the X-residues AX. Dimerization of the ferulate esters results in covalent crosslinking of AX chains and also linking AX to other cell wall constituents (Lazaridou et al., 2007).

Degree of polymerization of the xylan backbone, A/X ratio, and substitution pattern influences greatly the functional properties of AX (Glitsø et al., 2000) such as the solubility and viscosity properties and also susceptibility to enzymatic degradation (Bengtsson et al., 1992; Gruppen et al., 1993a, 1993b; Dervilly et al., 2000; Nilsson et al., 2000). AX is typically categorized to water-extractable (WE-AX) and water-unextractable AX (WU-AX). Generally most of the AX of cereal grains is WU-AX since AX is a structural component of cell walls crosslinking itself and other cell wall constituents via covalent and noncovalent interactions (Verwimp et al., 2007).

β-Glucans

Alongside with AX, β-glucans are an important DF constituent of grain, and in barley and oat they are the major structural components of cell walls (Ramakers et al., 2007). β-Glucans are carbohydrates consisting of linked glucose molecules with a molecular mass (MM) between 50 and 2,300 kDa. β-Glucans (Figure 10.3a) of cereal origin are linear homopolymers of D-glucopyranosyl residues linked mostly via two or three consecutive β-(1→4) linkages separated by a single β-(1→3) linkage (Izydorczyk and Dexter, 2008). The 1,3-linkages break up the uniform structure of the β-D-glucan molecule and make it soluble and flexible. β-Glucans from different genera of cereals share the same general molecular structure, but exhibit variation in features such as the ratios of β-(1→4)/β-(1→3) linkages, ratios of cellotriosyl/cellotetraosyl units, and molecular size (Lazaridou et al., 2007). Due to chain conformation, β-glucans are mostly soluble in water. The soluble β-glucans make viscous, shear thinning solutions even at low concentrations. The viscosity is related to the molecular weight and is strongly dependent on the concentration (Åman et al., 2005; Mäkeläinen et al., 2007). 2.3.

Cellulose

Cellulose is an essential structural component of cereal cell walls and it is mostly concentrated in the outermost layers of grain and grain husk. Cellulose is a polysaccharide consisting of a linear chain of several hundred to over 10,000 $\beta(1\rightarrow4)$ linked D-glucose units (Figure 10.3b). Hence, like β-glucan, cellulose is a homopolymer of glucose, but due to the linear $\beta(1\rightarrow4)$ structure of the polymer, cellulose is insoluble in water. Cellulose can form parallel packing of the polymer chains to three-dimensional microfibrillar aggregates that are stabilized by hydrogen bonding and van der Waals interactions (Stone and Morell, 2009). Other cell wall polysaccharides such as AX and β-glucan are believed to be associated to these cellulose microcrystalline structures.

Fructans

Fructans encompass the naturally occurring plant oligo- and polysaccharides in which one or more fructosyl–fructose linkages comprise the majority of glycosidic bonds; hence, they are primarily polymers of fructose units (Kelly, 2008). Structurally, fructans (Figure 10.3d) are linear or branched fructose (oligo)-polymers, which are either β-2,1-linked inulins (fructans of plant origin) or β-2,6-linked levans (fructans produced by some fungi and bacteria) (Gibson and Roberfroid, 1995). Inulin is a generic term that covers all linear fructans with β-(2−1) fructosyl–fructose glycosidic bonds.

Cereal fructans are of the inulin type with some branching via $(2\rightarrow6)$-β linkages. The beta configuration of the glycosidic bond gives inulin its unique structural and physiological properties, e.g., that they resist enzymatic hydrolysis by human salivary and small intestinal digestive enzymes (Roberfroid, 2007). Fructans are mostly water soluble, and their molecular weight is higher in the pericarp fraction than in the endosperm (Stone and Morell, 2009). Fructans may be hydrolyzed enzymatically or by acids to produce fructooligosaccharides containing x-y fructose molecules (FOS).

Colonic Fermentation of Whole Grain Carbohydrates

Large intestinal fermentation is a very complex process in which many metabolic pathways are carried out by different groups of bacteria. The rate and amount of fermentation products depends on different factors such as the species, amounts, and activity of the microbiota present in the colon, the substrate source, and the gut transit time. The simultaneous fermentation and metabolism of different types of DF and "copassengers" in whole grain foods influences the microbiota composition, and produces an extensive variety of metabolites with both local and systemic effects in human metabolism.

Production of Short-Chain Fatty Acids

Undigested carbohydrates from whole grains that reach the colon are anaerobically fermented by intestinal microbiota to gases and short-chain fatty acids (SCFAs) such as acetic, butyric, and propionic acids (Wong et al., 2006). SCFA are the main metabolites formed in the intestine because of fermentation. They are an important source of energy for the human gut (SCFA are readily absorbed and account for less than 10% of total daily energy) and can be transported to other tissues and organs of the human body.

Propionate is mainly metabolized by the liver, being a substrate for hepatic gluconeogenesis. Fermentation-derived propionate has been suggested to have a beneficial role on glucose and insulin metabolism (Nilsson et al., 2006, 2008). Acetate is metabolized partly by the liver and partly by the muscle and other peripheral tissues. Butyrate is the most important energy source for colonocytes, which should be constantly supplied to maintain a healthy status. It also plays a major role in the regulation of cell proliferation and differentiation (Topping et al., 2007). Fermentation of soluble DF may also produce potential anti-inflammatory effects via butyrate (Galisteo et al., 2008). Lactate has a short life in the gut, as it is a preferred electron sink product in anaerobic metabolism and may be quickly used by bacteria such as sulfate reducing bacteria (Kedia et al., 2009).

SCFAs act by lowering colon pH thus reducing potential pathogens, leading to ionization of potentially cytotoxic compounds like biogenic amines and ammonia (Topping, 2007). The reduction of the luminal pH also inhibits the conversion of primary bile acids to secondary bile acids and makes the free bile acids less available, thereby reducing carcinogenic activity (Christl et al., 1997). In addition, increased fecal bulk due to consumption of a diet high in fermentable fiber contributes to the lowered concentration of total and secondary bile acids (Gråsten et al., 2000).

Fermentation Rate of Cereal Carbohydrates

The fermentation rate of different fiber forms by the gut microbiota is physiologically important. There are several factors influencing fiber fermentability such as the chemical composition, solubility, particle size, the induction of bacterial enzymes, the time of residence in the large intestine, and the physical form (Karppinen et al., 2001). Extractable DF, resistant starch (RS), and oligosaccharides are more readily fermented than insoluble cell wall structures and NSP (Lebet et al., 1998).

Fiber fermentation might promote L-cell differentiation in the proximal colon, contributing to a higher endogenous GLP-1 production and suggesting a new mechanism by which DFs may lower food intake and fat mass development (Cani et al., 2008).

Fermentability of AX

Colonic degradation of AXs mainly depends on their structural characteristics (Glitsø and Bach Knudsen, 1999; Glitsø et al., 1999, 2000). Glitsø and Bach Knudsen (1999) studied the DF and AX characteristics of rye breads produced from different milling fractions enriched in pericarp/testa, aleurone, and endosperm tissue, respectively. The four types of rye bread differing in AX were fed to pigs, showing substituted WE-AX were slowly fermented, while increased complexity of the AX structure reduced degradation of AX (Glitsø et al., 1999). Endosperm AX were characterized by a large content of WE-AX and were extensively and readily degraded in the cecum, while the pericarp/testa AX, with a very high degree of mono- and double-xylose substitution, were undegraded in the intestinal tract of pigs. In contrast, aleurone AXs were degraded to a large extent but at a slower rate (Glitsø et al., 1999). The composition of the ileal effluents obtained after 3 and 48 hours in vitro fermentation correlated well with cecal and fecal degradation of AX respectively. The in vitro degradation of pericarp/testa carbohydrates was lower than in vivo (Glitsø et al., 2000). The pericarp/testa diet also led to the highest fecal bulking and the shortest transit time, whereas the aleurone diet resulted in the lowest pH values and the highest concentrations of SCFA, especially butyrate (Glitsø et al., 1998).

In oat bran fermentation, β-glucans and starch were consumed rapidly and the consumption of arabinose and xylose indicated a constant, but slow degradation of AX (Lebet et al., 1998). This slow fermentation of AX was also found in a study of fermentation of rye, wheat, and oat bran (Karppinen et al., 2001).

The importance of the AX fermentation in health promotion is therefore stressed by the different range in the AX fermentation rate through the whole gastrointestinal tract, mainly related to their different structural features.

Fermentability of β-Glucans

Water-extractable NSP like β-glucans may increase viscosity in the intestine, slowing intestinal transit, delaying gastric emptying, and slowing glucose and sterol absorption in the intestine. The ability of oat and barley β-glucans to increase viscosity is as well a factor contributing to explain the reduction of postprandial glucose and insulin responses after an oral load (Wood et al., 2000).

β-Glucans and β-glucan oligosaccharides were shown to selectively stimulate the growth of lactobacilli populations in a rat model suggesting a potential prebiotic activity of these DF compounds (Snart et al., 2006).

β-Glucans significantly modulated the microbial communities, and the resulting SCFA profile in a recent study focused on fermentation of β-glucan fractions from barley and oats by human fecal microbiota (Hughes et al., 2008). A considerable increase of propionate concentrations was observed, as reported in a previous in vitro study using rat cecal inocula to ferment oat and wheat brans (Monsma et al., 2000).

Enteral administration of β-glucans may also have biological effects. The addition of β-glucans to the diet may be used to modulate immune function and by that way might improve the resistance against invading pathogens in humans. Besides intraperitoneal or subcutaneous injection of oat β-glucans, also intragastric administration of oat β-glucans in mice enhanced resistance to bacterial and parasitic infections (Yun et al., 2003; Volman et al., 2008). Furthermore, a daily ingestion of oat β-glucan counteracted the decrease in macrophage antiviral resistance induced by exercise stress in mice (Davis et al., 2004).

β-Glucan, recognized as the main functional component of cereal fibers and also known as a prebiotic, stimulates the growth of some beneficial residential colon microorganisms such as bifidobacteria (Jaskari et al., 1998) and therefore has an interesting potential role in digestive health through microbiota-mediated effects.

Fermentability of Fructans

Fructans are not digested in the small intestine but reach the colon to be fully fermented by colon bacteria (Kelly, 2008). Degree of polymerization of fructans affects the site of fermentation in the colon. While fermentation of FOS seems to occur primarily in the proximal colon, inulin appears to be fermented in more distal parts of the colon.

Fructans have beneficial effects on stool production and fecal excretion (Roberfroid, 2007), which may relieve constipation. They also have beneficial effects on the immune system, and both animal and human studies show a lowering effect on blood triglycerides at the fasting and postprandial state (Roberfroid, 2005).

To our knowledge, however, there are no studies on the effects of cereal fructans on colon fermentation. Fermentability of inulin and FOS has been widely investigated but the fermentation of fructans in whole grains needs further research.

Bowel Function in Human Interventions on Whole Grain Carbohydrates

High-DF whole grain-based rye and wheat foods have been shown to be equally effective in increasing the weight of fecal output in a study with 28 Australian overweight middle-aged men (McIntosh et al., 2003) Two whole grain diets were shown to decrease the pH and fecal β-glucuronidase, secondary bile acids, and p-cresol concentrations as compared to the intake of low-fiber foods (McIntosh et al., 2003). These results agreed with those obtained from a randomized crossover trial in 17 healthy Finnish subjects. The ingestion of wholemeal rye bread for 4 weeks significantly increased fecal output and shortened intestinal transit time compared with white wheat bread (Gråsten et al., 2000). The rye foods were more effective in increasing plasma enterolactone, a mammalian lignan, and fecal butyrate concentrations than were whole-wheat and low-fiber foods, which did not differ from each other. Both high-fiber rye and wheat increased fecal bulking, but only rye foods significantly increased the fecal butyrate concentration. Oat bran has also shown to promote the production of propionate by human fecal bacteria in vitro, but produces less butyrate than FOS (Kedia et al., 2009). Consumption of rye bread and wheat bread with added AX has been also shown to increase the concentration of butyrate in feces, especially in men (Gråsten et al., 2000, 2003).

Influence of Whole Grain Carbohydrates on the Gut Microbiota

The intestinal microbiota of humans represent a complex ecosystem of a large number of metabolically active bacteria that strongly impact intestinal function and host health (Cani and Delzenne, 2009). In addition to the species belonging to the genera *Bifidobacterium* and *Lactobacillus*, maintenance of stable and diverse populations of commensal bacteria such as *Eubacterium* spp., *Atopobium* spp., and certain *Bacteroides* spp., has been shown to contribute to improved colonization resistance and protection against gastrointestinal disorders (Rakoff-Nahoum et al., 2004).

Prebiotics have been defined as carbohydrates stimulating the growth of human bacterial species belonging to the genera *Bifidobacterium* and *Lactobacillus* (Macfarlane et al., 2006). Inulin is by far the most studied prebiotic together with oligofructose and FOS (Meyer and Stasse-Wolthuis, 2009). However, the potential and synergic prebiotic effect of other carbohydrates present in whole grain foods deserves to be studied.

Evidence from In Vitro Studies

There are not many studies about gut effects of whole grain fiber in humans due to the invasive and expensive nature of colonic observation as well as the dynamic nature of the colon. Excreted colon contents do not necessarily represent colon contents in the proximal or even distal colon, due to continual fermentation of fiber and continual absorption of minerals and SCFA across the epithelium. In vitro fermentation is thus suggested as a noninvasive, time-efficient option to estimate fiber fermentability (Queenan et al., 2007). Despite the prebiotic effects of AX, AX-derived arabinoxylooligosaccharides (AXOS) have been studied far less than those of FOS. It is nowadays quite clear that they exert similar bifidogenic properties (Grootaert et al., 2006, 2007, 2009; Van Craeyveld et al., 2008; Vardakou et al., 2008; Pastell et al., 2009). New insights into the structure–activity relation of both AX and AXOS open up new perspectives for the production and application of AX and AXOS preparations with optimized prebiotic and fermentation properties.

A structure–function relationship was demonstrated between the MM of wheat-derived AX polysaccharide fractions and their fermentation properties in vitro by human fecal microbial communities (Hughes et al., 2007). The lowest MM AX fraction had the greatest selectivity for the *Bifidobacterium* genus and *Lactobacillus–Enterococcus* group, while the highest MM AX fraction was less selective for these groups. Therefore, the increased numbers and growth of bifidobacteria and lactobacilli groups from several in vitro fermentation studies have suggested prebiotic-like effects of AX fiber (Crittenden et al., 2002; Grootaert et al., 2006; Hughes et al., 2007; Vardakou et al., 2008). In a study investigating breakdown of FA-crosslinked AX and non-crosslinked AX in children's intestinal microbiotas, *Bacteroides* showed to play a major role in AX degradation in children's fecal microbiota. With the exception of enterobacterial metabolism, bacterial metabolism was not markedly affected by FA crosslinking between polymers (Hopkins et al., 2003).

However, components separated from the whole grain complex, or treated with different bioprocessing technologies, may have different effects in the gut than the whole grain food per se. In an in vitro study of separated bran and whole grain (WG) fiber fractions, oat bran increased bifidobacteria growth more than whole oat flour, being this prebiotic activity comparable to that found in commercial prebiotic FOS (Kedia et al., 2009). The growth of lactobacilli on the oat bran was also increased.

Fructans stimulate the growth of bifidobacteria, but they may also promote the growth of other undesirable microorganisms (Kelly, 2008). This is possibly due to the variance in the responses of individuals, and different types of fructans consumed. The effects of fructans on bacterial enzyme activities, production of SCFA, and end products of the fermentation of amino acids, mucins, and bile acids vary among the studies conducted, which is why clear conclusions of the influence of fructans on colon bacterial metabolism are difficult to make (Roberfroid, 2007).

Evidence from Human Studies

So far, there are only two human intervention studies investigating the effect of whole grain consumption on gut microbiota. This fact points out the need for designing controlled crossover studies to elucidate the bifidogenic effects of different whole grain carbohydrates in humans and their derived health outcomes.

In the first study, 17 healthy Finnish subjects consumed a diet supplemented with either white wheat or whole grain rye bread during a 4-week intervention period. The randomized crossover trial aimed to elucidate whether a high-fiber rye bread diet may positively influence the adhesion of several probiotics to the intestinal mucus (Ouwehand et al., 2000). Adhesion to the intestinal mucosa is one of the main selection criteria for probiotic microorganisms as it is considered a prerequisite for colonization, stimulation of the immune system, healing of damaged intestinal mucosa, and for antagonistic activity against enteropathogens (Mengheri, 2008). According to results of the study, *Lactobacillus rhamnosus* GG, *Bifidobacterium lactis* Bb12, and *Saccharomyces boulardii* adhered well to immobilized fecal mucus. Interestingly, men consuming the rye bread supplemented diet were found to excrete more fecal mucus per day than during the wheat bread diet period (Ouwehand et al., 2000).

The second study consisted of a randomized crossover trial in 31 healthy subjects given whole grain wheat and wheat bran in two 3-week study periods. The authors reported numbers of fecal bifidobacteria and lactobacilli significantly higher upon whole grain wheat ingestion compared with wheat bran (Costabile et al., 2008). Whole grain wheat can thus exert a pronounced prebiotic effect on gut microbiota composition and increase the FA concentration in fasting plasma. This fact suggests that whole grain intake might produce a continuous release of antioxidant into the bloodstream, which might explain the healthy benefits of its regular consumption.

Copassengers of Whole Grain Carbohydrates

The beneficial effects derived from whole grains might not be only explained by their characteristic types of carbohydrates, but may also be potentiated by other bioactive compounds such as micronutrients and phytochemicals (Figure 10.4) interacting with the fiber matrix (Figure 10.5).

The synergistic effects of constituents in the DF complex of whole grain foods are gaining interest in trying to give an explanation to the observed health benefits of whole grain intake (Jensen et al., 2006; Vitaglioni et al., 2008).

Fiber copassengers, mainly PC, might simultaneously influence not only the microflora, but also the mucosa and overall host health. Most of these compounds are concentrated in the outermost aleurone layers

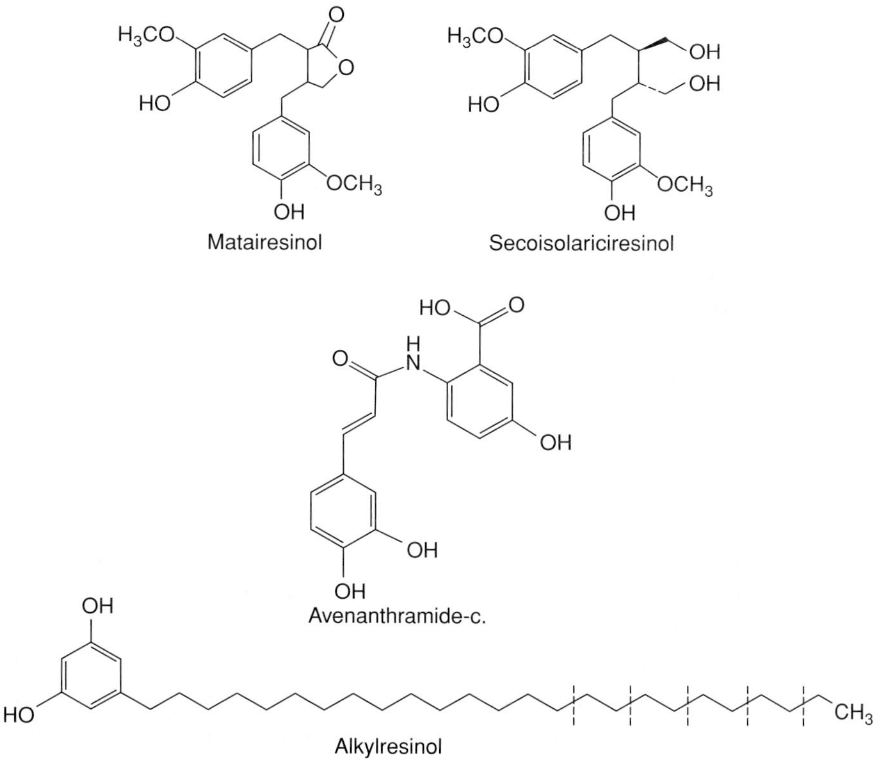

Figure 10.4. Chemical structures of some phytochemicals in whole grains.

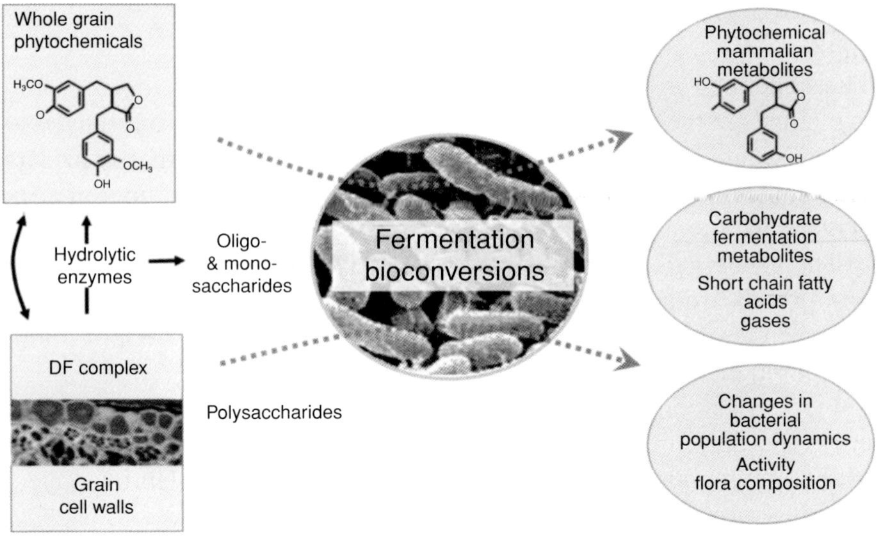

Figure 10.5. Fermentation bioconversions of the dietary fiber complex.

and bran of the grain, and are either covalently bound to the cell wall (phenolic acids) or closely associated and liberated during fiber fermentation. Grains are a major source of phytoestrogens such as lignans and sterols (β-sitosterol, andavenasterols), phytic acid, and PC (Kuhnle et al., 2009).

Lignans are metabolized to the mammalian lignans enterodiol and enterolactone by the intestinal microflora and may exert many biological activities (Adlercreutz et al., 2007).

There is a great variety of PC in whole grains, which range from flavonoids, anthocyanins, and tannins, to phenolic acids. The most abundant PCs in whole grain belong to the chemical class of hydroxycinnamic acids. About 95% of grain PC are linked to cell wall polysaccharides. FA is the most abundant phenolic acid in cereal grains, and is found mainly in the *trans* form, which is esterified to AX via the acid group acetylating the primary hydroxyl at the C5 position of α-L-arabinofuranosyl residues (Hatfield et al., 1999) and to hemicelluloses in the aleurone and pericarp of the grain. Several dimers of FA are also found in whole grains and form bridge structures between chains of hemicelluloses (Andreasen et al., 2000; Bunzel et al., 2000).

Unique phenolic lipids characteristic of the outer layers of wheat and rye grains are the so called alkylresorcinols (AR), a group of 1,3-dihydroxybenzene derivatives (Ross et al., 2004) localized in the external cuticle of testa/internal cuticle of pericarp (Landberg et al., 2008a). Their specific homolog profile makes ARs to be considered as potential biomarkers of rye and wheat intake (Landberg et al., 2008b). In addition, oats are a rich source of many other antioxidants (N-cinnamoylanthranilate alkaloids, apigenin, luteolin, and tricin) and unique polyphenols, called avenanthramides, with potential anti-inflammatory and anti-atherogenic properties (Nie et al., 2006; Guo et al., 2008).

Bioavailability of the Copassengers in the Whole Grain DF Complex

The "DF-antioxidant complex" is believed to mediate its biological activity on host health through the colonic microbiota (Scalbert et al., 2002; Wong et al., 2006). Once the new metabolites, resulting from the DF-antioxidant complex transformations, are formed in the colon, they circulate in plasma and are excreted via urine. The enterohepatic circulation plays an important role as it ensures that the residence time of these metabolites in plasma is extended compared to that of their parent compounds (Aura, 2008). Both, the physical structure and the soluble DF/insoluble DF ratio of the cereal fraction play a major role also in the bioavailability of the fiber copassengers (Napolitano et al., 2009). For instance, it seems the limited bioavailability of FA in the small intestine is due to its low bioaccessibility. An in vitro model mimicking digestion in the small intestine recently showed that the bioaccessibility of FA from aleurone, bran, flour, and bread with added aleurone is very low (Anson et al., 2009).

Human studies on the bioavailability of phenolic acids derived from whole grains are scarce (Manach et al., 2005). So far, only one study has provided evidence of FA absorption in humans. A significant increase was observed in the amount of FA excreted in the urine after a 6-week crossover intervention with rye bran bread (Harder et al., 2004). The estimated FA recovery after the rye bran bread intake was nevertheless ninefold higher than the recovery reported after a single serving of wheat bran breakfast cereal (Kern et al., 2003).

Dehydro-diferulic acids (DFA) are abundant components of insoluble cereal fiber. It has been hypothesized that the amount of DF diferulates is inversely correlated to the fermentability by intestinal microflora (Wang et al., 2004) but data about the potential breakdown and further DFA metabolism by intestinal bacteria is very limited. A recent study demonstrated the ability of human gut microbiota to transform two selected DFA regioisomers and identify their microbial metabolites (Braune et al., 2009). The cleavage of DFA by human intestinal microbiota, which depends on their coupling type, might affect not only the bioavailability of DFA but also the degradability of DFA-coupled fiber in the gut.

Conclusions and Future Prospects

Whole grain foods are the most important sources of nondigestible carbohydrates and associated phytochemicals in our diet. The complex architecture of the cereal cell walls and the diversity of chemical structures make whole grain carbohydrates a versatile substrate and also carrier for the intestinal microbiota. Whole grain foods contain nondigestible carbohydrates with very different intestinal fermentation rates, thus influencing the microbiota and its metabolism until the very distal parts of the colon. The associated phytochemicals also may act as selective antimicrobials, and their metabolism by microbiota is in many cases a prerequisite for their absorption into the circulation. Knowledge of their role in the events influencing the health of the gastrointestinal tract is thus far very limited.

The gastrointestinal microbiota is considered as a promising target to prevent and treat chronic diseases, such as type 2 diabetes. The current research focuses on the crosstalk between microbiota and intestinal epithelial cells, as well as immunomodulatory and anti-inflammatory responses. Of the whole grain carbohydrates, increasing interest has been directed towards AX which have been shown to be prebiotic, i.e., to promote the growth of bifidobacteria. The complex structures of AX and their binding of FA also makes them a diverse substrate for the gut bacteria. β-glucans and fructans are more rapidly fermented, and may also influence the microbiota composition.

The PC associated with the cell wall carbohydrates of whole grains possess antioxidative and other biological activities. They might influence both the digestibility of carbohydrates and the intestinal flora, but also locally exert protective reactions.

Whole grain foods have been shown to reduce colonic transit time and increase the fecal volume, as expected of DF-rich foods. It is surprising that the role of cereal fiber and whole grain foods has been studied so little in connection with gastrointestinal tract diseases. The whole grain DF matrix not only interacts with the microbiota but also may bind bacteria, metabolites, and harmful substances arising from food. Providing versatile substrates with different rates of fermentability in the digestive system, such as those in whole grain foods, seems a good rationale to maintain digestive health. Analysis of the many parallel phenomena taking place in the large intestine, and of their reflections in the peripheral circulation, will in the future cast light to the mechanisms of maintaining healthy gut function.

References

Adlercreutz, H. (2007) Lignans and human health. *Crit. Rev. Clin. Lab. Sci.* 44, 483–525.

Åman, P., Rimsten, L., Andersson, R. (2004) Molecular weight distribution of β-glucan in oat-based foods. *Cereal Chem.* 81(3), 356–360.

Andersson, A.A.M., Lampi, A-M., Nyström, L., et al. (2008) Phytochemical and dietary fiber components in barley varieties in the HEALTHGRAIN diversity screen. *J. Agric. Food Chem.* 56, 9767–9776.

Andreasen, M.F., Christensen, J., Meyer, A.S., Hansen, A. (2000) Ferulic acid dehydrodimersin rye (*Secale cereale* L.). *J. Cereal Sci.* 31, 303–307.

Anson, N.M., Selinheimo, E., Havenaar, R., et al. (2009) Bioprocessing of wheat bran improves in vitro bioaccessibility and colonic metabolism of phenolic compounds. *J. Agric. Food Chem.* 57(14), 6148–6155.

Antoine, C., Peyron, S., Mabille, F., Lapierre, C., Bouchet, B., Abecassis J. (2003) Individual contribution of grain outer layers and their cell wall structure to the mechanical properties of wheat bran. *J. Agric. Food Chem.* 51(7), 2026–2033.

Aura, A. (2008) Microbial metabolism of dietary phenolic compounds in the colon. *Phytochem. Rev.* 7, 407–429.

Barron, C., Surget, A., Rouau, X. (2007) Relative amounts of tissues in mature wheat (*Triticum aestivum* L) grain and their carbohydrate and phenolic acid composition. *J. Cereal Sci.* 45(1), 88–96.

Bengtsson, S., Andersson, R., Westerlund, E., Aman, P. (1992) *J. Sci. Food Agric.* 58, 331–337.

Bijkerk, C.J., de Wit, N.J., Muris, J.W., Whorwell, P.J., Knottnerus, J.A., Hoes, A.W. (2009) Soluble or insoluble fiber in irritable bowel syndrome in primary care? Randomised placebo controlled trial. *Br. Med. J.* 339(272), 1–7.

Bijkerk, C.J., Muris, J.W., Knottnerus, J.A., Hoes, A.W., de Wit, N.J. (2004) Systematic review: the role of different types of fiber in the treatment of irritable bowel syndrome. *Aliment. Pharmacol. Therap.* 19(3), 245–251.

Braune, A., Bunzel, M., Yonekura, R., Blaut, M. (2009) Conversion of dehydrodiferulic acids by human intestinal microbiota. *J. Agric. Food Chem.* 57(8), 3356–3362.

Bunzel M, Ralph J, Marita J, Steinhart H. (2000) Identification of 4-O-5'-coupled diferulic acid from insoluble cereal fiber. *J. Agric. Food Chem.* 48(8), 3166–3169.

Cani, P.D., Delzenne, N.M. (2009) The role of the gut microbiota in energy metabolism and metabolic disease. *Curr. Pharm. Des.* 15(13), 1546–1558.

Cani, P.D., Delzenne, N.M., Amar, J., Burcelin, R. (2008) Role of gut microflora in the development of obesity and insulin resistance following high-fat diet feeding. *Pathol. Biol.* 56(5), 305–309.

Christl, S.U., Bartram, H.P., Paul, A., Kelber, E., Scheppach, W., Kasper, H. (1997) Bile acid metabolism by colonic bacteria in continuous culture: effects of starch and pH. *Ann. Nutr. Metabol.* 41, 45–51.

Chudleigh, V.A., Hunter, J.O. (2005) Diseases of the gastrointestinal tract. In: Catherine, A. Geissler, Powers, Hilary J. (editors), *Human Nutrition*, pp. 429–441. Churchill Livingstone, Edinburgh.

Costabile, A., Klinder, A., Fava, F., et al. (2008) Whole-grain wheat breakfast cereal has a prebiotic effect on the human gut microbiota: a double-blind, placebo-controlled, crossover study. *Br. J. Nutr.* 99, 110–120.

Crittenden, R., Karppinen, S., Ojanen, S., et al. (2002) In vitro fermentation of cereal dietary fiber carbohydrates by probiotic and intestinal bacteria. *J. Sci. Food Agric.* 82, 781–789.

Davis, J.M., Murphy, E.A., Brown, A.S., Carmichael, M.D., Ghaffar, A., Mayer, E.P. (2004) Effects of oat beta-glucan on innate immunity and infection after exercise stress. *Med. Sci. Sports Eexerc.* 36(8), 1321–1327.

de Munter, J.S.L., Hu, F.B., Spiegelman, D., Franz, M., van Dam, R.M. (2007) Whole grain, bran, and germ intake and risk of type 2 diabetes: a prospective cohort study and systematic review. *PLoS Med.* 4, 1385–1395.

Englyst, H.N., Anderson, V., Cummings, J.H. (1983) Starch and non-starch polysaccharides in some cereal foods. *J. Sci. Food Agric.* 34(12), 1434–1440.

Ford, A.C., Talley, N.J., Spiegel, B.M., et al. (2008) Effect of fiber, antispasmodics, and peppermint oil in the treatment of irritable bowel syndrome: systematic review and meta-analysis. *Br. Med. J.* 337(132), 1–11.

Galisteo, M., Duarte, J., Zarzuelo, A. (2008) Effects of dietary fibers on disturbances clustered in the metabolic syndrome. *J. Nutr. Biochem.* 19(2), 71–84.

Gibson, G.R., Roberfroid, M.B. (1995) Dietary modulation 1 of the human colonic microbiota: introducing the concept of prebiotics. *J. Nutr.* 125, 1401–1412.

Glitsø, L.V., Bach Knudsen, K.E. (1999). Milling of whole grain rye to obtain fractions with different dietary fiber characteristics. *J. Cereal Sci.* 29, 89–97.

Glitsø, L.V., Brunsgaard, G., Hojsgaard, S., Sandstrom, B., Bach Knudsen, K.E. (1998) Intestinal degradation in pigs of rye dietary fiber with different structural characteristics. *Br. J. Nutr.* 80, 457–468.

Glitsø, L.V., Gruppen, H., Schols, H.A., Hoejsgaard, S., Sandstrom, B., Bach Knudsen, K.E. (1999) Degradation of rye arabinoxylans in the large intestine of pigs. *J. Sci. Food Agric.* 79, 961–969.

Glitsø, L.V., Jensen, B.B., Bach Knudsen, K.E. (2000) In vitro fermentation of rye carbohydrates including arabinoxylans of different structure. *J. Sci. Food Agric.* 80, 1211–1218.

Gråsten, S., Juntunen, K., Poutanen, K., Gylling H, Miettinen, T., Mykkänen, H. (2000) Rye bread improves bowel function and decreases the concentrations of some compounds that are putative colon cancer risk markers in middle-aged women and men. *J. Nutr.* 130(9), 2215–2221.

Gråsten, S., Liukkonen, K.H., Chrevatidis, A., El-Nezami, H., Poutanen, K., Mykkänen, H. (2003) Effects of wheat pentosan and inulin on the metabolic activity of fecal microbiota and on bowel function in healthy humans. *Nutr. Res.* 23, 1503–1514.

Grootaert, C., Delcour, J.A., Courtin, C.M., Broekaert, W.F., Verstraete, W., Van de Wiele, T. (2007) Microbial metabolism and prebiotic potency of arabinoxylan oligosaccharides in the human intestine. *Trends Food Sci Technol* 18(2), 64–71.

Grootaert, C., Van Den Abbeele, P., Marzorati, M., et al. (2009) Comparison of prebiotic effects of arabinoxylan oligosaccharides and inulin in a simulator of the human intestinal microbial ecosystem. *FEMS Microbiol. Ecol.* 69(2), 231–242.

Grootaert, C., Verstraete, W., Van de Wiele, T. (2006) Arabinoxylan oligosaccharides with different structures exert a bifidogenic effect in a mixed intestinal community. *Commun. Agric. Appl. Biol. Sci.* 71(1), 159–163.

Gruppen, H., Kormelink, F.J., Voragen, A.G.J. (1993a) Enzymatic degradation of water-unextractable cell wall material and arabinoxylans from wheat flour. *J. Cereal Sci.* 18, 129–143.

Gruppen, H., Kormelink, F.J., Voragen, A.G.J. (1993b) Water-unextractable cell wall material from wheat flour. 3. A Structural model for arabinoxylands. *J. Cereal Sci.* 18(2), 111–128.

Guo, W., Wise, M.L., Collins, F.W., Meydani, M. (2008) Avenanthramides, polyphenols from oats, inhibit IL-1beta-induced NF-kappaB activation in endothelial cells. *Free Radic. Biol. Med.* 44(3), 415–429.

Haboubi, N.Y., Taylor, S., Jones, S. (2006) Coeliac disease and oats: a systematic review. *Postgrad. Med. J.* 82, 672–678.

Hallfrisch, J., Facn, N., Behall, K.M. (2000) Mechanisms of the effects of grains on insulin and glucose responses. *J. Am. Coll. Nutr.* 19(3 Suppl), 320S–325S.

Harder, H., Tetens, I., Let, M.B., Meyer, A.S. (2004) Rye bread intake elevates urinary excretion of ferulic acid in humans, but does not affect the susceptibility of LDL to oxidation. *Eur. J. Nutr.* 43, 230–236.

Hatfield, R.D., Ralph, J., Grabber, J.H. (1999) Cell wall structural foundations: molecular basis for improving forage digestibilities. *Crop Science* 39(1), 27–37.

Henry, R.J. (1985) A comparison of the non-starch carbohydrates in cereal grains. *J. Sci. Food Agric.* 36, 1243–1253.

Hopkins, M.J., Englyst, H.N., Macfarlane, S., Furrie, E., Macfarlane, G.T., McBain, A.J. (2003) Degradation of cross-linked and non-cross-linked arabinoxylans by the intestinal microbiota in children. *Appl. Environ. Microbiol.* 69(11), 6354–6360.

Hughes, S.A., Shewry, P.R., Gibson, G.R., McCleary, B.V., Rastall, R.A. (2008) In vitro fermentation of oat and barley derived beta-glucans by human faecal microbiota. *FEMS Microbiol. Ecol.* 64(3), 482–493.

Hughes, S.A., Shewry, P.R., Li, L., Gibson, G.R., Sanz, M.L., Rastall, R.A. (2007) In vitro fermentation by human fecal microflora of wheat arabinoxylans. *J. Agric. Food Chem.* 55(11), 4589–4595.

Izydorczyk, M.S. and Dexter, J.E. (2008) Barley β-glucans and arabinoxylans: molecular structure, physicochemical properties, and uses in food products—a review. *Food Res. Int.* 41, 850–868.

Izydorczyk, M.S., Symons, S.J., Dexter, J.E. (2002) Fractionation of wheat and barley. In: Marquart, L., Slavin, J.L., Fulcher, R.G. (editors), *Whole-grain Foods in Health and Disease*, pp. 47–82. American Association of Cereal Chemists, Minnesota, USA.

Jacobs, D.R., Andersen, L.F., Blomhoff, R. (2007) Whole-grain consumption is associated with a reduced risk of noncardiovascular, noncancer death attributed to inflammatory diseases in the Iowa Women's Health Study. *Am. J. Clin. Nutr.* 85(6), 1606–1614.

Janatuinen, E.K., Kemppainen, T., Julkunen, R.J., et al. (2000) No harm from five year ingestion of oats in coeliac disease. *Gut* 50, 332–335.

Janatuinen, E.K., Pikkarainen, P.H., Kemppainen, T.A., et al. (1995) A comparison of diets with and without oats in adults with coeliac disease. *N. Engl. J. Med.* 333, 1033–1037.

Jang, Y., Lee, J.H., Kim, O.Y., Park, H.Y., Lee, S.Y. (2001) Consumption of whole grain and legume powder reduces insulin demand, lipid peroxidation, and plasma homocysteine concentrations in patients with coronary artery disease: randomized controlled clinical trial. *Arterioscler. Thromb. Vasc. Biol.* 21(12), 2065–2071.

Jaskari, J., Kontula, P., Siitonen, A., Jousimies-Somer, H., Mattila-Sandholm, T., Poutanen, K. (1998) Oat beta-glucan and xylan hydrolysates as selective substrates for *Bifidobacteruim* and *Lactobacillus* strains. *Appl. Microbiol. Biotechnol.* 49, 175–181.

Jensen, M.K., Koh-Banerjee, P., Franz, M., Sampson, L., Grønbaek, M., Rimm, E.B. (2006) Whole grains, bran, and germ in relation to homocysteine and markers of glycemic control, lipids, and inflammation 1. *Am. J. Clin. Nutr.* 83(2), 275–283.

Jensen, M.K., Koh-Banerjee, P., Hu, F.B., et al. (2004) Intakes of whole grains, bran, and germ and the risk of coronary heart disease in men. *Am. J. Clin. Nutr.* 80, 1492–1499.

Jones, J.M., Reicks, M., Adams, J., Fulcher, G., Marquart, L. (2004) Becoming proactive with the whole-grains message. *Nutr. Today* 39(1), 10–17.

Kamal-Eldin, A., Åman, P., Zhang, J.X., Bach Knudsen, K.E., Poutanen, K. (2008) High-fiber products: rye bread and other rye products. In: Hamaker, B.R. (editor), *Technology of Functional Cereal Products*, pp. 233–260. Woodhead Publishing Limited, Abington, Cambridge, England.

Kamal-Eldin, A., Lærke, H.N., Knudsen, K.E., et al. (2009) Physical, microscopic and chemical characterisation of industrial rye and wheat brans from the Nordic countries. *Food Nutr. Res.* 53, 1–11.

Karppinen, S., Liukkonen, K., Aura, A.M., Forssell, P., Poutanen, K. (2001) In vitro fermentation of polysaccharides of rye, wheat and oat brans and inulin by human faecal bacteria. *J. Sci. Food Agric.* 80, 1469–1476.

Karppinen, S., Myllymäki, O., Forssell, P., Poutanen, K. (2003) Fructan content of rye and rye products. *Cereal Chem.* 80(2), 168–171.

Kedia, G., Vázquez, J.A., Charalampopoulos, D., Pandiella, S.S. (2009) In vitro fermentation of oat bran obtained by debranning with a mixed culture of human fecal bacteria. *Curr. Microbiol.* 58(4), 338–342.

Kelly, G. (2008) Inulin-type prebiotics—a review: part 1. *Altern. Med. Rev.* 13(4), 315–329.

Kern, S.M., Bennett, R.N., Mellon, F.A., Kroon, P.A., Garcia-Conesa, M.T. (2003) Absorption of hydroxycinnamates in humans after high-bran cereal consumption. *J. Agric. Food Chem.* 51(20), 6050–6055.

Koh-Banerjee, P., Rimm, E.B. (2003) Whole grain consumption and weight gain: a review of the epidemiological evidence, potential mechanisms and opportunities for future research. *Proc. Nutr. Soc.* 62(1), 25–29.

Kuhnle, G.G., Dell'aquila, C., Aspinall, S.M., Runswick, S.A., Mulligan, A.A., Bingham, S.A. (2009) Phytoestrogen content of cereals and cereal-based foods consumed in the UK. *Nutr. Cancer* 61(3), 302–309.

Landberg, R., Kamal-Eldin, A., Andersson, A., Vessby, B., Aman, P. (2008a) Alkylresorcinols as biomarkers of whole-grain wheat and rye intake, plasma concentration and intake estimated from dietary records. *Am. J. Clin. Nutr.* 87, 832–838.

Landberg, R., Kamal-Eldin, A., Salmenkallio-Martilla, M., Rouau, X., Aman, P. (2008b) Localization of alkylresorcinols in wheat, rye and barley kernels. *Cereal Chem.* 48, 401–406.

Lazaridou, A., Biliaderis, C.G., Izydorczyk, M.S. (2007) Cereal β-glucans: structures, physical properties, and physiological functions. In: Biliaderis, C.G., Izydorczyk, M.S. (editors), *Functional Food Carbohydrates*, pp. 1–72. CRC Press, Taylor & Francis Group, Boca Raton, FL.

Lebet, V., Arrigoni, E., Amado, R. (1998) Measurement of fermentation products and substrate disappearance during incubation of dietary fiber sources with human faecal flora. *Lebensm. Wiss. Technol.* 31, 473–479.

Macfarlane, S., Macfarlane, G.T., Cummings, J.H. (2006) Review article: prebiotics in the gastrointestinal tract. *Aliment. Pharmacol. Ther.* 24(5), 701–714.

Mäkeläinen, H., Anttila, H., Sihvonen, J., et al. (2007) The effect of beta-glucan on the glycemic and insulin index. *Eur. J. Clin. Nutr.* 61(6), 779–785.

Manach, C., Williamson, G., Morand, C., Scalbert, A., Rémésy, C. (2005) Bioavailability and bioefficacy of polyphenols in humans. I. Review of 97 bioavailability studies. *Am. J. Clin. Nutr*. 81(1 Suppl.), 230S–242S.

Marquardt, L., Asp, N.G., Richardson, P. (2004) Whole grain health claims in the United States, United Kingdom and Sweden. In: Kamp, J.W., Asp, N.G., Miller Jones, J., Schaafsma, G. (editors), *Dietary Fiber—Bioactive Carbohydrates in Food and Feed*, pp. 39–57. Wageningen Academic Publishers, The Netherlands.

McIntosh, G.H., Noakes, M., Royle, P.J., Foster, P.R. (2003) Whole-grain rye and wheat foods and markers of bowel health in overweight middle-aged men. *Am. J. Clin. Nutr.* 77(4), 967–974.

Mellen, P.B., Liese, A.D., Tooze, J.A., Vitolins, M.Z., Wagenknecht, L.E., Herrington DM. (2007) Whole-grain intake and carotenoid artery atherosclerosis in a multiethnic cohort, the Insulin Resistance Atherosclerosis Study. *Am. J. Clin. Nutr.* 85, 1495–1502.

Mengheri, E. 2008 Health, probiotics, and inflammation. *J. Clin. Gastroenterol.* 42(Suppl 3), S177–S188.

Meyer, D., Stasse-Wolthuis, M. (2009) The bifidogenic effect of inulin and oligofructose and its consequences for gut health. *Eur. J. Clin. Nutr.* 63(11), 1277–1289.

Monsma, D.J., Thorsen, P.T., Vollendorf, N.W., Crenshaw, T.D., Marlett, J.A. (2000) In vitro fermentation of swine ileal digesta containing oat bran dietary fiber by rat caecal inocula adapted to the test fiber increases propionate production but fermentation of wheat bran ileal digesta does not produce more butyrate. *J. Nutr.* 130, 585–593.

Napolitano A, Costabile A, Martin-Pelaez, S., et al. (2009) Potential prebiotic activity of oligosaccharides obtained by enzymatic conversion of durum wheat insoluble dietary fiber into soluble dietary fiber. *Nutr., Metab. Cardiovasc. Dis.*, 19, 283–290.

Nie, L., Wise, M.L., Peterson, D.M., Meydani, M. (2006) Avenanthramide, a polyphenol from oats, inhibits vascular smooth muscle cell proliferation and enhances nitric oxide production. *Atherosclerosis*. 186(2), 260–266.

Nilsson, A., Grandfeldt, Y., Östman, E., Preston, T., Björk, I. (2006) Effects of GI and content of indigestible carbohydrates of cereal-based evening meals on glucose tolerance at a subsequent standardised breakfast. *Eur. J. Clin. Nutr.* 60, 1092–1099.

Nilsson, A., Ostman, E., Preston, T., Björck, I. (2008) Effects of GI vs content of cereal fiber of the evening meal on glucose tolerance at a subsequent standardized breakfast. *Eur. J. Clin. Nutr.* 62(6), 712–720.

Nilsson, M., Andersson, R. Andersson, R.E., Autio, K., Aman, P. (2000) Heterogeneity in a water-extractable rye arabinoxylan with a low degree of disubstitution. *Carbohyd. Polym.* 41, 397–405.

Nilsson, M., Åman, P., Härkönen, H., et al. (1997) Nutrient and lignan content, dough properties and baking performance of rye samples used in Scandinavia. *Acta Agric. Scand.* 47, 26–34.

Nyström, L., Lampi, A.M., Andersson, A., et al. (2008) Phytochemicals and dietary fiber components in rye varieties in the HEALTHGRAIN diversity screen. *J. Agric. Food Chem.* 56, 9758–9766.

Ouwehand, A.C., Gråsten, S., Niemi, P., Mykkänen, H., Salminen, S. (2000) Wheat or rye supplemented diets do not affect faecal mucus concentration or the adhesion of probiotic micro-organisms to faecal mucus. *Lett. Appl. Microbiol.* 31(1), 30–33.

Pastell, H., Westermann, P., Meyer, A.S., Tuomainen, P., Tenkanen, M. (2009) In vitro fermentation of arabinoxylan-derived carbohydrates by bifidobacteria and mixed fecal microbiota. *J. Agric. Food Chem.* 57(18), 8598–8606.

Peräaho, M., Collin, P., Kaukinen, K., Kekkonen, L., Miettinen, S., Mäki M. (2004) Oats can diversify a gluten-free diet in celiac diasease and dermatitis herpetiformis. *J. Am. Diet. Assoc.* 104(7), 1144–1150.

Pereira, M.A., Jacobs, DR Jr, Pins, J.J., et al. (2002) Effect of whole grains on insulin sensitivity in overweight hyperinsulinemic adults. *Am. J. Clin. Nutr.* 75, 848–855.

Poutanen, K., Flander, L., Katina, K. (2009) Sourdough and cereal fermentation in a nutritional perspective. *Food Microbiol.* 26(7), 693–699.

Queenan, K.M., Stewart, M.L., Smith, K.N., Thomas W, Fulcher, R.G., Slavin JL. (2007) Concentrated oat beta-glucan, a fermentable fiber, lowers serum cholesterol in hypercholesterolemic adults in a randomized controlled trial. *Nutr. J.* 6, 6.

Rakoff-Nahoum S, Paglino J, Eslami-Varzaneh F, Edberg S, Medzhitov R. (2004) Recognition of commensal microflora by toll-like receptors is required for intestinal homeostasis. *Cell* 118(2), 229–241.

Ramakers, J.D., Volman, J.J., Birklund M, Önning G, Mensink, R.P., Plat J. (2007) Fecal water from ileostomic patients consuming oat β-glucan enhances immune responses in enterocytes. *Mol. Nutr. Food Res.* 51, 211–220.

Riccardi G, Rivellese, A.A., Giacco R. (2008) Role of glycemic index and glycemic load in the healthy state, in prediabetes, and in diabetes. *Am. J. Clin. Nutr.* 87(1), 269S–274S.

Richardson, D.P. (2003) Wholegrain health claims in Europe. *Proc. Nutr. Soc.* 62, 161–169.

Roberfroid, M. (2007) Prebiotics: the concept revisited. *J. Nutr.* 137(3 Suppl 2), 830S–837S.

Roberfroid, M.B. (2005) Introducing inulin-type fructans. *Br. J. Nutr*. 93(Suppl 1), S13–S25.

Ross, A.B., Kamal-Eldin, A., Åman, P., (2004) Dietary alkylresorcinols, absorption, bioactivities, and possible use as biomarkers of whole-grain wheat- and rye rich foods. *Nutr. Rev.* 62, 81–95.

Scalbert, A., Morand, C., Manach, C., Remesy, C. (2002) Absorption and metabolism of polyphenols in the gut and impact on health. *Biomed. Pharmacother.* 56, 276–282.

Shewry, P.R., Li, L., Piironen, V., et al. (2008) Phytochemicals and fiber components in oat varieties in the HEALTHGRAIN diversity screen. *J. Agric. Food Chem.* 56, 9777–9784.

Slavin, J. (2003) Why whole grains are protective: biological mechanisms. *Proc. Nutr. Soc.* 62(1), 129–134.

Snart J, Bibiloni R, Grayson T. (2006) Supplementation of the diet with high-viscosity beta-glucan results in enrichment for lactobacilli in the rat cecum. *Appl. Environ. Microbiol.* 72, 1925–1931.

Stone, B. and Morell, M.K. (2009) Carbohydrates. In: Khan, K. and Shewry, P.R. (editors), *Wheat Chemistry and Technology*, 4th ed, pp. 299–362. AACC International Inc., St. Paul, MN.

Storsrud, S., Hulthen, L.R., Lenner, R.A. (2003) Beneficial effects of oats in the gluten-free diet of adults with special reference to nutrient status, symptoms and subjective experiences. *Br. J. Nutr*. 90(1), 101–107.

Topping, D. (2007) Cereal complex carbohydrates and their contribution to human health. *J. Cereal Sci.* 46, 220–229.

Van Craeyveld, V., Swennen, K., Dornez, E., et al. (2008) Structurally different wheat-derived arabinoxylooligosaccharides have different prebiotic and fermentation properties in rats. *J. Nutr.* 138(12), 2348–2355.

Vardakou, M., Palop, C.N., Christakopoulos, P., Faulds, C.B., Gasson, M.A., Narbad, A. (2008) Evaluation of the prebiotic properties of wheat arabinoxylan fractions and induction of hydrolase activity in gut microflora. *Int. J. Food Microbiol.* 123(1–2), 166–170.

Venn, B.J., Mann, J.I. (2004) Cereal grains, legumes and diabetes. *Eur. J. Clin. Nutr.* 58, 1443–1461.

Verwimp, T., Van Craeyveld, V., Courtin, C.M., Delcour, J.A. (2007) Variability in the structure of rye flour alkali-extractable arabinoxylans. *J. Agric. Food Chem.* 55(5), 1985–1992.

Vitaglioni, P., Napolitano, A., Fogliano, V. (2008) Cereal dietary fiber: a natural functional ingredient to deliver phenolic compounds into the gut. *Trends Food Sci. Technol.* 19, 451–463.

Volman, J.J., Ramakers, J.D., Plat, J. (2008) Dietary modulation of immune function by beta-glucans. *Physiol. Behav.* 94(2), 276–284.

Wang, X., Geng, X., Egashira, Y., Sanada, H. (2004) Purification and characterization of a feruloyl esterase from the intestinal bacterium Lactobacillus acidophilus. *Appl. Environ. Microbiol.* 70, 2367–2372.

Ward, J.L., Poutanen, K., Gebruers, K., et al. (2008) The HEALTHGRAIN cereal diversity screen: concept, results, and prospects *J. Agric. Food Chem.* 56, 9699–9709.

Wong, J.M., de Souza, R., Kendall, C.W., Emam, A., Jenkins, D.J. (2006) Colonic health: fermentation and short chain fatty acids. *J. Clin. Gastroenterol.* 40(3), 235–243.

Wood, P.J., Beer, M.U., Butler, G. (2000) Evaluation of role of concentration and molecular weight of oat beta-glucan in determining effect of viscosity on plasma glucose and insulin following an oral glucose load. *Br. J. Nutr.* 84(1), 19–23.

Yun, C.H., Estrada, A., Van Kessel, A., Park, B.C., Laarveld, B. (2003) Beta-glucan, extracted from oat, enhances disease resistance against bacterial and parasitic infections. *FEMS Immunol. Med. Microbiol.* 35, 67–75.

Chapter 11

Fermentability of Polydextrose, Resistant Maltodextrin, and Other Soluble Fibers: Prebiotic Potential

Maria Stewart

Introduction

Evidence of the prebiotic properties of polydextrose (PDX), resistant maltodextrin (RMD), and other soluble fiber sources are slowly emerging, but the body of evidence is small compared to the gold standard of fructooligosaccharides (FOS) and inulin. This chapter reviews the current evidence for prebiotic function of PDX, RMD as well as other comparable soluble fibers.

Dietary fiber can produce many desirable physiological effects in humans including attenuating blood glucose, lowering blood cholesterol, and improving laxation. Some dietary fibers, particularly soluble dietary fibers, are fermented by gut microorganisms to produce short-chain fatty acids (SCFAs) and other metabolites. The three major SCFAs produced in the colon are acetate, propionate, and butyrate. Butyrate usually receives the most attention of the three as the preferred substrate for colonocyte metabolism, but all SCFAs lower intestinal pH, which increases solubility of some minerals and inhibits pathogenic bacteria growth. Acetate can be used as a substrate by specific types of bacteria to produce butyrate. Both acetate and propionate enter the circulating bloodstream providing energy for the host among other emerging functions. Propionate may have cholesterol lowering properties through inhibition of cholesterol production in the liver (Delzenne

Nondigestible Carbohydrates and Digestive Health. Edited by T. M. Paeschke and W. R. Aimutis

et al., 2002). SCFA production is necessary for maintaining the health of the colonic epithelium. Although simple fermentation of fiber by gut microbes is beneficial to the host and can be accomplished with most soluble fibers, not all fibers ferment to the same extent or to the same metabolites or end products. Differences in fermentation products are shown as a function of the monosaccharide composition, bond linkages between the monosaccharides, purity of the fiber (e.g., locked in cellulosic plant cell wall material or extracted), and tertiary structure of fibers and nondigestible carbohydrates (NDCs). The term prebiotic was developed to qualify and classify NDCs or fibers through specific criteria. "Prebiotic" was coined by Gibson and Roberfroid in 1995 to describe a "*nondigestible food ingredient that beneficially affects the host by selectively stimulating the growth and/or activity of one or a limited number of bacteria in the colon and thus improves host health* (Gibson and Roberfroid, 1995)." This definition was updated in 2004 to contain three specific criteria (Gibson et al., 2004). Prebiotic food constituents "(1) *resist gastric acidity, hydrolysis by mammalian enzymes and gastrointestinal absorption;* (2) *are fermented by intestinal microflora;* (3) *stimulate selectively the growth and/or activity of intestinal bacteria associated with health and well-being.*" Using these specific criteria, it is insufficient to cite fermentability alone as a determinant of prebiotic activity. While many fibers are fermentable, only a few types of fiber have produced a consistent, positive effect on the gut microbiota; inulin and FOS are the most well-known prebiotics. Trans-galactooligosaccharides and lactulose have also been recognized as prebiotic substances (Gibson et al., 2004).

The aim of this chapter is to examine the data regarding fermentability and prebiotic potential of resistant starch (RS), PDX, gum arabic (acacia gum), β-glucan, arabinoxylans (AX) and arabinoxylooligosaccharides, guar gum, and konjac glucomannan (KGM). These fibers have not been designated as prebiotics at the current time.

Methods for Determining Fermentability and Establishing a Prebiotic Effect

Fermentability of fibers is largely dependent on the host's colonic microflora, as well as the molecular composition as previously mentioned. Fermentation products are difficult to assess in vivo because colonic observation in humans is highly invasive. The colon is a very dynamic organ and metabolites of fermentation are continually absorbed by the host. Absorption and utilization of SCFAs and other metabolites is not well understood at the present time, and thus difficult to measure. In vitro measures of SCFA production using either batch fermentation system or continuous culture fermentation system reflect results in vivo (Macfarlane et al., 1998; Wisker et al., 1998; Wisker et al., 2000). Data for exact correlation of in vitro versus in vivo systems is limited, but these methods provide estimates of fermentation products and differences between NDOs at lower cost than human clinical trials.

In vitro fermentation methods are generally classified as batch or dynamic methods. The batch method is a method in which minimal equipment is required (McBurney and Thompson, 1987; Amrein et al., 2003). Fermentations are carried out in sealed flasks at 37°C using fecal inoculum in a buffered media. Batch fermentation does not allow for the removal of SCFAs, gases, or other metabolites. In some respects, batch fermentation provides a reasonable measurement of end products because SCFAs and gases are trapped until the samples are collected for analysis. However, when SCFAs are not removed from the system, pH can drop dramatically which affects bacterial metabolism (Louis et al., 2007). Increased gas pressure as fermentation progresses may also affect the fermentation process and influence the outcome.

Multiple stage or continuous culture fermentation systems are simple dynamic systems that allow for each segment of the colon (ascending, transverse, and descending) to be represented, which is an improvement over batch fermentation (Macfarlane et al., 1998). In each stage or flask, pH is adjusted to best replicate colonic conditions. The lowest pH occurs in the proximal colon (first flask) and the highest pH occurs in the distal colon (last flask).

Automated dynamic digestion and fermentation requires complex instrumentation than batch fermentation and is hence more expensive and time consuming. The two most cited dynamic fermentation systems are the TNO Intestinal Models 1&2 (TIM-1, TIM-2) (TNO Netherlands) (Venema et al., 2000) and the Simulated Human Intestinal Microbial Ecosystem model (SHIME) (De Boever et al., 2000). These anaerobic systems allow for absorption of SCFAs, maintenance of pH, and mimicry of physical conditions such as peristaltic movements.

The major caveat of using in vitro systems is the variability of microbial communities among systems, which may influence the fermentation profiles. This is largely due to the variability of the microbes in the fecal inoculum used and it is also true of human clinical studies as well. Although it has been shown that pooling three fecal donors creates a replicable system (McBurney and Thompson, 1987), it has been suggested that a several unpooled, single donors may be more representative of the general population. The theory is that each individual's microbial community is in homeostasis, and pooling the microbial communities disrupts the homeostasis.

In vivo SCFA measurement is generally measured through cecal contents, colonic contents, or fecal samples in animals, or fecal samples in humans. Fecal SCFA ratio is approximately 60:20:20 (acetate:propionate:butyrate) in humans (Bergman, 1990). This ratio is altered by introducing fermentable fibers or changing the microbial community; both are actions of prebiotics. Fecal SCFA analysis provides data on the remaining SCFAs in the colon, but it cannot capture how much fermentation occurred in the more proximal parts of the colon, nor how much was absorbed by the host or utilized by gut bacteria.

Fermentability is one of the three criteria for establishing a prebiotic effect. Prebiotic substances must also resist gastric acidity and hydrolysis by mammalian enzymes, and several approaches may be taken to establish these characteristics. Dietary fiber analysis methods (AOAC 991.43 and similar methods) do not effectively demonstrate in vitro resistance to gastric acidity and hydrolysis by mammalian enzymes (Marlett, 1990). The current dietary fiber method does not expose substrates to low pH, which is necessary to demonstrate resistance to gastric acidity. Additionally, this method requires heat-resistant enzymes from fungal sources, thus not representing the actions of mammalian enzymes. In vivo data is required to demonstrate resistance to gastrointestinal (GI) absorption and it is difficult to obtain such data. An alternative method is an in vitro digestion, which typically encompasses both resistance to gastric acidity and hydrolysis by mammalian enzymes (Amrein et al., 2003). In most cases, however, it is assumed that a substrate that analyzes as dietary fiber meets these prebiotic criteria, although this assumption may not be accurate.

The third, and most difficult, criterion to meet is to stimulate selectively the growth and/or activity of intestinal bacteria associated with health and well-being. Ideally, growth of saccharolytic bacteria such as Bifidobacteria and/or Lactobacillus or decreased counts of potentially pathogenic bacteria such as *Clostridium perfringens* is documented in a mixed microbial community. Molecular techniques utilizing 16S rDNA such as fluorescent in situ hybridization (FISH) or PCR-denaturing gel gradient electrophoresis/temperature gradient gel electrophoresis (DGGE/TGGE) are becoming the standard techniques for microbial analysis, although selective media plating is still used. Activity of the bacteria may be altered to render a substrate as a prebiotic, as detected in increased enzyme activity. However this method is less frequently used to satisfy the third criteria. Gibson et al. discusses the appropriate methods for establishing a prebiotic effect in greater detail (Gibson et al., 2004) and other methods are being continually developed. In the following sections of this chapter, a summary of data supporting a prebiotic effect of selected dietary fiber is given.

Review of Prebiotic Potential for Specific Nondigestible Carbohydrates

Resistant Starch and Resistant Maltodextrins

RS is present in almost every food containing starch, but the quantity of RS depends on the type and amount of starch, food processing techniques, storage, and ingestion (Brown, 2004). RS can be classified into following four different subtypes: RS1, RS2, RS3, and RS4. RS1 type of RS is frequently found in whole grains and seeds (Topping and Clifton, 2001; Brown, 2004). This type of RS is physically separated from the digestive tract due to its food matrix. Milling and chewing of grains and seeds reduces the amount of RS1 in a food product by simply breaking up the matrix (Topping and Clifton, 2001). RS2 type of RS is characterized by its granular structure. The conformations of native starch granules in the food render the starch nondigestible (Brown, 2004). Some examples of RS2 sources include green bananas and raw potatoes (Topping and Clifton, 2001). RS3 type of RSs is commonly termed as retrograded starches. These types of starches are formed when a starch-containing product (e.g., potatoes) is cooked and then cooled. During the cooking process, the starch granules become disrupted and lose their structure (Brown, 2004), becoming amorphous. Upon cooling, the starch molecules reorient into more crystalline structures that inhibit digestion of the starch in the upper GI tract. RS4 is the final classification of RS. RS4 consists of starch that is chemically modified to decrease digestibility (Brown, 2004). The RS classification system describes why RS is resistant to digestion. However, the RS1–4 classification may not adequately describe the physiological effects within the colon. Multiple RS3 preparations have exhibited different fermentation profiles (Brouns et al., 2007), while other researchers found that RS2 and RS3 preparations have shown no difference in fermentability (Fassler et al., 2006).

Topping and Clifton (2001) summarized the fermentation of RS in a comprehensive review of RS and other nondigestible oligosaccharides. In vitro studies have generated inconsistent data regarding the fermentability of RS. Starch structure may affect the enzyme accessibility necessary for fermentation (Robertson et al., 2001). Additionally, the methods used to assess fermentation affect the results. In vitro digestion (batch vs dynamic) and in vitro fermentation (batch vs dynamic) methods produced different SCFA concentrations for RS (type 2 and type 3) (Fassler et al., 2006). Additionally, RS in ileostomy effluents (in vivo digestion*)* produced significantly lower SCFA concentrations and gas production in an in vitro batch fermentation compared to RS subjected to in vitro digestion and batch fermentation. This supports a previous study showing that lower RS recovery in ileal residues were compared to RS measured via Englyst method for RS analysis (Danjo et al., 2003).

Discrepancies reported between in vitro and in vivo methods for assessing fermentability and digestibility may be due to individual variation of human gut bacteria. Fermentability of RS is dependent on the individual and increased fermentation may be an adaptation (Christl et al., 1997). As the fields of metabolomics and genetic nutrition are developed, it may be possible to identify which individuals can more completely digest RS, and which individuals' microbial communities can better ferment RS.

Most human studies assessing SCFA production have used RS-containing baked goods as the dietary source of RS. Heating or cooking RS may result in altered digestibility of the RS products. Four human studies have reported increased fecal butyrate concentrations after RS consumption (21–75 g RS/day) (Phillips et al., 1995; Cummings et al., 1996; Noakes et al., 1996; Jenkins et al., 1998). Fecal acetate and propionate concentrations after RS consumption have been inconsistent among the studies. Cummings et al. found that potato RS2 was more butyrogenic than maize RS3 (Cummings et al., 1996). Heijnen et al. (1998) reported no differences in fecal SCFA concentrations after RS consumption (32 g/day).

Animal studies have confirmed RS's prebiotic characteristic, but evidence in humans is still weak. Rodent studies have documented stimulation of *Bifidobacterium* (Silvi et al., 1999; Wang et al., 2002; Le Leu et al., 2005), *Lactobacillus* (Silvi et al., 1999; Wang et al., 2002; Le Leu et al., 2005), *Bacteroides* (Kleessen et al., 1997; Silvi et al., 1999; Wang et al., 2002), *Enterobacter* (Kleessen et al., 1997; Silvi et al., 1999), *Staphylococcus* (Silvi et al., 1999), *Streptococcus* (Kleessen et al., 1997), and total anaerobes (Le Leu et al., 2005). RS may have a synbiotic effect (synergistic effect when administered with a probiotic bacteria). Pigs fed high-amylose maize starch and also supplemented with *Bifidobacterium longum* had significantly greater *B. longum* colonization in the feces compared to low-amylose maize starch (Brown et al., 1997). Animals that did not receive the *B. longum* supplement did not have *B. longum* in the feces.

One human study investigated changes in fecal microbes after RS consumption. A short-term intervention (7 days), showed no changes in fecal microflora as a result of consuming 10 g of RS3 type starch per day. However, fecal microbe profiles were altered significantly in response to FOS, soy oligosaccharides, and galactooligosaccharides consumption (Bouhnik et al., 2004).

RMD is considered by some to be a low-degree of polymerization (DP) RS, while others prefer to classify it as its own substrate. SCFA production by RMD has been inconsistent. In vitro fermentation of different preparations of RMD (Actistar® tapioca, Actistar® potato, and Fibersol®, DP = 10–35) indicated that total SCFA production was greatest for Fibersol® and butyrate production was greatest for both of the Actistar® products, regardless of source (Brouns et al., 2007). The fermentation kinetics for the Actistar® products were slow over the first 4 hours and very rapid from 4 to 6 hours of fermentation, while the Fibersol® was more slowly degraded over the entire 24-hour fermentation. The fermentation of these three preparations of RMD is likely influenced by microbial communities and how easily the carbohydrate is accessed by microbial enzymes. It is also influenced by the molecular structure of these products as they are made with different processes. Fecal SCFA concentrations (mmol/g dry matter) in humans increased more for the control group consuming 0 g/day RMD compared to groups consuming 7.5 or 15 g/day RMD for 3 weeks (Fastinger et al., 2008). The authors attribute this to increased stool volume in the treatment groups, thereby diluting the SCFA concentrations. RS supplementation (30 g/day or 45 g/day) for 5 weeks did not affect fecal SCFA concentrations in healthy males (Pasman et al., 2006).

Evidence for a prebiotic effect of RMD in humans is weak. In human subjects, Pasman et al. saw a significant increase in Lactobacillus after a 35-day treatment of 45 g Nutriose® consumed in yogurt or beverages (Pasman et al., 2006). The 30-g dose of Nutriose® and control treatment (Glucidex®) did not differ from baseline. Using Dice's similarity coefficient from DGGE analysis of fecal bacteria, doses of 7.5 g/day or 15 g/day RMD for 3 weeks both increased the diversity of the microbial communities more than the control group (0 g/day RMD) (Fastinger et al., 2008). However, no significant changes were detected in total bacteria, Bifidobacteria, or Lactobacilli.

RS and RMDs are a diverse group of NDCs, and although many preparations are fermentable, it is not prudent to state that all RSs or RMDs are well fermented in the colon or that RS or RMD has a prebiotic effect. Although they are all glucose polymers, the molecular bonding arrangement and three-dimensional structure cause differences in fermentability and they are in no way equivalent. Focused research on specific RS and RMD preparations will be necessary to establish a prebiotic effect and better understand fermentation patterns.

Polydextrose

PDX is a low molecular weight (162–10,000 kDa) glucose polymer created by mixing and heating glucose with sorbitol and a food grade acid (Mitchell et al., 2001), creating a randomly bonded glucose that has limited digestibility. PDX is commonly used in food manufacturing to create low-calorie or low-carbohydrate foods because it contributes the same texture and functional attributes as caloric sweeteners.

PDX is degraded in model colon systems, suggesting that it is fermented by colonic bacteria (Makelainen et al., 2007). Gas production and SCFA production in vitro confirms fermentability. Individual gas production ($n = 3$) was highly variable in vitro. PDX produced high gas volumes for two fecal donors and inulin produced high gas volumes for one fecal donor (Ghoddusi et al., 2007). When PDX and FOS were fermented using fecal inoculum from the same donors in a continuous culture system, PDX produced slightly higher total SCFA, acetate, and propionate concentrations in the three vessels compared to FOS (Probert et al., 2004). FOS produced slightly higher butyrate concentrations than PDX. PDX increased butyrate concentrations more than control in all fermentation vessels using a four-stage fermentation system (Makelainen et al., 2007). PDX elevated acetate, propionate, and total SCFA concentrations in the 2nd, 3rd, and 4th vessels more than control. Current in vitro data suggests that PDX is fermented, but the extent of fermentation depends on the microbial community.

Animal studies support the in vitro studies. In pigs, PDX was recovered in the distal colon, showing that it's a slowly and partially fermented fiber (Fava et al., 2007). Luminal concentrations of SCFAs from the distal small intestine to the distal colon were lower for the PDX-supplemented group compared to control, but plasma acetate and lactate concentrations were high for the PDX-supplemented group, suggesting that lower SCFA concentrations in the lumen were due to colonic absorption.

In humans, fecal SCFA excretion did not change with 8 g/day PDX for 3 weeks (Hengst et al., 2008). PDX at 8 g/day and 12g/day significantly increased acetate and butyrate excretion after 4 weeks administration compared to the control (0g/day) (Jie et al., 2000). Gas production due to PDX in humans (10 g/day) was higher than isphagula husk (psyllium) (Tomlin and Read, 1988). The results in vivo confirm what was shown in vitro; PDX is a fermentable fiber, but the extent to which it is fermented depends on the individual.

The evidence for a prebiotic effect is strong for PDX. PDX supported growth of five *Bifidobacteria* strains in pure culture, but the growth rate was slow compared to other NDCs such as galactooligosaccharide (Vernazza et al., 2006). PDX increased *Bifidobacteria* counts in a continuous-culture fermentation system using human fecal inoculum (Probert et al., 2004). *Bifidobacteria infantis* increased in the presence of PDX but not oligofructose. Ghoddusi et al. reported a stronger prebiotic effect, based on a prebiotic index, for inulin (0.9) and FOS (2.4) than PDX (0.2) (Ghoddusi et al., 2007). Bifidobacteria counts were unchanged as a result of PDX administration in a four-stage fermentation system (Makelainen et al., 2007). In pigs, the composition of the microbial community shifted with PDX treatment based on %G+C analysis (where G, guanine; C, cytosine), but FISH analysis of individual bacterial groups showed no difference with treatment (Fava et al.,

2007). Fecal counts of eubacteria decreased during PDX treatment (8 g/day for 3 weeks) while placebo did not change the eubacteria counts (Hengst et al., 2008). PDX measuring 4 g/day, 8 g/day, and 12 g/day significantly decreased *Bacteroides fragilis*, *Bacteroides vulgatus*, and *Bacteroides intermedius*, and significantly increased *Lactobacillus* and *Bifidobacterium* after 4 weeks administration compared to control (0 g/day) (Jie et al., 2000). PDX may have a synbiotic effect in humans with low initial bifidobacteria counts, but this work needs to be replicated before conclusions can be drawn (Tiihonen et al., 2008). Some evidence supports the prebiotic effects, but more studies are necessary to confirm this effect.

Gum Arabic (Acacia Gum)

Gum arabic is a soluble fiber from the acacia tree comprised of a galactose backbone, with rhamnose, arabinose, and glucuronic acid sidechains (Ali et al., 2009). It is a highly branched polysaccharide and therefore exhibits low viscosity at higher doses necessary for use as a fermentable fiber. Gum arabic was completely fermented by gut bacteria in a batch in vitro system based on monosaccharide (arabinose, rhamnose, galactose, and uronic acid) disappearance and SCFA production (Titgemeyer et al., 1991). Gum arabic produced more butyrate and propionate at 24 and 48 hours compared to pectin, oat fiber, and corn bran (Titgemeyer et al., 1991). Fermentability using a batch in vitro system was confirmed for gum arabic from two different species of acacia trees (Michel et al., 1998). Tree species slightly influenced total SCFA production, acetate production, and pH. Absolute concentrations and molar percentages of propionate and butyrate were similar between the two gum arabic samples and FOS. When the same fibers were fermented in a continuous culture system, tree species and culture medium pH (6.5 or 5.8) influenced the absolute amount and proportions of SCFAs produced. Gum arabic significantly lowered *Clostridium* sp. counts at pH 5.8 but not pH 6.5. Increased counts of *Lactobacillus* sp. and lactic acid bacteria were modestly, but not significantly increased by gum arabic in vitro compared to control. Tree species only influenced lactic acid bacteria counts at pH 6.5.

The prebiotic nature of gum arabic was demonstrated in humans consuming 10 g/day or 15 g/day for 10 days (compared with 10 g/day or 15 g/day sucrose) (Cherbut et al., 2003). Bifidobacteria and total lactic acid-producing bacteria increased with gum arabic treatment in a dose-dependent manner, while *Clostridia* counts remained unchanged. Gum arabic is a potential prebiotic, but further studies need to confirm these results and establish an effective dose.

Guar Gum

Guar gum is a viscous soluble fiber isolated from the guar bean (*Cyamoposis tetragonolobus*). Guar gum contains a mannose backbone with galactose side chains in a ratio of 1:1.5–1.8. Partially hydrolyzed guar gum is produced from the enzymatic cleavage of native guar gum. Numerous studies have examined the fermentability of guar gum and partially hydrolyzed guar gum in vitro. In most cases, acetate was the major SCFA produced at 24 hours (42–52% total SCFA) with propionate being produced in the second highest concentration (29–37% total SCFA) (McBurney and Thompson, 1987; McBurney and Thompson, 1989a, 1989b; Adiotomre et al., 1990; Velazquez et al., 2000; Khan and Edwards, 2005). Butyrate was the least abundant SCFA in these studies, comprising 9–23% of total SCFA. One study reported low acetate production and high butyrate production (molar percents (Ac:Pr:Bu)—25:13:62 for 15 kDa PHGG and 28:1:71 for 10 kDa partially hydrolyzed guar gum (PHGG)) (Pylkas et al., 2005). Differences in molar ratios of SCFA may be a result of microbes present in the inoculum. A comparison of SCFA production in vitro between FOS and guar gum (MW 370 kDa), showed that guar gum was less fermentable than FOS over 48 hours (Khan and Edwards, 2005). Molecular weight of guar gum plays a role in SCFA production, with 400 kDa being a near optimal MW for maximal SCFA production (Stewart and Slavin, 2006).

Guar gum is not currently an accepted prebiotic fiber. Early studies showed a minimal prebiotic effect in vitro and a modest prebiotic effect with increased Bifidobacteria after 1 week and increased Lactobacillus after 2 weeks of treatment (21 g/day partially hydrolyzed guar gum) (Okubo et al., 1994). However, these results have not been replicated by other groups. FISH analysis of prebiotic effect of guar gum and partially hydrolyzed guar gum indicated that both products increased microbial growth in vitro, but the increases were not species-specific (Vulevic et al., 2004). The present data confirms the fermentability of guar gum and partially hydrolyzed guar gum, but not a prebiotic effect.

β-*Glucan*

β-Glucan is the primary soluble fiber found in oats and barley. Oat products, in particular, contain 2.5–8.5 g β-glucan/100 g whole grain while total dietary fiber constitutes approximated 10% of the entire grain (Sadiq Butt et al., 2008). The remainder of the dietary fiber component is lignin and other nonstarch polysaccharides that may contribute to colonic fermentation. Fermentation of oat bran or "oat fiber" may not accurately represent fermentation of β-glucan because of the presence of other fermentable carbohydrates.

Early work with concentrated β-glucan in pigs suggested that β-glucan and oat bran affected fermentation differently, (Knudsen et al., 1993). β-glucan produced greater acetate concentrations than oat bran throughout the colon, but oat bran produced greater propionate and butyrate concentrations throughout the colon. The authors suggest that the arabinoxylan fraction of oat bran is responsible for the elevated butyrate concentrations, as shown with the "insoluble fraction" diet that contained primarily oat AX.

Barley β-glucan (average MW 130, 172, 243 kDa) produced 41–47 mM total SCFA, of which 41–51% was acetate, 31–37% was propionate, and 18–24% was butyrate after 48 hours of batch in vitro fermentation (Hughes et al., 2008). Oat β-glucan (average MW 150, 230 kDa) produced 45 mM total SCFA, of which 49–54% was acetate, 30–31% was propionate, and 16–20% was butyrate. The fermentation patterns of 130 kDa and 243 kDa barley β-glucan and 150 kDa oat β-glucan preparations and inulin (for reference) were similar over 48 hours. Barley β-glucan (172 kDa) exhibited consistently high SCFA production and oat β-glucan (230 kDa), exhibited consistently low SCFA production.

SCFA production of oat flour from oat cultivars with varying amounts of β-glucan (4.7–8.1%) were compared to purified β-glucan (Sayar et al., 2007). The oat flours contained varying amounts of arabinose, xylose, glucose, and other neutral sugars that were not present in the purified β-glucan. Significantly lower total SCFA concentrations were produced by the oat flours compared with the purified β-glucan. The oat flours were significantly more acetogenic, at the expense of propionate and butyrate production, compared to purified β-glucan. A subsequent in vitro fermentation with three oat flours (5.5, 5.3, and 7.7% β-glucan), indicated that fecal donor, not β-glucan content of the oats, influenced SCFA production the most (Kim and White, 2009).

Animal data slightly favors the selective stimulation of bacteria by β-glucan, but in vitro data is mixed. Oat products (oat flour, oat bran, oat bran + Novelose® RS, or autoclaved oat flour) exhibited a bifidogenic effect in rats when consumed for 6 weeks (Drzikova et al., 2005). The bifidogenic diets contained 39–92 g β-glucan/kg chow. The average molecular weight of the β-glucan in the autoclaved oat flour was roughly half of the original oat flour (49.1 vs 94.6 kDa), which did not impact the bifidogenic effect. Cecal acetate, propionate, and butyrate concentrations were greater for the above diets and also oat flour + Novelose® RS compared to control. Oat β-glucan significantly altered cecal microbes in rats; high viscosity and low viscosity β-glucan produced 30 and 20% dissimilar DGGE profiles, respectively, compared to oat flour, barley flour, or cellulose (Snart et al., 2006). High viscosity oat β-glucan increased cecal *Lactobacillus* counts in rats. This study did not investigate how low viscosity β-glucan affected cecal *Lactobacillus* counts (Snart et al., 2006). This data, along with the fermentation data presented earlier, suggests that molecular weight or viscosity of β-glucan may influence the potential prebiotic effect. Barley and oat β-glucans, regardless of grain origin, showed minimal selectivity in increasing bacterial counts in vitro (Hughes et al., 2008). Hughes et al. found that barley β-glucan hydrosylates (average DP 3–9) significantly increased *Bifidobacterium* group and *Lactobacillus–Enterococcus* group counts, and significantly decreased Clostridia cluster IX in a batch in vitro system over 12 hours compared to 0 hours (Hughes et al., 2008). Despite these differences, the authors conclude that the increases in microbe groups were not specific enough to render a prebiotic effect. Evidence supporting a prebiotic effect of glucooligosaccharides, both hydrosylates and enzymatically synthesized substrates, is growing, and this class of substrates may be accepted prebiotics in the future (Sanz et al., 2006; Wichienchot et al., 2006). β-glucan is a fermentable fiber. However, data supporting a prebiotic effect of β-glucan is limited and no human studies have been published to date.

Arabinoxylans and Arabinoxylooligosaccharides

AX are common cereal fibers consisting of a xylan backbone with arabinofuranose sidechains (Izydorczyk and Biliaderis, 1995). Molecular weight of AX is related to the branching pattern, and depending on the method of analysis, ranges from 65 to 5,000 kDa. Enzyme treatment of AX yields lower molecular weight arabinoxylooligosaccharides (AXOS).

Wheat-derived AX have the potential to act as prebiotic substrates. In vitro fermentation of three AX preparations (354 kDa, 278 kDa, and 66 kDa) were compared to inulin (Hughes et al., 2007). Total SCFA production over 24 hours was not significantly different among any of the AX preparations or inulin. All AXs increased *Bifidobacterium* by the same magnitude as inulin (~0.7 log increase). Effect of AX on *Clostridium* and *Lactobacillus* were inconsistent among preparations. AXOS (average DP = 15) did not significantly alter the microbial community residing in the SHIME system (simulated human intestinal microbial ecosystem) (Grootaert et al., 2009). Fermentation of AXOS was unexpected with total SCFA and acetate concentrations in the model ascending colon decreasing during treatment, but SCFA concentrations were elevated in the model transverse colon. In a three-vessel fermentation system, wheat arabinoxylan increased xylanase activity of the microbial slurry, suggesting a mechanism by which AXOS could be degraded further along the colon (Vardakou et al., 2007).

Rats fed AX had elevated cecal total SCFA and acetate concentrations compared to the no fiber control, and lower propionate and butyrate concentrations (Lu et al., 2000). Cecal propionate and butyrate concentrations were significantly lower for the AX-fed rats compared to guar gum and wheat bran. Fecal SCFA concentrations after AX diet were not different than the control, except for fecal propionate, which was slightly higher with the AX diet. Guar gum and wheat bran resulted in higher fecal SCFA concentrations than AX. Five AXOS preparations were developed with differing DP (3, 5, 12, 15, and 61) and fed to rats for 6 weeks (Van Craeyveld et al., 2008). Inulin, FOS, and xylooligosaccharides were fed as control substrates. DP significantly affected cecal acetate and propionate concentrations and fecal acetate and butyrate concentrations. All treatments except AXOS-DP-61 significantly increased cecal *Bifidobacterium* counts. Total anaerobes and *Lactobacillus* counts did not change with treatment. AX and AXOS are fermentable fibers with the potential to act as prebiotics. More research is needed to confirm a prebiotic effect and establish an effective dose.

Konjac Glucomannan

KGM is a type of dietary fiber derived from the tuber of *Amophophallus konjac*, and is a β-D-(1–4) linked linear polymer of glucose and mannose with O-acetate being substituted every 9–19 units (Nishinari et al., 1992). Molecular weight of KGM ranges from 200 to 2,000 kDa and averages 1,000 kDa (cited in Keithley and Swanson (2005)). KGM is a viscous, fermentable fiber based on incubation with fecal enzymes (extracted from fecal material) and anaerobic fermentation with fecal inoculum (Matsuura, 1998; Vuksan et al., 2001). However, SCFA production over 24 hours, the typical colonic retention time, has not been published. It is unclear how rapidly KGM is degraded in colon. The presence of KGM in the colon may influence bacterial enzyme activity or expression, but the mechanisms and physiological effects are unclear (Fujiwara et al., 1991).

Using a mouse model (Balb/C), Chen et al. suggested that hydrolyzed KGM was more fermentable than conventional KGM (Chen et al., 2005). After 2 weeks of a KGM supplemented diet (50 g/kg food), KGM and hydrolyzed KGM produced greater fecal acetate concentrations than control. After 4 weeks, KGM and hydrolyzed KGM produced greater acetate, propionate, and *n*-butyrate concentrations than control. Hydrolyzed KGM produced greater SCFA concentrations than control for all SCFAs and for 2 and 4 weeks. Fermentation of KGM was documented in humans on the basis of decreased fecal pH, and increased total SCFA, acetate, propionate, and *i*-butyrate concentrations (Chen et al., 2008).

A prebiotic effect of KGM (10% of diet by weight) was shown in mice inoculated with a human microbiota. The KGM treatment reduced the counts of *Streptococcus* bacteria in the stool compared to control (Fujiwara et al., 1991). Mice consuming the KGM were more likely to have detectable Bifidobacteria in the stool (8 out of 8) than the control mice (3 out of 10). However, this study lacked appropriate controls, as no baseline Bifidobacteria counts were given for the KGM or control groups. The prebiotic effect of KGM was confirmed in Balb/C mice that were fed varying doses of KGM (25, 50, or 75 g/kg food) (Chen et al., 2005). Fecal microbiota was assessed after 2 and 4 weeks of diet. Bifidobacteria counts increased at 2 and 4 weeks in a dose-dependent manner. *C. perfringens* counts decreased at 4 weeks. In humans, KGM exhibited a prebiotic effect after consuming 4.5 g/day for 3 weeks (Chen et al., 2008). *Lactobacillus* counts increased with treatment.

The prebiotic effect of KGM could possibly stem from its alkaline nature (from processing with calcinated calcium), although acidic environment is typically associated with prebiotic substances (Kitamoto et al., 2003). The pH of konjac fluid ranges from 11.4 to 12.5, and can inhibit the growth of microorganisms in food such as *Escherichia coli* O157:H7, *Salmonella enteriditis*, etc. It is unlikely, however, that konjac fluid maintains its alkalinity in the colon due to the presence of gastric acid and bicarbonate in the stomach and small intestine, respectively. Present data show that KGM is fermentable and has prebiotic potential. Future work should focus on fermentation patterns and confirming a replicable prebiotic effect in humans.

Discussion

Few studies have compared more than one of the fibers discussed in this chapter. One such in vitro batch fermentation with human fecal inoculum compared partially hydrolyzed guar gum, guar gum, RMD, and gum arabic (Flickinger et al., 2000). Total SCFA and butyrate concentrations over the 24 hours were highest for the guar gum and partially hydrolyzed guar gum. The RMD exhibited similar fermentation patterns during the first 3 or 6 hours, and then fermentation rates declined or plateaued. Gum arabic produced the lowest concentrations of butyrate and produced low total SCFA concentrations from 0 to 12 hours.

SCFAs have many beneficial effects on the human body, from providing colonocytes with a preferred fuel to modulating gene expression. Fermentation is a desirable trait, but the fermentation profile is important to consider. Rapidly fermentable fibers such as lactulose or FOS create high SCFA concentrations in the ascending colon which are readily absorbed. The transverse and descending colon are exposed to very low SCFA concentrations, because the fermentable substrate has been completely metabolized. Contrastingly, slowly fermented fibers allow for SCFA production to occur beyond the proximal colon. Rose et al. reviewed how fermentation pattern of dietary fibers may influence inflammatory bowel diseases and colorectal cancer (Rose et al., 2007). Most cases of colorectal cancer and ulcerative colitis occur in the distal colon, where, typically, SCFA concentrations are low.

The proportions of acetate, propionate, and butyrate produced by dietary fiber also elicit varying physiological effects. Acetate may be absorbed and transported to the liver for lipogenesis or may be further metabolized by gut microbes into butyrate (Wong et al., 2006; Duncan et al., 2004). Propionate influences human metabolism in a variety of manners: substrate for gluconeogenesis, inhibition of gluconeogenesis (via pyruvate carboxylase inhibition), glycolysis enhancer (via modulation of phosphofructokinase), and lipid metabolism modulator (Wong et al., 2006). The hypolipidemic effect of propionate has been inconsistent in humans. Butyrate is the preferred substrate for colonocyte metabolism (Wong et al., 2006). Other actions of butyrate are derived from its ability to modulate signal transduction; however, in vitro effects and in vivo effects have been contradictory (Hamer et al., 2008).

A recent review of prebiotics, immunity, and inflammation concluded that inulin has the potential to improve human immune function and reduce inflammation on the basis of its prebiotic properties (Lomax and Calder, 2009). As more evidence amasses regarding the prebiotic effect of the fiber discussed in this chapter, we may see other fibers exhibiting these therapeutic properties.

References

Adiotomre, J., Eastwood, M.A., Edwards, C.A., Brydon, W.G. (1990) Dietary fiber: in vitro methods that anticipate nutrition and metabolic activity in humans. *Am. J. Clin. Nutr.* 52(1), 128–134.

Ali, B.H., Ziada, A., Blunden, G. (2009) Biological effects of gum arabic: a review of some recent research. *Food Chem. Toxicol.* 47(1), 1–8.

Amrein, T.M., Granicher, P., Arrigoni, E., Amado, R. (2003) In vitro digestiblilty and colonic fermentability of aleurone isolated from wheat bran. *Int. J. Food Sci.* 36(4), 451.

Bergman, E.N. (1990) Energy contribution of volatile fatty acids from the gastrointestinal tract in various species. *Phys. Rev.* 70(2), 567–590.

Bouhnik, Y., Raskine, L., Simoneau, G., et al. (2004) The capacity of nondigestible carbohydrates to stimulate fecal bifidobacteria in healthy humans: a double-blind, randomized, placebo-controlled, parallel-group, dose-response relation study. *Am. J. Clin. Nutr.* 80(6), 1658–1664.

Brouns, F., Arrigoni, E., Langkilde, A.M., et al. (2007) Physiological and metabolic properties of a digestion-resistant maltodextrin, classified as type 3 retrograded resistant starch. *J. Agric. Food Chem.* 55(4), 1574–1581.

Brown, I., Warhurst, M., Arcot, J., Playne, M., Illman, R.J., Topping, D.L. (1997) Fecal numbers of bifidobacteria are higher in pigs fed *Bifidobacterium longum* with a high amylose cornstarch than with a low amylose cornstarch. *J. Nutr.* 127(9), 1822–1827.

Brown, I.L. (2004) Applications and uses of resistant starch. *J. AOAC Int.* 87(3), 727–732.

Chen, H.L., Cheng, H.C., Wu, W.T., Liu, Y.J., Liu, S.Y. (2008) Supplementation of konjac glucomannan into a low-fiber Chinese diet promoted bowel movement and improved colonic ecology in constipated adults: a placebo-controlled, diet-controlled trial. *J. Am. Coll. Nutr.* 27(1), 102–108.

Chen, H.L., Fan, Y.H., Chen, M.E., Chan, Y. (2005) Unhydrolyzed and hydrolyzed konjac glucomannans modulated cecal and fecal microflora in Balb/c mice. *Nutrition* 21(10), 1059–1064.

Cherbut, C., Michel, C., Raison, V., Kavtchenko, T., Severine, M. (2003) Acacia gum is a bifidogenic dietary fibre with high digestive tolerance in healthy humans. *Microb. Ecol. Health Dis.* 15(1), 43–50.

Christl, S.U., Katzenmaier, U., Hylla, S., Kasper, H., Scheppach, W. (1997) In vitro fermentation of high-amylose cornstarch by a mixed population of colonic bacteria. *JPEN* 21(5), 290–295.

Cummings, J.H., Beatty, E.R., Kingman, S.M., Bingham, S.A., Englyst, H.A. (1996) Digestion and physiological properties of resistant starch in the human large bowel. *Br. J. Nutr.* 75733–75747.

Danjo, K., Nakaji, S., Fukuda, S., Shimoyama, T., Sakamoto, J., Sugawara, K. (2003) The resistant starch level of heat moisture-treated high amylose cornstarch is much lower when measured in the human terminal ileum than when estimated in vitro. *J. Nutr.* 133(7), 2218–2221.

De Boever, P., Deplancke, B., Verstraete, W. (2000) Fermentation by gut microbiota cultured in a simulator of the human intestinal microbial ecosystem is improved by supplementing a soygerm powder. *J. Nutr.* 130(10), 2599–2606.

Delzenne, N.M., Daubioul, C., Neyrinck, A., Lasa, M., Taper, H.S. (2002) Inulin and oligofructose modulate lipid metabolism in animals: review of biochemical events and future prospects. *Br. J. Nutr.* (87 Suppl 2), S255–S259.

Drzikova, B., Dongowski, G., Gebhardt, E. (2005) Dietary fibre-rich oat-based products affect serum lipids, microbiota, formation of short-chain fatty acids and steroids in rats. *Br. J. Nutr.* 94(6), 1012–1025.

Duncan, S.H., Holtrop, G., Lobley, G.E., Calder, A.G., Stewart, C.S., Flint, H.J. (2004) Contribution of acetate to butyrate formation by human faecal bacteria. *Br. J. Nutr.* 91(6), 915–923.

Fassler, C., Arrigoni, E., Venema, K., Brouns, F., Amado, R. (2006) In vitro fermentability of differently digested resistant starch preparations. *Mol. Nutr. Food Res.* 50(12), 1220–1228.

Fastinger, N.D., Karr-Lilienthal, L.K., Spears, J.K., et al. (2008) A novel resistant maltodextrin alters gastrointestinal tolerance factors, fecal characteristics, and fecal microbiota in healthy adult humans. *J. Am. Coll. Nutr.* 27(2), 356–366.

Fava, F., Makivuokko, H., Siljander-Rasi, H., et al. (2007) Effect of polydextrose on intestinal microbes and immune functions in pigs. *Br. J. Nutr.* 98(1), 123–133.

Flickinger, E.A., Wolf, B.W., Garleb, K.A., et al. (2000) Glucose-based oligosaccharides exhibit different in vitro fermentation patterns and affect in vivo apparent nutrient digestibility and microbial populations in dogs. *J. Nutr.* 130(5), 1267–1273.

Fujiwara, S., Hirota, T., Nakazato, H., Muzutani, T., Mitsuoka, T. (1991) Effect of konjac mannan on intestinal microbial metabolism in mice bearing human flora and in conventional F344 rats. *Food Chem. Toxicol.* 29(9), 601–606.

Ghoddusi, H.B., Grandison, M.A., Grandison, A.S., Tuohy, K.M. (2007) In vitro study on gas generation and prebiotic effects of some carbohydrates and their mixtures. *Anaerobe* 13(5–6), 193–199.

Gibson, G., Probert, H., van Loo, J., Rastall, R., Roberfroid, M. (2004) Dietary modulation of the human colonic microbiota: updating the concept of prebiotics. *Nutr. Res. Rev.* 17259–17275.

Gibson, G.R., Roberfroid, M.B. (1995) Dietary modulation of the human colonic microbiota: introducing the concept of prebiotics. *J. Nutr.* 125(6), 1401–1412.

Grootaert, C., Van Den Abbeele, P., Marzorati, M., et al. (2009) Comparison of prebiotic effects of arabinoxylan oligosaccharides and inulin in a simulator of the human intestinal microbial ecosystem. *FEMS Microbiol. Ecol.* 69(2), 231–242.

Hamer, H.M., Jonkers, D., Venema, K., Vanhoutvin, S., Troost, F.J., Brummer, R.J. (2008) Review article: the role of butyrate on colonic function. *Aliment. Pharmacol. Ther.* 27(2), 104–119.

Heijnen, M.L., van Amelsvoort, J.M., Deurenberg, P., Beynen, A.C. (1998) Limited effect of consumption of uncooked (RS2) or retrograded (RS3) resistant starch on putative risk factors for colon cancer in healthy men. *Am. J. Clin. Nutr.* 67(2), 322–331.

Hengst, C., Ptok, S., Roessler, A., Fechner, A., Jahreis, G. (2008) Effects of polydextrose supplementation on different faecal parameters in healthy volunteers. *Int. J. Food Sci. Nutr.* 1–10.

Hughes, S.A., Shewry, P.R., Gibson, G.R., McCleary, B.V., Rastall, R.A. (2008) In vitro fermentation of oat and barley derived beta-glucans by human faecal microbiota. *FEMS Microbiol. Ecol.* 64(3), 482–493.

Hughes, S.A., Shewry, P.R., Li, L., Gibson, G.R., Sanz, M.L., Rastall, R.A. (2007) In vitro fermentation by human fecal microflora of wheat arabinoxylans. *J. Agric. Food Chem.* 55(11), 4589–4595.

Izydorczyk, M.S., Biliaderis, C.G. (1995) Cereal arabinoxylans: advances in structure and physicochemical properties. *Carbohydr. Polym.* 28(1), 2833–2848.

Jenkins, D.J., Vuksan, V., Kendall, C.W., et al. (1998) Physiological effects of resistant starches on fecal bulk, short chain fatty acids, blood lipids and glycemic index. *J. Am. Coll. Nutr.* 17(6), 609–616.

Jie, Z., Bang-Yao, L., Ming-Jie, X., et al. (2000) Studies on the effects of polydextrose intake on physiologic functions in Chinese people. *Am. J. Clin. Nutr.* 72(6), 1503–1509.

Keithley, J., Swanson, B. (2005) Glucomannan and obesity: a critical review. *Altern. Ther. Health Med.* 11(6), 30–34.

Khan, K.M., Edwards, C.A. (2005) In vitro fermentation characteristics of a mixture of Raftilose and guar gum by human faecal bacteria. *Eur. J. Nutr.* 44(6), 371–376.

Kim, H.J., White, P.J. (2009) In vitro fermentation of oat flours from typical and high beta-glucan oat lines. *J. Agric. Food Chem.* 57(16), 7529–7536.

Kitamoto, N., Kato, Y., Ohnaka, T., Yokota, M., Tanaka, T., Tsuji, K. (2003) Bactericidal effects of konjac fluid on several food-poisoning bacteria. *J. Food Prot.* 66(10), 1822–1831.

Kleessen, B., Stoof, G., Proll, J., Schmiedl, D., Noack, J., Blaut, M. (1997) Feeding resistant starch affects fecal and cecal microflora and short-chain fatty acids in rats. *J. Anim. Sci.* 75(9), 2453–2462.

Knudsen, K.E., Jensen, B.B., Hansen, I. (1993) Oat bran but not a beta-glucan-enriched oat fraction enhances butyrate production in the large intestine of pigs. *J. Nutr.* 123(7), 1235–1247.

Le Leu, R.K., Brown, I.L., Hu, Y., et al. (2005) A synbiotic combination of resistant starch and *Bifidobacterium lactis* facilitates apoptotic deletion of carcinogen-damaged cells in rat colon. *J. Nutr.* 135(5), 996–1001.

Lomax, A.R., Calder, P.C. (2009) Prebiotics, immune function, infection and inflammation: a review of the evidence. *Br. J. Nutr.* 101(5), 633–658.

Louis, P., Scott, K.P., Duncan, S.H., Flint, H.J. (2007) Understanding the effects of diet on bacterial metabolism in the large intestine. *J. Appl. Microbiol.* 102, 1197–1208.

Lu, Z.X., Gibson, P.R., Muir, J.G., Fielding, M., O'Dea, K. (2000) Arabinoxylan fiber from a by-product of wheat flour processing behaves physiologically like a soluble, fermentable fiber in the large bowel of rats. *J. Nutr.* 130(8), 1984–1990.

Macfarlane, G.T., Macfarlane, S., Gibson, G.R. (1998) Validation of a three-stage compound continuous culture system for investigating the effect of retention time on ecology and metabolism of bacteria in the human colon. *Microb. Ecol.* 35, 180–187.

Makelainen, H.S., Makivuokko, H.A., Salminen, S.J., Rautonen, N.E., Ouwehand, A.C. (2007) The effects of polydextrose and xylitol on microbial community and activity in a 4-stage colon simulator. *J. Food Sci.* 72(5), M153–M159.

Marlett, J.A. (1990) Issues in dietary fiber analysis. *Adv. Exp. Med. Biol.* 270, 183–192.

Matsuura Y. (1998) Degradation of konjac glucomannan by enzymes in human feces and formation of short-chain fatty acids by intestinal anaerobic bacteria. *J. Nutr. Sci. Vitaminol. (Tokyo)* 44(3), 423–436.

McBurney, M.I., Thompson, L.U. (1987) Effect of human faecal inoculum on in vitro fermentation variables. *Br. J. Nutr.* 58(2), 233–243.

McBurney, M.I., Thompson, L.U. (1989a) In vitro fermentabilities of purified fiber supplements. *J. Food Sci.* 54(2), 347–350.

McBurney, M.I., Thompson, L.U. (1989b) Effect of human faecal donor on in vitro fermentation variables. *Scand. J. Gastroenterol.* 24(3), 359–367.

Michel, C., Kravtchenko, T.P., David, A., Gueneau, S., Kozlowski, F., Cherbut, C. (1998) In vitro prebiotic effects of acacia gums onto the human intestinal microbiota depends on both botanical origin and environmental pH. *Anaerobe* 4(6), 257–266.

Mitchell, H., Auerbach, M.H., Moppett, F.K. (2001) Polydextrose. In: O'Brien Nabors, L. (editor), Alternative Sweeteners, pp. 499–517. Marcel Dekker, New York.

Nishinari, K., Williams, P., Phillips, G. (1992) Review of the physico-chemical characteristics and properties of konjac mannan. *Food Hydrocolloid.*6, 199–222.

Noakes, M., Clifton, P.M., Nestel, P.J., Le Leu, R., McIntosh, G. (1996) Effect of high-amylose starch and oat bran on metabolic variables and bowel function in subjects with hypertriglyceridemia. *Am. J. Clin. Nutr.* 64(6), 944–951.

Okubo, T., Ishihara, N., Takahashi, H., et al. (1994) Effects of partially hydrolyzed guar gum intake on human intestinal microflora and its metabolism. *Biosci. Biotech. Biochem.* 58, 1364.

Pasman, W., Wils, D., Saniez, M.H., Kardinaal, A. (2006) Long-term gastrointestinal tolerance of NUTRIOSE FB in healthy men. *Eur. J. Clin. Nutr.* 60(8), 1024–1034.

Phillips, J., Muir, J.G., Birkett, A., et al. (1995) Effect of resistant starch on fecal bulk and fermentation-dependent events in humans. *Am. J. Clin. Nutr.* 62(1), 121–130.

Probert, H.M., Apajalahti, J.H., Rautonen, N., Stowell, J., Gibson, G.R. (2004) Polydextrose, lactitol, and fructo-oligosaccharide fermentation by colonic bacteria in a three-stage continuous culture system. *Appl. Environ. Microbiol.* 70(8), 4505–4511.

Pylkas, A.M., Juneja, L.R., Slavin, J.L. (2005) Comparison of different fibers for in vitro production of short-chain fatty acids by intestinal microflora. *J. Med. Foods* 8(Spring), 113–116.

Robertson, J.A., Ryden, P., Botham, R.L., Gibson, G., Ring, S.G. (2001) Structural properties of diet-derived polysaccharides and their influence on butyrate production during fermentation. *Lebensm. Wiss. Technol.* 34, 567–573.

Rose, D.J., DeMeo, M.T., Keshavarzian, A., Hamaker, B.R. (2007) Influence of dietary fiber on inflammatory bowel disease and colon cancer: importance of fermentation pattern. *Nutr. Rev.* 65(2), 51–62.

Sadiq Butt, M., Tahir-Nadeem, M., Khan, M.K., Shabir, R., Butt, M.S. (2008) Oat: unique among the cereals. *Eur. J. Nutr.* 47(2), 68–79.

Sanz, M.L., Cote, G.L., Gibson, G.R., Rastall, R.A. (2006) Influences of glycosidic linkages and molecular weight on the fermentation of maltose-based oligosaccharides by human gut bacteria. *J. Agri. Food Chem.* 54, 9779–9984.

Sayar, S., Jannink, J.L., White, P.J. (2007) Digestion residues of typical and high-beta-glucan oat flours provide substrates for in vitro fermentation. *J. Agric. Food Chem.* 55(13), 5306–5311.

Silvi, S., Rumney, C.J., Cresci, A., Rowland, I.R. (1999) Resistant starch modifies gut microflora and microbial metabolism in human flora-associated rats inoculated with faeces from Italian and UK donors. *J. Appl. Microbiol.* 86(3), 521–530.

Snart, J., Bibiloni, R., Grayson, T., et al. (2006) Supplementation of the diet with high-viscosity beta-glucan results in enrichment for lactobacilli in the rat cecum. *Appl. Environ. Microbiol.* 72(3), 1925–1931.

Stewart, M.L., Slavin, J.L. (2006) Molecular weight of guar gum affects short-chain fatty acid profile in model intestinal fermentation. *Mol. Nutr. Food Res.* 50(10), 971–976.

Tiihonen, K., Suomalainen, T., Tynkkynen, S., Rautonen, N. (2008) Effect of prebiotic supplementation on a probiotic bacteria mixture: comparison between a rat model and clinical trials. *Br. J. Nutr.* 99(4), 826–831.

Titgemeyer, E.C., Bourquin, L.D., Fahey, G.C., Jr., Garleb, K.A. (1991) Fermentability of various fiber sources by human fecal bacteria in vitro. *Am. J. Clin. Nutr.* 53(6), 1418–1424.

Tomlin, J., Read, N.W. (1988) A comparative study of the effects on colon function caused by feeding ispaghula husk and polydextrose. *Aliment. Pharmacol. Ther.* 2(6), 513–519.

Topping, D.L., Clifton, P.M. (2001) Short-chain fatty acids and human colonic function: roles of resistant starch and nonstarch polysaccharides. *Physiol. Rev.* 81(3), 1031–1064.

Van Craeyveld, V., Swennen, K., Dornez, E., Van de Wiele, T., et al. (2008) Structurally different wheat-derived arabinoxylooligosaccharides have different prebiotic and fermentation properties in rats. *J. Nutr.* 138(12), 2348–2355.

Vardakou, M., Nueno Palop, C., Gasson, M., Narbad, A., Christakopoulos, P. (2007) In vitro three-stage continuous fermentation of wheat arabinoxylan fractions and induction of hydrolase activity by the gut microflora. *Int. J. Biol. Macromol.* 41(5), 584–589.

Velazquez, M., Davies, C., Marett, R., Slavin, J.L., Feirtag, J.M. (2000) Effect of oligosaccharides and fibre substitutes on short-chain fatty acid production by human faecal microflora. *Anaerobe* 6, 87–92.

Venema, K., van Nuenen, M.H., Smeets-Peeters, M.J.E., Minekus, M., Havenaar, R. (2000) TNO's in vitro large intestinal model: an excellent screening tool for functional food and pharmaceutical research. *Ernaehrung/Nutrition* 24, 558–564.

Vernazza, C.L., Gibson, G.R., Rastall, R.A. (2006) Carbohydrate preference, acid tolerance and bile tolerance in five strains of *Bifidobacterium*. *J. Appl. Microbiol.* 100(4), 846–853.

Vuksan, V., Sievenpiper, J.L., Xu, Z., et al. (2001) Konjac-mannan and American ginsing: emerging alternative therapies for type 2 diabetes mellitus. *J. Am. Coll. Nutr.* 20(5 Suppl), 370S–380S; discussion 381S–383S.

Vulevic, J., Rastall, R.A., Gibson, G.R. (2004) Developing a quantitative approach for determining the in vitro prebiotic potential of dietary oligosaccharides. *FEMS Microbiol. Lett.* 236(1), 153–159.

Wang, X., Brown, I.L., Khaled, D., Mahoney, M.C., Evans, A.J., Conway, P.L. (2002) Manipulation of colonic bacteria and volatile fatty acid production by dietary high amylose maize (amylomaize) starch granules. *J. Appl. Microbiol.* 93(3), 390–397.

Wichienchot, S., Prasertsan, P., Hongpattarakere, T., Gibson, G.R., Rastall, R.A. (2006) In vitro fermentation of mixed linkage glucooligosaccharides produced by Gluconobacter oxydans NCIMB 4943 by the human colonic microflora. *Curr. Issues Intest. Microbiol.* 7(1), 7–12.

Wisker, E., Daniel, M., Rave, G., Feldheim, W. (1998) Fermentation of non-starch polysaccharides in mixed diets and single fibre sources: comparative studies in human subjects and in vitro. *Br. J. Nutr.* 80(3), 253–261.

Wisker, E., Daniel, M., Rave, G., Feldheim, W. (2000) Short-chain fatty acids produced in vitro from fibre residues obtained from mixed diets containing different breads and in human faeces during the ingestion of the diets. *Br. J. Nutr.* 84(1), 31–37.

Wong, J.M., de Souza, R., Kendall, C.W., Emam, A., Jenkins, D.J. (2006) Colonic health: fermentation and short chain fatty acids. *J. Clin. Gastroenterol.* 40(3), 235–243.

Chapter 12

Development and Evaluation Bimuno®, a Novel Second-Generation Prebiotic Galactooligosaccharide Mixture

George Tzortzis

Introduction

In recent years, the science of nutrition and consumers has begun to look at food not only for basic nutrition, but for health benefits as well. The introduction of functional foods that provide demonstrated physiological benefits or reduce the risk of chronic disease, above and beyond their basic nutritional function, is being driven by a growing consumer understanding of diet/disease links, ageing populations, rising health care costs, and advances in food technology and nutrition. The targets of functional foods include cholesterol reduction, cardiovascular disease, osteoporosis, child development, and gastrointestinal (GI) disorders with the aim being prevention rather than curing of conditions.

The GI tract is one of the most metabolically active organs in the body, mainly due to the high number of resident bacteria that can reach 10^{12} cells/mL of colonic content (Holzapfel and Veld, 1998). This colonic microflora is made up of a very complex consortium of microorganisms of more than 400 different bacterial species. The main role of this resident microflora is to ferment substances provided by the diet, that cannot be digested by the host in the upper GI tract, such as resistant starch, dietary fiber, oligosaccharides, protein, etc., to end products

Nondigestible Carbohydrates and Digestive Health. Edited by T. M. Paeschke and W. R. Aimutis

such as short-chain fatty acids (SCFA) that contribute positively toward host daily energy requirements. However, under certain circumstances, the fermentative process may produce undesirable metabolites that may cause the onset of gut disorders that are manifested through both acute and chronic conditions (Rowland, 1995).

At the same time, the integrity of the existing colonic microflora, that normally presents a barrier to invading organisms, can be impaired through stress, illness, antibiotic treatment, changes in diet, or physiological alterations in the gut and general living conditions allowing pathogenic or detrimental organisms to become established (Gibson and Macfarlane, 1999). Knowledge of the gut microflora and its interactions with the host has led to the development of dietary strategies that serve to sustain, or even improve normal GI microflora. Such interventions include probiotics and prebiotics.

Prebiotics are nondigestible food ingredients that beneficially affect the host by selectively stimulating the growth and/or activity of health-promoting bacteria already resident in the GI tract (Gibson and Roberfroid, 1995). To be effective, prebiotics must escape digestion in the upper gut, reach the large intestine, and be utilized selectively by a restricted group of microorganisms that have clearly identified health promoting properties. The prebiotics most often used in foods are oligosaccharides and their preferred targets are microorganisms belonging to the genera of *Lactobacillus* and *Bifidobacterium*. Since they require neither sophisticated temperature settings nor major adjustments during storage and handling, prebiotics are becoming a very popular ingredient in the functional food market. Applications of prebiotics include beverages, dairy products, bakery products, infant formulae, and weaning foods, as well as breakfast cereals and confectionery.

The health benefits of prebiotics are mainly related to their ability to rebalance the colonic microflora by increasing the numbers of bifidobacteria and lactobacilli, although specific prebiotics have also been reported to bind to specific receptors on the host epithelium suggesting that a direct interaction between prebiotics and the host is in place (Vos et al., 2007).

There are two main types of ingredients with proven prebiotic properties: fructan-type saccharides (inulin and short-chain fructooligosaccharides (sc-FOS)) and galactooligosaccharides (GOS) (Gibson et al., 2004).

GOS Production

GOS are defined as a mixture of those substances produced from lactose which comprise between two and eight saccharide units, with one of these units being a terminal glucose and the remaining saccharide units being galactose, and disaccharides comprising two units of galactose. They can be synthesized by classical chemical synthesis methods from simple sugars, but the preferred mode for their synthesis is by enzymatic catalysis from lactose using an appropriate β-galactosidase enzyme. The two types of enzymes that can be used in GOS preparation are the glycosyltransferases (E.C. 2.4) and the glycohydrolases (E.C. 3.2.1).

The most extensively studied β-galactosidases for GOS synthesis are of microbial origin (β-galactohydrolase, EC 3.2.1.23). Enzymes from species belonging to *Kluyveromyces*, *Aspergillus*, *Bacillus*, *Streptococcus*, and *Cryptococcus* have been used for the synthesis of GOS from lactose showing differing requirements for reaction conditions in terms of pH and temperature and differing product formation in terms of the glycoside bonds formed between the galactose moieties and the degree of polymerization (DP) of the synthesized oligosaccharides. Usually 55% of the initial lactose is converted into a mixture of products containing glucose and galactose due to the hydrolytic activity of the enzyme, unreacted lactose, disaccharides of galactose, and glucose with different β-glycoside bonds from lactose due to direct internal transfer, and transgalactosylation products such as galactobiose, galactotriose, galactosyl lactose, tetra- to octasaccharides of similar rearrangement and/or side chain formations (Playne, 2002). The enzymatic conversion of lactose into GOS by β-galactosidases is a kinetically controlled reaction during which the thermodynamically favored hydrolysis of the substrate generates D-galactose and D-glucose in competition to the transferase activity that generates a complex mixture of various galactose based di- and oligosaccharides of different structures. In addition to this transferase activity, another mechanism of GOS synthesis that leads directly to formation of the disaccharide allolactose is by the direct internal transfer of galactose from the position 4, found in lactose, to the position 6 of the glucose moiety without release of the glucose moiety from the active site. Quantitatively, allolactose is one of the major oligosaccharides formed by neutral pH β-galactosidases and although this mechanism has been demonstrated only for β-galactosidase enzymes from *Escherichia coli*, it has been proposed also for other β-galactosidases with similarities to the LacZ enzyme. 6′-galactosyl lactose is the main product when the yeast β-galactosidase from *Kluyveromyces* (*K. marxianus* ssp. *lactis*, *K. fragilis*) is used, whilst 3′- and 6′- galactosyl lactose are formed when the fungal lactase of *Aspergillus oryzae* is used. Enzymes from *Bacillus circulans* or *Cryptococcus laurentii* form mainly 4′-galactosyl lactose and enzymes from *Streprococcus thermophilus* 3′-galactosyl lactose.

During this enzymatic reaction, the amount and nature of the formed oligosaccharide mixture is affected by the ratio of hydrolytic and transferase activities of the enzyme. This ratio depends on the enzyme source, the concentration and nature of the substrate, and the reaction conditions (pH, temperature, and time) following the general principle that the transferase activity is favored at high lactose concentration, elevated reaction temperature, and lower water activity. The source of the enzyme is directly influencing the type of glycoside bond formed between the galactose moieties of the produced GOS, and is also setting the range of pH and temperature conditions available for the synthesis reaction.

Current forms of those prebiotic GOS are efficient at stimulating the beneficial bacteria of the colonic microbiota at genus level, but the emerging interest focuses on the development of prebiotics with more specific functional properties. This functionality can derive either by increasing the selectivity of the prebiotics from the genus to species level, thus taking advantage of the biological functional properties of specific bacterial species and/or manufacturing prebiotic oligosaccharides that have a direct effect on the host beyond the indirect effect through the microbiota modulation.

An interesting approach for sourcing microbial β-galactosidase for the production of oligosaccharides for species selectivity has been explored in the use of probiotic bacteria. The rationale behind the use of β-galactosidases from probiotic bacteria is that since the origin of the enzyme used in this type of manufacturing is important in the final GOS mixture composition and therefore functionality, the use of enzymes originating from probiotic bacteria as synthetic catalyst will produce oligosaccharide mixtures that will be more readily metabolized by the producing organism, resulting so in higher selectivity toward that organism. Following this approach, enzymes from various *Bifidobacterium bifidum* strains have been used to produce mixtures of linear 3′-galactosyl lactose as well as branched oligosaccharides.

Table 12.1. Bimuno composition on dry matter.

Ingredient	Grams per 100 g
Monosaccharide	<15.3
Lactose	<28.0
B-GOS	50.0
DP 2	52%
DP 3	26%
DP 4	14%
DP 5	8%
Minerals	<5.0
Protein	<1.0
Fat	0.5

DP, degree of polymerization; B-GOS, galactooligosaccharides in Bimuno

Bimuno® Production

Enzymes from *B. bifidum* NCIMB 41171 are used to produce the Bimuno® mixture (Clasado Ltd, UK) (Table 12.1).

B. bifidum is an attractive probiotic target based on its functional properties. It is reported as the main *Bifidobacterium* species identified in the colonic microbiota of healthy breast-fed infants, but is absent in infants with atopic dermatitis (Ouwehand et al., 2001). The specific *B. bifidum* strain, NCIMB 41171, used for the manufacturing of Bimuno® was isolated from a healthy human volunteer (Tzortzis et al., 2005) and carries four galactosidase enzymes, three β- and one α-galactosidase, all contributing to the unique properties of the GOS mixture produced (Goulas et al., 2007).

For Bimuno® production, 40–50% (w/w) lactose solution is incubated with permeated *B. bifidum* cells carrying the β-galactosidase enzymes in a batch reactor until the optimum of lactose conversion into oligosaccharides is reached. The solution containing 48% (w/w) B-GOS is then diluted to 28% (w/w) solids and ultrafiltrated to remove bacterial cells and free β-galactosidase enzymes. After nanofiltration to remove monosaccharides, the solution containing 60–65% (w/w) B-GOS is concentrated by evaporation usually to 67–75% total solids to produce the syrup format or spray dried to produce the powder format (Figure 12.1).

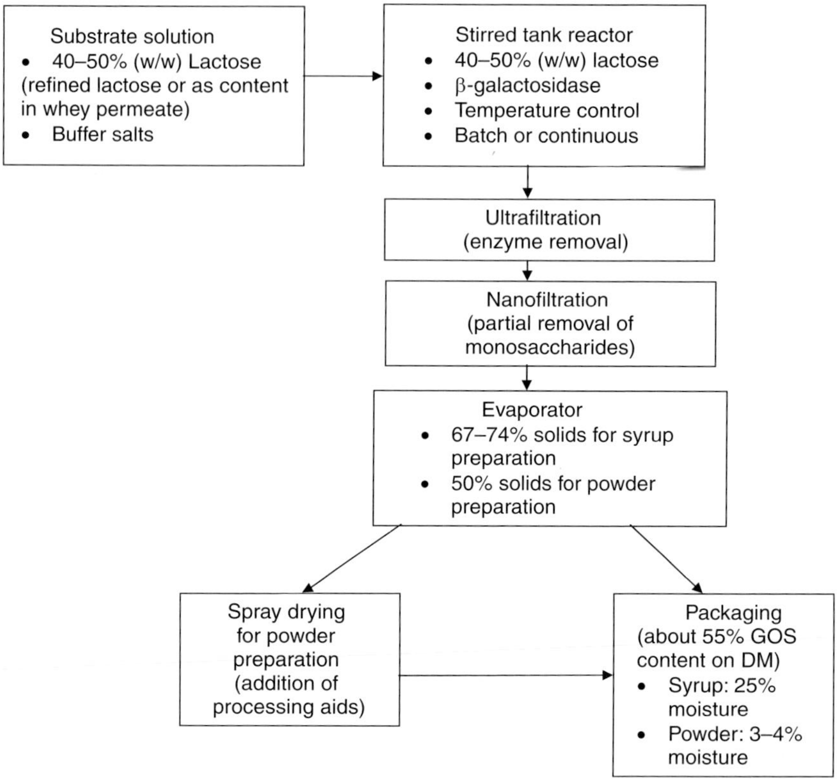

Figure 12.1. Process flow schematic for the manufacture of Bimuno® from lactose

Physicochemical Properties of Bimuno®

The GOS in Bimuno® (B-GOS) are water soluble with a viscosity similar to high fructose syrup leading to improved body and mouthfeel. They are stable to pH 2 at 37°C for several months (Figure 12.2) making their application in nonrefrigerated fruit juice matrices possible, while the presence of β-type linkages makes them ingredients of increased resistance to high temperature in acidic medium. They remain unchanged after treatment at 160°C for 10 minutes at neutral pH and after treatment at 120°C for 10 minutes at pH 3, or 100°C for 10 minutes at pH 2, offering potential for a wide range of food applications. They are mildly sweet, typically 0.3–0.6 times the sweetness of sucrose and can be used in very sweet foods as bulking agent to enhance other food flavors. B-GOS are resistant to salivary degradation and are not utilized

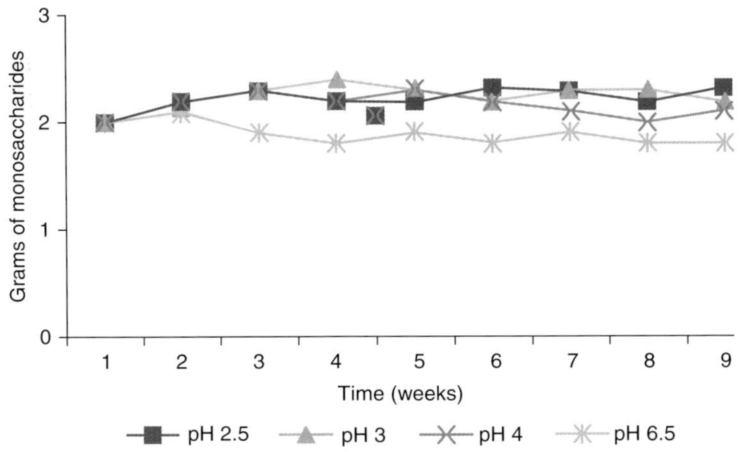

Figure 12.2. Stability of B-GOS in phosphate buffer solution (pH 2.5–6.5) at 37°C for a period of 9 weeks. Monosaccharide content measured by high performance liquid chromatography.

by the mouth microbiota and can therefore be used as low cariogenic sugar substitutes in chewing gums and confectionery. They are not hydrolyzed by pancreatic enzymes and gastric juice passing the small intestine offering reduced glycemic index and a calorific value lower than 50% that of sucrose, which makes them suitable for low-calorie diet food and for consumption by individuals with diabetes. Overall, GOS have a high moisture retaining capacity preventing excessive drying that can be useful in baked goods especially bread where B-GOS are not broken down during fermentation with yeast and the baking process, providing the bread with excellent taste and texture. Their low water activity can help in controlling microbial contamination and depending on the molecular weight of the oligosaccharide content, they can alter the freezing temperature of frozen foods and reduce the amount of coloring, due to Maillard reactions, in heat processed food as relatively fewer reducing moieties are available.

Functional Properties of Bimuno®

Bimuno® has been extensively studied in vitro and in vivo in terms of its functional properties including prebiotic, immunomodulatory, and protection from GI pathogen invasion (Table 12.2).

Table 12.2. Effect of B-GOS on the colonic microbiota in vitro and in vivo.

Method	Substrate	Effect
Three-stage continuous culture system	Bimuno®, GOS, inulin	Significantly higher increase in the levels of bifidobacteria in the first vessel in the presence of Bimuno® compared to GOS and inulin
Piglet study (10 piglets per group, 4-week treatment, parallel design)	Control, Bimuno®, inulin	Significantly higher levels of bifidobacterial population in the proximal colonic content in the presence of Bimuno®
		Significantly lower pH in the proximal colonic content in the presence of Bimuno®
BalbC Mice oral challenge study with *Salmonella enterica* typhimurium	Control, Bimuno®	Significant reduction in the invasion of the pathogen in the presence of Bimuno®. Damage in tissue histopathology and clinical symptoms of Salmonellosis prevented in the presence of Bimuno®
Healthy adults (20 subject, 1-week treatment, crossover design)	Placebo, Bimuno® (2 levels), GOS	Significantly higher increase of the bifidobacterial population in the presence of Bimuno®
		Dose response effect between B-GOS and bifidobacterial population increase
Healthy elderly (44 subjects, 10-week treatment, crossover design)	Placebo, Bimuno®	Significant increase in the bifidobacterial numbers compared to baseline and control in the presence of Bimuno®. Effect significant after 5 and 10 weeks of treatment
		Significant increase in phagocytic and NK cell activity when Bimuno® had been consumed
		Significant reduction in pro-inflammatory and increase in anti-inflammatory cytokines after consumption of Bimuno®

Table 12.2. (*Continued*)

Method	Substrate	Effect
IBS sufferers (44 subjects, 4-week treatment, parallel design)	Placebo, Bimuno® (2 levels)	Significant increase in the bifidobacterial population for both levels of Bimuno® administration. Change in microflora was followed by the alleviation of IBS symptoms
Travellers to medium and high risk for diarrhea countries	Placebo, Bimuno®	Significant reduction in the incidence and duration of travelers' diarrhea during consumption of Bimuno®

Placebo: Maltodextrin.
IBS, inflammatory bowel syndrome; GOS, galactooligosaccharides; NK, natural killer; B-GOS, galactooligosaccharides in Bimuno®.

Bifidogenic Properties of Bimuno®

From the very early stages of bifidobacteria research, it has been recognized that these bacteria express higher activity of β-galactosidase than many other members of the colonic microbiota (Desjardins and Roy, 1990). This has been further supported by isolation and characterization of various β-galactosidases from different *Bifidobacterium* species and subsequently confirmed at the genome level through the sequencing and annotation of numerous β-galactosidase encoding genes present in many *Bifidobacterium* strains of human origin. Within the available literature, most bifidobacterial strains make use of more than one β-galactosidase isoenzyme for their growth. Amongst those bifidobacterial β-galactosidases many have been proven to be lactases, hydrolyzing mainly lactose into its simpler glucose and galactose moieties, but some of them, the ones belonging to the GH 42 family seem to preferentially hydrolyze GOS other than lactose, as demonstrated with a β-galactosidase extracted from *Bifidobacterium adolescentis* DSM20083. The presence of β-galactosidases with different activity within the same cell of *Bifidobacterium* strains enables them to better utilize GOS mixtures by working in a synergistic fashion.

The bifidogenic properties of Bimuno® have been confirmed in vitro and in vivo in animal models and human intervention studies.

In vitro Bimuno® has been studied in a three-stage continuous culture system (Tzortzis et al., 2005), which simulates various physicochemical conditions within anatomically distinct regions of the colon (Macfarlane et al., 1998). An increase of 0.9 log10 cells in the bifidobacterial population numbers was observed in the first vessel, which represents the proximal colon. The effect was reduced in the second vessel and almost maintained in the third, suggesting that due to their low molecular weight, the synthesized oligosaccharides were mainly fermented in the earlier parts of the gut model. The overall increase in the *Bifidobacterium* population numbers in the presence of Bimuno® was higher than in the presence of either inulin or a GOS mixture produced using β-galactosidase from a *Bacillus* strain.

The prebiotic capacity of Bimuno® has been further investigated in a pig feeding trial (Tzortzis et al., 2005). During this parallel blinded study, 40 piglets were divided in four groups and their commercial diet was supplemented either with 2% (w/w) Bimuno® 4% (w/w) Bimuno®, or 4% (w/w) inulin. The reduced pH, compared to the control group, in the proximal colon of the animals combined with significant increases in SCFA production has shown that the Bimuno® mixture is not absorbed in the small intestine, but arrives in the colon where it becomes available for colonic fermentation. The increased concentration of acetic acid (the main fermentation product of bifidobacteria) in the proximal colon together with the significant increase in the bifidobacterial population numbers, have shown that fermentation of Bimuno® is selective by bifidobacteria and higher compared to control and inulin supplementation.

Bimuno® has been used in human volunteer studies. In a double blinded, placebo-controlled crossover design study, Depeint et al. (2008) showed that administration of Bimuno® mixture (2.75 g of B-GOS per day) for 1 week resulted in a better bifidogenic effect than an equal amount of another commercially available GOS mixture produced using enzymes from a *Bacillus* strain. Similar effect on the bifidobacterial population, after the B-GOS treatment, was observed at a lower dose of 1.33 g/day, and this bifidogenic effect was found to be dose-dependent. In addition, a different composition of bifidobacteria was suggested after the two GOS treatments, which was indicated through the isolation of different fecal *Bifidobacterium* species. Difference in the magnitude and quality of the effect on bifidobacteria between the two GOS mixtures was attributed to differences in structures of the oligosaccharide present in the mixtures due to the different enzymes used for their production. This was supported by the fact that the *Bifidobacterium* species isolated from the volunteers consuming Bimuno® were ones that shared high (>75%) similarity in their β-galacosidase genes with the genes of the β-galacosidase enzymes used in the production of the specific GOS mixture. Not all GOS mixtures necessarily result in the same effect on the microbiota, since differences in their DP as well as structure composition can become significant when it comes to their assimilation by bifidobacteria in the complex colonic ecosystem.

Vulevic et al. (2008) reported in a double blinded, placebo-controlled crossover design the effect of Bimuno® administration over a period of 5 and 10 weeks in the aging population (>60 years). Aging is associated with changes in the function of many organs and tissues, including the GI tract. Elderly intestinal microflora composition differs from that of younger adults and studies investigating these differences report increases in putrefactive bacteria, such as clostridia (especially *C. perfringens*) and enterobacteria at the expense of more beneficial groups such as *Eubacterium* spp., and more importantly, bifidobacteria. Administration of 2.75 g B-GOS/day significantly improved the colonic bifidobacterial numbers of the elderly and restored the microflora toward one similar to the younger adult population after 5 weeks.

Another area of great interest, in terms of microflora modulation, is functional GI disorders. Inflammatory bowel syndrome (IBS) specifically continues to represent a significant therapeutic challenge. Current interest has focused on the role of gut microflora–mucosa interactions linked to inflammatory and immune processes likely to contribute toward enteric neuromuscular dysfunction in these patients (Shanahan, 2000). The development of IBS following bacteriologically proven gastroenteritis, abnormal fermentation, and evidence of small bowel bacterial overgrowth (Silk et al., 2009) are some of the indicators the indigenous microflora have pointed to a pivotal role in IBS pathogenesis. In a blinded, placebo-controlled, parallel design study, Silk et al. (2009) reported significant improvement in the propensity of bifidobacterial population after 4 weeks of Bimuno® (1.33 or 2.75 g B-GOS/day) administration. The treatment with lower B-GOS intake (1.33 g/day) not only induced qualitative changes in the fecal flora, but also was associated with significant positive changes in stool consistency, flatulence, and composite scores (abdominal pain/discomfort, bloating/distension, and bowel movement difficulty). In addition, subjective global assessment values in the patients were positively influenced.

Immunomodulatory Properties of Bimuno®

The idea prebiotic ingredients could help intestinal defense system originated from observations that newborn babies, who have an underdeveloped intestinal host defense system, lack an appropriate capacity to defend themselves against intestinal infections. Furthermore, infants consuming their mother's milk were found to have a greatly reduced risk of diarrheal diseases, and a lower risk of respiratory and other infections. Human milk contains various protective components and active ingredients, including nondigestible oligosaccharides, which represent the third largest component of human milk and have been identified as the main factors involved in development of an appropriate colonization process in infants, which in turn stimulates the maturation of intestinal host defenses.

Although it is known that human milk oligosaccharides can exert a prebiotic effect, research into the immunomodulatory actions of prebiotics is very recent, with most data originating from animal models and in relation to FOS and its prebiotic effect (i.e., bifidogenicity and SCFA production). However, there are a few studies and lines of evidence that either suggest or demonstrate the effect of GOS on the immune system. This effect could either be direct in the form of interactions with immune, mucosal, or epithelial cells and/or indirect through the species or even strain selective modulation of the microbiota and their metabolic products.

The potential direct immunomodulatory properties of Bimuno® have been investigated in vitro using adult human colonic NCM460 cells (Dubert-Ferrandon et al., 2008). Confluent NCM460 cells stimulated with TNFα in the presence or absence of B-GOS were assayed for IL-8 and MIP-3α stimulation, mRNA induction, and NF-κB nuclear translocation. It was shown that in the presence of B-GOS, TNFα-induced IL-8, and MIP-3α secretions were significantly decreased compared to those stimulated with TNFα alone. The expression of pro-inflammatory cytokines mRNA levels revealed IL-8 and MIP-3α mRNA expression in the presence of B-GOS was significantly decreased by B-GOS compared to TNFα only stimulation. By immunofluorescence NF-κB was shown to translocate into the nuclei when cells were stimulated with TNFα, but this translocation was partially inhibited by B-GOS. These data showed for the first time that B-GOS has direct immunomodulatory properties in TNFα-mediated inflammation suggesting supplementation of B-GOS in the diet may reduce the inflammatory response of the intestinal innate immune system.

A recent study investigated the effect on immune function (natural killer (NK) cells, phagocytosis, and cytokines) of feeding daily 5.5 g of Bimuno® (2.75 g B-GOS) to 44 healthy elderly volunteers (Vulevic et al., 2008). This study found significant positive effects upon the immune response, evidenced by an improvement in NK cell activity and phagocytosis, increased secretion of the anti-inflammatory cytokine, IL-10, and decreased secretion of pro-inflammatory (IL-6, IL-1β, and TNF-α) cytokines by stimulated peripheral blood mononuclear cell (PBMC). This pattern of cytokine response (increased IL-10 followed by reduction in IL-6, IL-1β, and TNF-α) was previously reported to be regulated by NF-κB (Baldwin, 1996). Additionally, a clear positive correlation between the number of bifidobacteria and both NK cell activity and phagocytosis was demonstrated. Several studies have shown that Bimuno® increases the population numbers of bifidobacteria, lactobacilli, and subsequent SCFA production that are known to have immunomodulatory effect. Furthermore, specific probiotic species belonging to either bifidobacteria or lactobacilli isolated from volunteers after administration of Bimuno® (Depeint et al., 2008), are known to increase secretory immunoglobulin A (sIgA) secretion in the small intestine and feces, and to stimulate Peyer's patches (PP) B lymphocyte IgA production (Hoyles and Vulevic, 2008). These in vitro and in vivo data suggest Bimuno® immunomodulatory properties are not only restricted to the effect through microflora modulation but also to a direct effect on gut associated lymphoid tissue.

Protection from Pathogens

Previous studies have suggested that oligosaccharides, besides their prebiotic properties, could directly be involved in protecting the gut from infection and inflammation through inhibition of the attachment and/or invasion of pathogenic bacteria or their toxins to the colonic epithelium (Karlsson, 1989). This attachment is necessary before pathogens can colonize and cause disease, and is mediated by glycoconjugates on proteins and lipids present on the microvillus membrane. Protection is accomplished because certain oligosaccharides contain structures, similar to those found on the microvillus membrane, interfere with bacterial receptors by binding to them, and thus preventing bacterial attachment to the same sugar moiety of the microvillus glycoconjugates. For example, α-linked galactobioses are known to have antiadhesive properties and be capable of toxin neutralization. Bimuno® contains an oligosaccharide in alpha anomeric configuration, and has been shown to significantly decrease attachment of enteropathogenic *E. coli* (EPEC) and *Salmonella enterica* serovar typhimurium to HT-29 epithelial cell line (Tzortzis et al., 2005). This functional property of Bimuno® was further studied in infection caused by *S.* typhimurium (SL1344nalr). A murine ligated ileal gut loop model and a murine oral challenge model were used to assess the efficacy of *S.* typhimurium (SL1344nalr) to cause salmonellosis, in the presence or absence of Bimuno® (Searle et al., 2009). In tissue culture adherence and invasion assays, with HT-29–16E cells, the presence of 2 mM B-GOS significantly reduced *S.* typhimurium (SL1344nalr) invasion. In the murine ligated ileal gut loops, Bimuno® presence prevented colonization and associated pathology of *S.* typhimurium, while in the BalbC mouse model, the oral delivery of Bimuno® prior to challenge with *S.* typhimurium resulted in significant reductions in colonization in all five organs sampled (spleen, liver, ileum, colon, and cecum), with highly significant reductions being observed in the spleen at 72 and 96 hours post challenge. Overall, the results showed that Bimuno® can significantly reduce colonization and pathology associated with *S.* typhimurium infection in a murine model system by reducing pathogen invasion into host cells. The exact mechanisms of action by which oligosaccharides, such as the ones present in B-GOS, mitigate infection is still to be elucidated. However, it is evident from the studies conducted so far, a reduction in adherence to the gut mucosa may play an important role in preventing subsequent pathogen invasion.

Following these findings, a placebo-controlled, randomized, double-blind parallel design human study with travelers at high risk of pathogenic infections was recently completed (Drakoularakou et al., 2010). Travelers' diarrhea (TD) is one of the most common conditions experienced by individuals traveling to a developing country (Kamat and Mathur, 2006). It has an attack rate of 20–50% and studies have shown that bacteria and some viruses are the most common causative agents of TD (Black, 1990; Paredes et al., 2000) with enterotoxigenic *E. coli* (ETEC) being reported as the bacterial cause in 50% of TD cases (Vila et al., 2000). In this study, 159 healthy volunteers traveling to medium- or high-risk countries, were divided in two groups consuming either Bimuno® or placebo for a period of 7 days prior to their traveling and during their traveling abroad. Diarrhea was reported in 31% of persons who took part in the study during the holiday period and of those 62% were in the placebo group and 38% in the Bimuno® group showing a statistically significant difference between the two treatments. Moreover the duration of diarrhea for the people in the Bimuno® group was 2.4 days compared to the 4.6 days in the placebo group even though 34% of volunteers in the placebo group followed a "relief" treatment (such as loperamide or ciprofloxamin) versus 22% in the Bimuno® group. Additionally, Bimuno® did not induce any flatulence symptoms, a common side effect of prebiotics. These results were similar to a previously reported clinical study with irritable bowel syndrome sufferers, during which administration of B-GOS, at similar levels as used in this study significantly reduced the reported flatulence and bloating scores of the patients. This property of Bimuno® has been attributed to the selective fermentation of the oligosaccharide mixture by *Bifidobacterium* spp. which does not produce gas, as a result of its mode of preparation. Finally, analysis regarding the overall quality of life of the volunteers, as based on a scoring system, showed that although there were no differences between the treatments for the subjects that did not experience diarrhea, the people enrolled in the Bimuno® group reported better overall feelings concerning their health than those in the placebo group during the duration of acute diarrhea.

Conclusion

Although further studies are required to understand the exact mechanism of action, the beneficial effects of Bimuno®, of which not all have been reported for other prebiotic oligosaccharides, are attributed to the increased functionality of the specific mixture. This functionality includes selective bifidogenic properties in the large intestine, direct protection from the adhesion, and invasion of pathogenic organism in the small intestine, and stimulation of the immune function resulting in increased levels of phagocytic and NK cell activities. In addition, regulation of the pro- and anti-inflammatory cytokine profile of the host has been observed.

References

Baldwin, A.S. (1996) The NF kappa B and I kappa B proteins: new discoveries and insights. *Ann. Rev. Immunol.* 14, 649–683.

Black, R.E. (1990) Epidemiology of travelers' diarrhea and relative importance of various pathogens. *Rev. Infect. Dis.* 12, S73–S79.

Depeint, F., Tzortzis, G., Vulevic, J., I'Anson, K., Gibson, G.R. (2008) Prebiotic evaluation of a novel galactooligosaccharide mixture produced by the enzymatic activity of *Bifidobacterium bifidum* NCIMB 41171, in healthy humans: a randomized, double-blind, crossover, placebo-controlled intervention study. *Am. J. Clin. Nutr.* 87, 785–791.

Desjardins, M.L., Roy, D. (1990) Uncoupling of growth and acids production in *Bifidobacterium* ssp. *J. Dairy Sci.* 73, 299–307.

Drakoularakou, A., Tzortzis, G., Rastall, R.A. et al. (2010) A double blind, placebo controlled, randomised human study assessing the capacity of a novel galacto-oligosaccharide mixture in reducing travelers' diarrhea. *Eur. J. Clin. Nutr.* 64(2), 146–152.

Dubert-Ferrandon, A., Nanthakumar, N., Rautava, S. et al. (2008) Direct immunomodulatory properties of galacto-oligosaccharides on TNFalpha mediated inflammation in human intestinal epithelial cells. WCPGHAN 2008 Iguassu Falls, Brazil.

Gibson, G.R., Macfarlane, G.T. (1999) The human colonic microbiota. In: Gibson, G.R., Roberfroid M.B. (editors), *Colonic Microbiota, Nutrition and Health*, pp. 1–27. Kluwer Academic Publishers, Massachusetts.

Gibson, G.R., Probert, H.M., van Loo, J. et al (2004). Dietary modulation of the human colonic microbiota: updating the concept of prebiotics. *Nutr. Res. Rev.* 17, 259–275.

Gibson, G. R., Roberfroid, M. B. (1995) Dietary modulation of the human colonic microflora—introducing the concept of prebiotics. *J. Nutr.* 125, 1402–1412.

Goulas, T.K., Goulas, A.K., Tzortzis, G. et al. (2007) Molecular cloning and comparative analysis of four beta-galactosidase genes from *Bifidobacterium bifidum* NCIMB41171. *Appl. Microbiol. Biotechnol.* 76, 1365–1372.

Holzapfel, W.H., Veld, H.I. (1998) Overview of gut flora and probiotics. *Int. J. Food Microbiol.*, 41, 85–101.

Hoyles, L., Vulevic, J. (2008) Diet, immunity and functional foods. In: Huffnagle, G.B., Noverr, M.C. (editors), *GI Microbiota and Regulation of the Immune System*, pp. 79–92. Landes Biosciences, Texas.

Kamat, D., Mathur, A. (2006) Prevention and management of travelers' diarrhea. *Travel Med.* 52, 289–302.

Karlsson, K.A. (1989) Meaning and therapeutic potential of microbial recognition of host glycoconjugates. *Mol. Microbiol.* 29, 1–11.

Macfarlane, G.T., Macfarlane, S., Gibson, G.R. (1998) Validation of a three-stage compound continuous culture system for investigating the effect of retention time on the ecology and metabolism of bacteria in the human colon. *Microb. Ecol.* 35, 180–187.

Ouwehand, A.C., Salminen, S., He, F. et al. (2001) Differences in *Bifidobacterium* flora composition in allergic and healthy infants. *J. Allergy Clin. Immunol.* 108, 144–145.

Paredes, P., Campbell-Forrester, S., Mathewson, J.J. et al. (2000) Etiology of travelers' diarrhea on a Caribbean Island. *J. Travel Med.* 7, 15–18.

Playne, M.J. (2002) The health benefits of probiotics. In: *Encyclopaedia of Dairy Sciences,* pp. 1151–1158. Academic Press, London.

Rowland, I.R. (1995). Toxicology of the colon: role of the intestinal flora. In: Gibson, G.R., MacFarlane, G.T. (editors), *Human Colonic Bacteria: Role in Nutrition, Physiology and Pathology*, pp. 155–174. Boca Raton, FL.

Searle, L.E.J., Best, A., Nunez, A., et al. (2009) A mixture containing galactooligosaccharide, produced by the enzymic activity of *Bifidobacterium bifidum*, reduces *Salmonella enterica* serovar typhimurium infection in mice. *J. Med. Microbiol.* 58, 37–48.

Shanahan, F. (2000) Therapeutic manipulation in gut flora. *Science* 289, 1211–1312.

Silk, D.B, Davis, A., Vulevic, J., et al. (2009) Clinical trial: the effects of a trans-galactooligosaccharide prebiotic on faecal microbiota and symptoms in irritable bowel syndrome. *Aliment. Pharmacol. Ther.* 29(5), 508–518.

Tzortzis, G., Goulas, A.K., Gee, J.M., et al. (2005) A novel galactooligosaccharide mixture increases the bifidobacterial population numbers in a continuous in vitro fermentation system and in the proximal colonic contents of pigs in vivo. *J. Nutr.* 135, 1726–1731.

Vila, J., Vargas, M., Ruiz, J., et al. (2000) Quinolone resistance in enterotoxigenic *Escherichia coli* causing diarrhea in travellers to India in comparison with other geographical areas. *Antimicrob. Agents Chemother.* 44, 1731–1733.

Vos, A.P., M'Rabert, L., Stahl, B., et al. (2007) Immunomodulatory effects and potential working mechanisms of orally applied nondigestible carbohydrates. *Crit. Rev. Immunol.* 27, 97–140.

Vulevic, J., Drakoularakou, A., Yaqoob, P., et al. (2008) Modulation of the fecal microflora profile and immune function by a novel trans-galactooligosaccharide mixture (B-GOS) in healthy elderly volunteers. *Am. J. Clin. Nutr.* 88(5), 1438–1446.

Chapter 13

Concluding Remarks: Gastrointestinal Health and Nondigestible Carbohydrates

William R. Aimutis and Teri M. Paeschke

Most living organisms require food for survival. Generally, the diet is composed of a healthy balance of proteins, carbohydrates, and lipids to maintain an individual's sustenance. Complex carbohydrates such as starches, fibers, hydrocolloids, and other oligosaccharides impact the consuming host by modulating molecular mechanisms of the normal microflora and epithelial cells in the intestinal tract. These interactions cascade a complex series of biochemical and physical activities that impact many physiological responses. While the majority of the population can take for granted that the gastrointestinal tract will function normally, an increasing number are finding difficulty originating from their intestinal tract. Diseases such as type 2 diabetes, irritable bowel syndrome (IBS), inflammatory bowel disease (IBD), obesity, and constipation have all been linked to a malfeasant intestinal tract. The fact that the diseases are all on the rise is puzzling, and is leading more researchers to study interactions of nutrients with the intestinal tract. However, the intestinal environment is very complex and often the knowledge reported can only be generalized and not applicable to many individuals given our unique genetic characteristics. Likewise, access to massive amounts of information by the consuming public has caused many misperceptions about the correct value of a nutrient versus hearsay.

Nondigestible Carbohydrates and Digestive Health. Edited by T. M. Paeschke and W. R. Aimutis

Public health care messages have confused the public as they tried to clarify the physiological role for dietary carbohydrates (Buttriss and Stokes, 2008). The consumer is faced daily with messages about carbohydrate from both a positive and negative context. The challenge for scientists and other public health communicators is to relate the growing body of scientific evidence about the physiological role of carbohydrates into a consumer-friendly format. Carbohydrates are an essential component of the human diet responsible for supplying energy to the host and its complex gastrointestinal microflora, but also for having an overall role in physiology including maintaining gastric function, hepatic metabolism, and glycosylation of lipids and proteins necessary for other physiological functions.

Consumption of dietary carbohydrates is under increasing scrutiny in recent years by professional health care personnel and nutritionists. The problematic carbohydrates have generally been mono- and disaccharides, but some attention is also given to complex carbohydrates digested in upper regions of the gastrointestinal tract. Therefore, carbohydrates absorbed predominantly in the small intestine are blamed as culprits for diseases such as obesity, type 2 diabetes, and ultimately cardiovascular disease.

Nutrigenomic methods are allowing researchers to gain a better understanding of the interaction of human genetics to nutritional components. The science behind nutrigenomics is to understand how dietary components influence gene transcription, protein expression, and metabolism. The human genome responds to carbohydrates by activating 25–30 genes when a sugar as simple as glucose is ingested. However, not all of us will respond identically to glucose because the insulin-signaling pathways inside our cells are responding at slightly different rates in all individuals. Thus, we now know that even if we all have identical diets (the same amounts of protein, carbohydrate, and lipids), we will all still potentially be different in our response to utilizing the potential energy from those nutrients and gain, lose, or maintain our weight differently.

We as scientists should not generalize whether carbohydrates are good or evil to individuals. The discipline of nutrigenetics will allow scientists to better predict how individuals will respond to carbohydrates on the basis of individual genetic disposition, manifesting as single-nucleotide- and copy-number polymorphisms. Our long-term goal will be to use nutrigenomics and nutrigenetics to personalize nutrition for maintenance of individual health and prevent disease.

Mother Nature has furnished a human genome to efficiently digest carbohydrates right from parturition. Consumption of carbohydrates from human or bovine milk by the neonate offers protection from pathogens, toxins, and other environmental stresses the infant faces in its young life. Galactooligosaccharides present in colostrum and milk are important to the rapidly establishing gastrointestinal microflora, and these maintain their importance as the human matures into adult life. Galactooligosaccharides isolated or enriched from dairy sources offer protection to adults by stimulating immune response and influencing gut microflora composition. The importance of specific carbohydrate ingestion to elderly humans has not been extensively explored. However, science has shown that the gut microflora population in geriatric individuals is more comparative to an infant's microflora than a middle age adult.

Intact fruits, legumes, vegetables, milk, and whole grains (including the bran, germ, and endosperm) contain appreciable amounts of nonstarch polysaccharides, and are considered important dietary components for individuals who already have established diseases such as obesity, type 2 diabetes, and cardiovascular disease (Mann, 2007). Epidemiological data and feeding studies indicate that complex carbohydrates such as those derived from these foods are important in preventing these diseases. Many of these foods reduce the risk of cardiovascular disease by lowering total and low-density lipoprotein cholesterol. Additionally, diabetics benefit from improved blood sugar control from these low glycemic index foodstuffs. More recent attention is looking beyond these diseases to other conditions that complex carbohydrates may impinge such as colon cancer, IBD, and disordered laxation. The perplexing view on the mechanism for these foodstuffs impacting disease is whether it is the low glycemic index value they have or the fiber content impacting the gut microflora and other physiological factors. These same foodstuffs often contain appreciable levels of resistant starches and other oligosaccharides that are increasingly being demonstrated to have an impact on the gut microflora.

The importance of carbohydrates stimulating immune response is an emerging science. Proteins have long been identified as antigenic stimulants to the immune response, but more recently the sugars attached to these protein backbones are being shown to be involved in immune recognition as T-cell independent antigens (Brown and Mayer, 2007). Carbohydrate recognition by the immune system is important for both foreign and self-recognition. Research advances have demonstrated a positive immunostimulatory response to various oligo- and polysaccharides. While respectable data has reported the sugar composition of these complex carbohydrates in immune response, little information is available, to date, to explain the role of specific molecular conformation or carbohydrate linkage types in eliciting these responses. We look for more interest in this area as researchers try to explain the modus operandi of carbohydrates as immune stimulants.

Inflammation is present in many digestive symptoms and diseases. Biomarkers to study gut inflammation are still lacking. The marker most studied has been C-reactive protein (CRP), and correlates well with disease activity in Crohn's patients but is less definitive in individuals with ulcerative colitis (Vermeire et al., 2006). Another promising, but less studied, biomarker linked to IBD pathogenesis is a protein derived from neutrophils. These proteins function as damage-associated molecular pattern molecules (DAMPs) released from the neutrophils and are followed by pro-inflammatory agents that activate the pattern recognition receptors (PRRs) in the gut lymphoid tissue (Foell et al., 2009). The significance of this particular family of biomarkers is that they are assayed for from patients' feces and thus a less invasive means of studying gut inflammation. Furthermore, they distinguish IBD from IBS (Kaiser et al., 2007).

Individuals suffering from IBS and IBD often have elevated levels of CRP and interleukin (IL)-6. Diets high in carbohydrates and fiber decreased IL-18 (another inflammation biomarker) in diabetic patients, but had little effect on CRP and other inflammation biomarkers (Esposito et al., 2004). Similarly, King et al. (2008) reported psyllium fiber supplementation had no significant effect on the inflammatory biomarkers CRP, fibrinogen, or IL-6 in overweight or obese individuals. Other studies (King et al., 2007; Estruch et al., 2009) have shown that diets naturally rich in fiber were effective at lowering CRP. In another study, it was shown that participants receiving whole grains in their diets had lower CRP than individuals receiving a hypocaloric diet with refined grains (Katcher et al., 2008). The impact of carbohydrates, and specifically fiber, on levels of DAMPs has not been investigated. Although progress has been made to identify genes that increase susceptibility to IBDs, more information is needed on the mechanistic processes associated with these genes. Much further research is needed to define the role of dietary factors on the biomarkers of inflammation and especially the mechanism of anti-inflammatory effects from complex carbohydrates.

Dietary fiber consumption affects expression of various gut hormones that can act as satiety signals. The linkage to fermentation and expression of volatile fatty acids by the lower gut microflora has long been considered part of the overall process for satiety. However recent evidence indicates that increasing viscosity in the stomach can slow gastric motility and lead to a prolonged feeling of satiety in individuals consuming viscous complex carbohydrates, and therefore fermentation and subsequent volatile fatty acid production is not an inherent signal to satiety. Incretin response to simple carbohydrate ingestion has been well studied, but the response to complex carbohydrates such as those found in fiber and other polysaccharides is not as developed. Many studies have focused on insulin-signaling after glucose absorption, but far fewer studies have used nutrigenomic tools to follow the impact of fiber ingestion and subsequent physiological response.

Numerous in vitro, cell culture, and in vivo models have further advanced our understanding of the physiological effects of complex carbohydrates. All of these methods are complex, time consuming, and expensive, but established critical criteria needed to build computer models as an alternative to quickly and quantitatively study the effect of dietary components to physiological response. *In silico* modeling is becoming more widely used especially by drug companies as they capitalize on this method to translate observations made in preclinical models to human response. Similarly, *in silico* models in the future will be utilized to predict physiological responses to consumption of various food ingredients and foods. Much of the success of predicting how complex carbohydrates impact incretin hormone response and gastric motility will be credited to *in silico* models built to study diabetes utilizing data from diabetic mice.

Researchers have been successful at showing the complexity of the gut microflora and its overall importance in digestion and maintaining host health by utilization of modern genetic techniques. We have shown that the gut contains thousands of microbial species, many of which have not been cultured or identified. We are at the threshold of better understanding the importance of the complex community in healthy and diseased individuals and what can be done to manipulate the microflora to improve an individual's well-being. The impact of carbohydrates on the microflora has generally focused on genera and species thought to contribute to the well-being of most individuals predominantly the *Bifidobacterium*, *Lactobacillus*, and some *Bacteriodes* species. Given the large population of other bacteria present, complex carbohydrates surely have influence on other bacteria present that also impact host physiology by as yet unknown processes whose functional consequences have not been appreciated yet. Discovery of other bacterial species which impact the host can lead to new generations of probiotics, and likewise the carbohydrates that influence their growth and metabolism will lead to new generations of prebiotics.

Interactions occurring at the mucosal surface are less well understood. This results partly from a lack of understanding of the mucosal surface and the mucin layer covering it. Major breakthroughs are occurring at understanding how colonizing bacteria interact with intestinal tissues to stimulate defense mechanisms. Improved understanding of the role of colonizing bacteria microbial-associated molecular patterns (MMAPs) with PRRs on epithelial cells will provide more insight as to how the host recognizes friend from foe in the intestinal microbial milieu, and modulates inflammatory response. Increasing our understanding of the mucosal immune response and interactions of commensal gut flora with the mucosa will enable us to develop new vaccines or other approaches such as dietary intervention to prevent or treat diseases that affect the digestive tract.

The enteric nervous system connects the digestive organs with the brain in a bidirectional communication system to control gastric motility. There is also evidence beginning to emerge that the gut–brain axis (GBA) has a role in inflammatory disorders of the gut including IBD (Collins and Bercik, 2009). It would be important to consider the impact of viscous carbohydrates on brain and gastric motility. Research should be conducted with particular emphasis on studying the influence of genetic differences, age, ethnicity, and gender on complex carbohydrate metabolism and gastric motility also. While one could speculate that decreased motility may be advantageous, there remains the possibility that it could also be deleterious in individuals with IBD by causing an overgrowth of unwanted microbes. Slowed gastric motility may also lead to more efficient carbohydrate metabolism and increased weight gain by the host (Tilg et al., 2009).

The connection of the nervous system to the gut microbiome is beginning to be looked at more seriously. The normal microflora influences mucosal immune function, epithelial cell integrity, motility, and nutrient absorption in healthy individuals. However, disruptions to the normal flora by external factors including infections, antibiotics, stress, laxatives, or other environmental influences increases the likelihood of an intestinal inflammatory response. Studies with mice have shown that psychological stress can negatively impact the microflora population by decreasing normal indigenous bacteria and increasing potential microbial pathogens (Tannock and Savage, 1974). Later research showed that the microbial flora could be restored to more normality after stress was removed and a probiotic was administered (Eutamene and Bueno, 2007). Few studies have looked at the effect of carbohydrates on gut microflora after a stressful event(s); most have studied the impact of carbohydrates on gut flora populations in healthy or diseased individuals. However, Sink and Lochmann (2008) showed that a prebiotic from dried yeast fermentation product could reduce mortality in fish if administered before a psychological stressful event. Also influenced by changes in the gut microflora is a change in host behavior. A study was conducted to determine if shifts in bacterial diversity due to dietary manipulation could be correlated with changes in memory and learning (Li et al., 2009). Test mice were fed ground beef in addition to their standard chow. These animals showed improved working and reference memory in standardized testing protocols (hole-board open field apparatus). The conclusion was that dietary-induced shifts in bacterial diversity might indicate a role in memory and learning. The connection to higher animals including the human would be of future interest.

In conclusion, carbohydrates are becoming increasingly well recognized by consumers as important to their health and well-being. This is not always a positive relationship, as we know that some carbohydrates have deleterious effects on the ingesting host. The interaction of carbohydrates (both simple and complex) with the gastrointestinal tract occurs at many interfaces and influences numerous physiological effects. Future researchers still have many opportunistic areas to study digestive health and carbohydrates.

References

Brown, S.J., Mayer, L. (2007) The immune response in inflammatory bowel disease. *Am. J. Gastroenterol.* 102, 2058–2069.

Buttriss, J.L., Stokes, C.S. (2008) Dietary fibre and health: an overview. *South. Med. J.* 33, 186–200.

Collins, S.M., Bercik, P. (2009) The relationship between the intestinal microbiota and the central nervous system in normal gastrointestinal function and disease. *Gastroenterology.* 136, 2003–2014.

Esposito, K., Marfella, R., Ciotola, M., et al. (2004) Effect of a Mediterranean-style diet on endothelial dysfunction and markers of vascular inflammation in the metabolic syndrome. *JAMA.* 292, 1440–1446.

Estruch, R., Martinez-Gonzalez, M.A., Corella, D., et al. (2009) Effects of dietary fibre intake on risk factors for cardiovascular disease in subjects at high risk. *J. Epidemiol. Community Health* 63, 582–588.

Eutamene, H., Bueno, L. (2007) Role of probiotics in correcting abnormalities of colonic flora induced by stress. *Gut.* 56, 1495–1497.

Foell, D., Wittkowski, H., Roth, J. (2009) Monitoring disease activity by stool analyses: from occult blood to molecular markers of intestinal inflammation and damage. *Gut.* 58, 859–868.

Kaiser, T., Langhorst, J., Wittkowski, H., et al. (2007) Faecal S100A12 as a non-invasive marker distinguishing inflammatory bowel disease from irritable bowel syndrome. *Gut.* 56, 1706–1713.

Katcher, H.I., Legro, R.S., Kunselman, A.R., et al. (2008) The effects of a whole grain-enriched hypocaloric diet on cardiovascular disease risk factors in men and women with metabolic syndrome. *Am. J. Clin. Nutr.* 87, 79–90.

King, D.E., Egan, B.M., Woolson, R.F., et al. (2007) Effect of a high-fiber diet vs a fiber-supplemented diet on C-reactive protein level. *Arch. Intern. Med.* 167, 502–506.

King, D.E., Mainous, A.G., Eagen, B.M., et al. (2008) Effect of psyllium fiber supplementation on C-reactive protein: the trial to reduce inflammatory markers (TRIM). *Ann. Fam. Med.* 6, 100–106.

Li, W., Dowd, S.E., Scurlock, B., et al. (2009) Memory and learning behavior in mice is temporarily associated with diet-induced alterations in gut bacteria. *Physiol. Behav.* 96, 557–567.

Mann, J. (2007) Dietary carbohydrate: relationship to cardiovascular disease and disorders of carbohydrate metabolism. *Eur. J. Clin. Nutr.* 61(Suppl), S100–S111.

Sink, T.D., Lochmann, R.T. (2008) Preliminary observations of mortality reduction in stressed, *Flavobacterium columnare*-challenged golden shiners after treatment with a diary-yeast prebiotic. *N. Am. J. Aquac.* 70, 192–194.

Tannock, G., Savage, D.C. (1974) Influences of dietary and environmental stress on microbial populations in the murine gastrointestinal tract. *Infect. Immun.* 95, 91–598.

Tilg, H., Moschen, A.R., Kaser, A. (2009) Obesity and the microbiota. *Gastroenterology.* 136, 1476–1483.

Vermeire, S., Van Assche, G., Rutgeerts, P. (2006) Laboratory markers in IBD: useful, magic, or unnecessary toys? *Gut.* 55, 426–431.

Appendix

Nondigestible Carbohydrates: Structure and Sources

A primer on carbohydrate structure

Carbohydrate Numbering System: Carbon molecules are numbered successively from the carbon group containing the aldehyde group (before cyclization) counterclockwise. In the case of fructose, the second carbon is the carbon extending from the anomeric carbon.

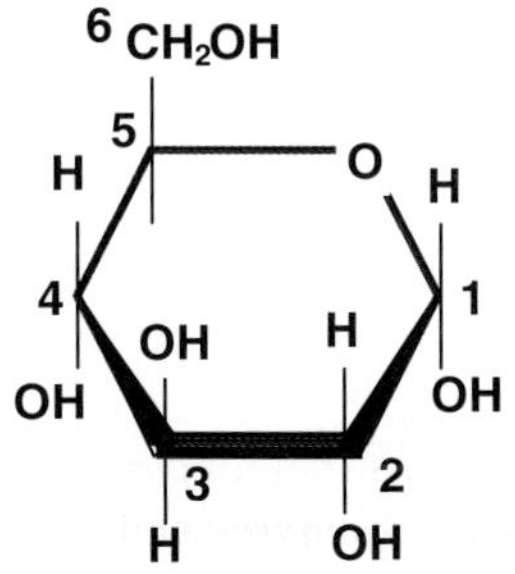

Nondigestible Carbohydrates and Digestive Health. Edited by T. M. Paeschke and W. R. Aimutis

Alpha and Beta Convention: The alpha–glycosidic linkage is below the plane of the rings and the beta glycosidic linkages is above the plane of the rings.

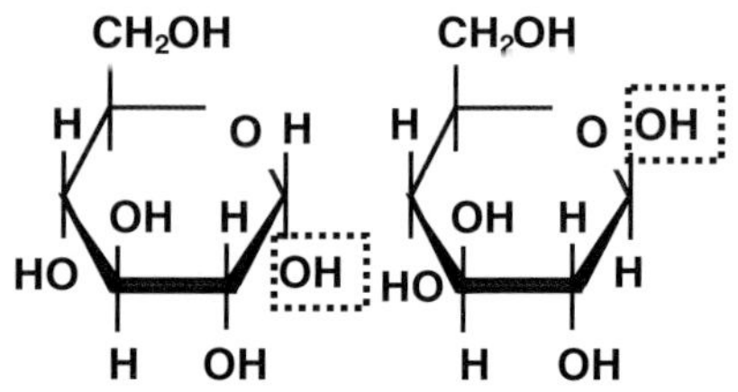

Alpha (α) Beta (β)

D and L Convention: The designation of D or L is based on the configuration around the next to the last carbon atom, counting the carbon with the aldehyde group as C1. For sugars with more than one chiral center, the D or L refers to chiral carbon farthest from the aldehyde or keto group.

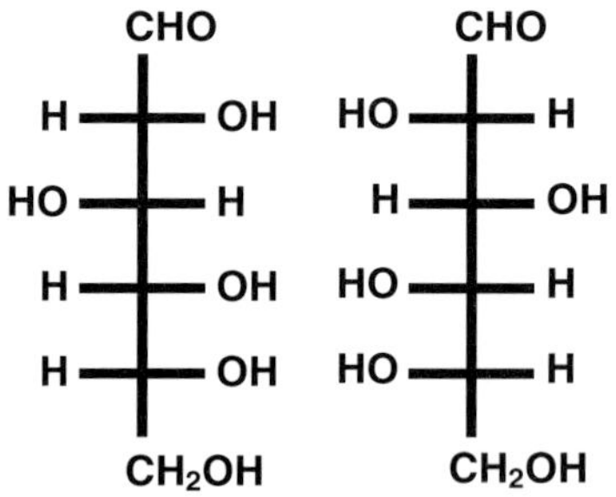

D-Glucose L-Glucose

Monosaccharide Building Blocks of Non-digestible Carbohydrates

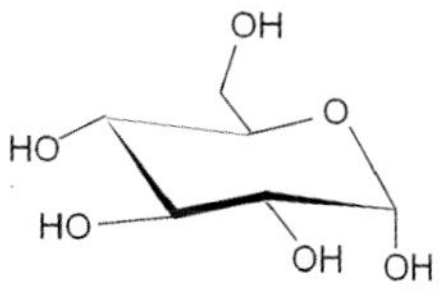

α-D-Glc*p*

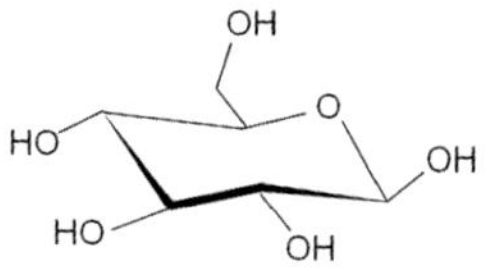

β-D-Glc*p*

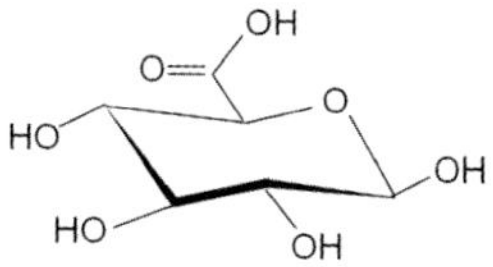

β-D-Glc*p*A

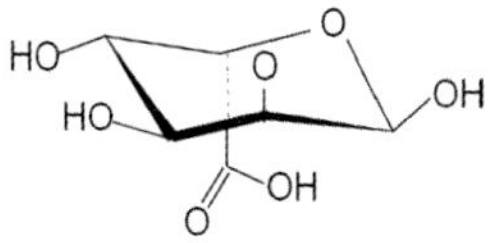

α-L-Gul*p*A

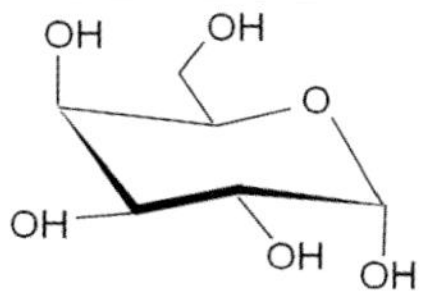

α-D-Gal*p*

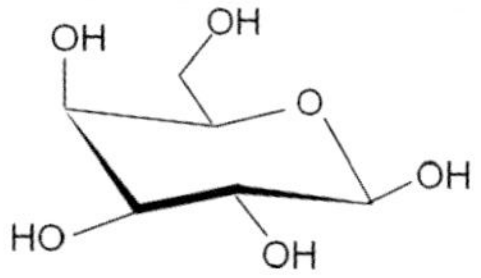

β-D-Gal*p*

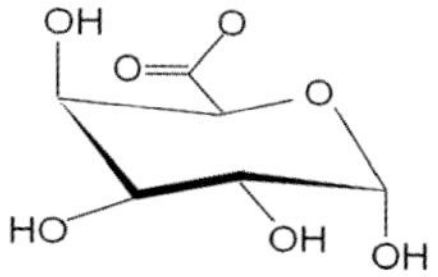

α-D-Gal*p*A

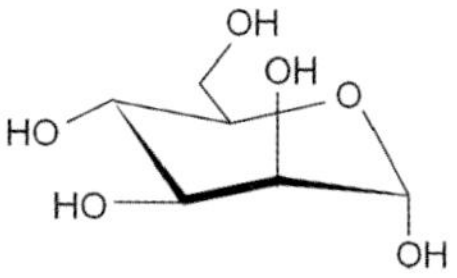

α-D-Man*p*

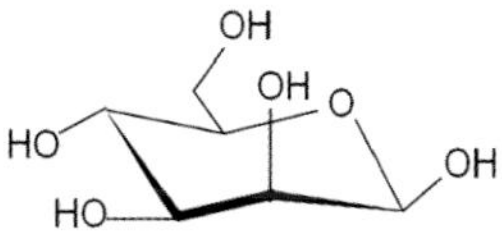

β-D-Man*p*

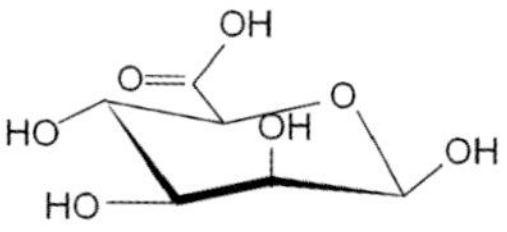

β-D-Man*p*A

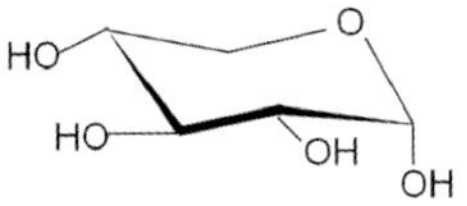

α-D-Xyl*p*

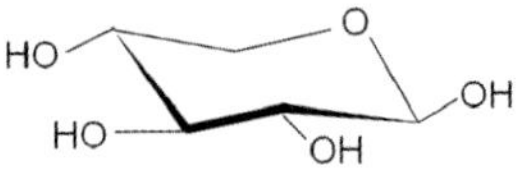

β-D-Xyl*p*

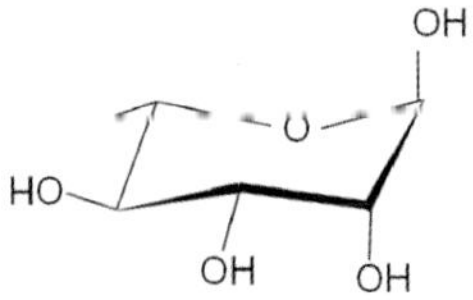

α-L-Rha*p*

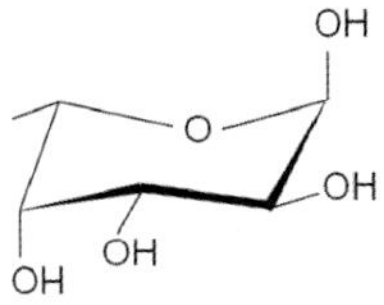

α-L-Fuc*p*

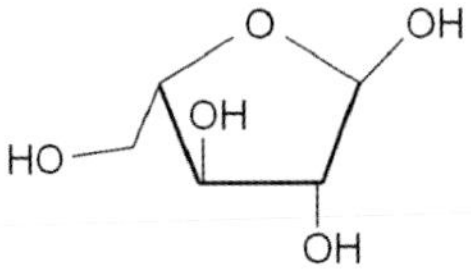

α-L-Ara*f*

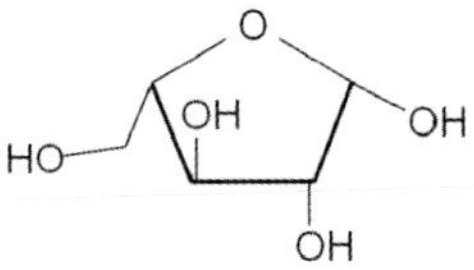

β-L-Ara*f*

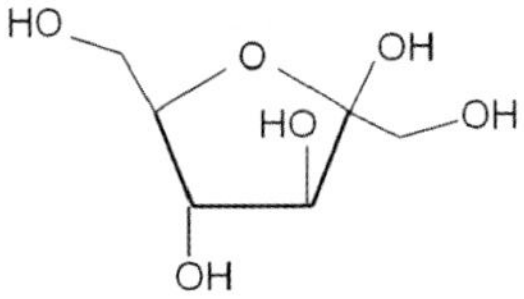

β-D-Fru*f*

Abbreviations of Common Monosaccharide Building Blocks:

Saccharide Monomer	Abbreviation
α-D-Glucose	α-D-Glc*p*
β-D-Glucose	β-D-Glc*p*
α-D-Galactose	α-D-Gal*p*
β-D-Galactose	β-D-Gal*p*
α-D-Mannose	α-D-Man*p*
β-D-Mannose	β-D-Man*p*
α-L-Fucose	α-L-Fuc*p*
β-D-Fructose	β-D-Fru*f*
α-L-Rhamnose	α-L-Rha*p*
α-D-Xylose	α-D-Xyl*p*
β-D-xylose	β-D-Xyl*p*
α-L-Arabinose	α-L-Ara*f*
β-L-Arabinose	β-L-Ara*f*
α-D-Galacturonic Acid	α-D-Gal*p*A
α-D-Guluronic Acid	α-D-Gul*p*A
β-D-Mannuronic Acid	β-D-Man*p*A
β-D-Glucuronic Acid	β-D-Glc*p*A

Dietary Fiber Sources, Monosaccharide Building Blocks, Structures, and Monomer Linkage

Nondigestible Carbohydrate	Source	Monomer Building Blocks	Structure, Monomer Linkage
Fructans			
Inulin	Chicory root, jerusalem artichoke	Fru	β(2→1) linked fructose, often with a terminal glucose
Fructooligosaccharide (FOS), oligofructose (OFS)	Hydrolyzed chicory root extract and enzymic synthesis from sucrose	Fru	β(2→1) linked fructose, often with a terminal glucose

(*Continued*)

Nondigestible Carbohydrate	Source	Monomer Building Blocks	Structure, Monomer Linkage
Levan	Enzymic synthesis from sucrose, plants	Fru	$\beta(2\rightarrow6)$ linked fructose, with branches of $\beta(2\rightarrow1)$ linked fructose
Galactans			
Galactooligosaccharide	Enzymic synthesis from lactose obtained from whey	Gal	Varies widely by manufacturer and enzyme choice: Galactose linked $\beta(1\rightarrow4)$, and/or $\beta(1\rightarrow6)$, and/or $\beta(1\rightarrow3)$
Synthetic Nondigestible Carbohydrates			
Polydextrose	Catalytic synthesis from glucose, sorbitol, and citric or phosphoric acid	Glc	Randomly linked glucose units, α and β $(1\rightarrow2)$, $(1\rightarrow3)$, $(1\rightarrow4)$, $(1\rightarrow6)$ linkages, with $(1\rightarrow6)$ predominating
Resistant maltodextrin	Corn or wheat starch pyrolysis, with subsequent enzyme digestion	Glc	Randomly linked glucose units α and β $(1\rightarrow2)$, $(1\rightarrow3)$, $(1\rightarrow4)$, $(1\rightarrow6)$
Grain, fruits and vegetable oligo and polysaccharides			
Pectin	Most often citrus based, sugar beet, fruits, and vegetables	Gal, Man, Rha, Glc, Xyl, Fuc, GlcA	Very complex heteropolymer, backbone of galacturonic acid linked $\alpha(1\rightarrow4)$, various side chains and substitutions on the backbone w/rhamnose, glucose, xylose, mannose, fucose, glucuronic acid

Nondigestible Carbohydrate	Source	Monomer Building Blocks	Structure, Monomer Linkage
β-Glucan	Oats, barley	Glc	Repeating tetraose and triose units of β(1→4) and β(1→3) glucose
Cellulose	Cell wall structural material in plants	Glc	β(1→4) linked glucose
Hemicellulose	Cell wall structural material in plants		Very complex heteropolymer of xyloglucans, arabinoxylans, etc. often linked with ferulic acid
Arabinoxylan	Grain-based materials, most often wheat or corn	Xyl, Ara	Heteropolymer with a backbone of β(1→4) linked xylose and side chains of arabinose
Arabino-xylooligos-accharides	Hydrolyzed grain arabinoxylan: wheat and corn most common sources	Xyl, Ara, Glc	Oligosaccharides from arabinoxylan, see above
Psyllium	Husk from Plantago	Xyl, Ara, Gal, Rha, GalA	Complex heteropolymer, xylose backbone linked β(1→4) and β(1→3) sidechains of Ara, Xyl, Gal, Rha, GalA also present
Larch gum	Larch tree	Gal, Ara	Highly branched galactose backbone linked β(1→3)
Resistant starches			
Type 1	Physically inaccessible, seeds and unprocessed grains	Glc	Crystalline α(1–4) linkages

(*Continued*)

Nondigestible Carbohydrate	Source	Monomer Building Blocks	Structure, Monomer Linkage
Type 2	Ungelatinized, crystalline high amylose starches	Glc	Crystalline α(1–4) linkages
Type 3	Retrograded, crystalline, corn starch	Glc	Crystalline α(1–4) linkages
Type 4	Chemically crosslinked starch, corn, wheat, tapioca, etc.	Glc	Varies by process and manufacturer
Glactomannan polysaccharides			β(1→4)-linked mannan backbone with α(1–6) linked galactose side chains (substitution ratio) depends on source
Guar gum	Legume/seed	Gal, Man	Ratio of mannose to galactose (M:G) ratio ~2:1
Hydrolyzed guar gum	Legume	Gal, Man	see guar above, low viscosity
Locust bean gum, aka carob gum	Legume/seed	Gal, Man	M:G ratio ~4:1
Fenugreek gum	Legume/seed	Gal, Man	M:G ratio ~1:1
Tara gum	Legume/seed	Gal, Man	M:G ratio ~3:1
Microbial polysaccharides			
Xanthan gum	Exocellular polysaccharide from xanthamonus campestrus	Glc, Man, GlcA	β(1→4) cellulose backbone w/side chains of (3→1) α-linked trisaccharide chain of (β-D-Man-(2→1)-β-D-GlcA acid (4→1)-β-D-Man) on alternating glucose residues

Nondigestible Carbohydrate	Source	Monomer Building Blocks	Structure, Monomer Linkage
Gellan gum	Exocellular polysaccharide from Sphingomonas elodea	Glc, Rha, GlcA	Repeating tetrasaccharide [Glc $\beta(1\rightarrow4)$ GlcA $\beta(1\rightarrow4)$-D-Glc $\beta(1\rightarrow4)$-L-Rha $\alpha(1\rightarrow3)]_n$
Seaweed polysaccharides			
Alginate	Brown seaweed	GlcA, ManA	Linear block copolymer of glucuronic and mannuronic acids
Glucomannan polysaccharides			
Konjac	Tuber/root	Glc, Man	$\beta(1\rightarrow4)$-linked mannose and glucose
Tree exudate polysaccharides			
Gum arabic, gum acacia	Tree resin	Gal, Ara, Rha, GlcA	Highly branched heteropolymer with some protein moeities, galactose backbone linked $\beta(1\rightarrow3)$, numerous and extensive branches of galactose, arabinose, rhamnose
Karaya, tragacanth, and ghatti gums	Tree resin	Gal, Xyl, Rha, Man, GlcA, GalA	Highly branched and very complex heteropolymers varying greatly by species, climate, etc. Pectic components may be present.

Index

Le développement des affaires

Membre de L'ASSOCIATION NATIONALE DES ÉDITEURS DE LIVRES

Presses de l'Université du Québec
Édifice Fleurie, 480, rue de la Chapelle, bureau F015, Québec (Québec) G1K 0B6
Téléphone : 418 657-4399 - Télécopieur : 418 657-2096
Courriel : puq@puq.ca - Internet : www.puq.ca

Diffusion / Distribution :

CANADA Prologue inc., 1650, boulevard Lionel-Bertrand
Boisbriand (Québec) J7H 1N7 - Tél. : 450 434-0306 / 1 800 363-2864

FRANCE ET BELGIQUE Sofédis, 11, rue Soufflot
75005 Paris, France - Tél. : 01 53 10 25 25

Sodis, 128, avenue du Maréchal de Lattre de Tassigny
77403 Lagny, France - Tél. : 01 60 07 82 99

SUISSE Servidis SA, chemin des Chalets 7
1279 Chavannes-de-Bogis, Suisse - Tél. : 022 960.95.25

Le développement des affaires

Comment accroître ses ventes
dans un monde numérique

Benoit Bourguignon

Presses de
l'Université du Québec

Catalogage avant publication de Bibliothèque et Archives nationales du Québec et Bibliothèque et Archives Canada

Titre: Le développement des affaires: comment accroître ses ventes dans un monde numérique / Benoit Bourguignon.

Noms: Bourguignon, Benoit, auteur.

Description: Comprend des références bibliographiques.

Identifiants: Canadiana (livre imprimé) 20250027690 | Canadiana (livre numérique) 20250027704 | ISBN 9782760562202 | ISBN 9782760562219 (PDF) | ISBN 9782760562226 (EPUB)

Vedettes-matière: RVM: Vente—Gestion. | RVM: Vente. | RVM: Entreprises—Croissance. | RVM: Planification stratégique. | RVM: Réseaux d'affaires.

Classification: LCC HF5438.4.B68 2025 | CDD 658.8/1—dc23

Financé par le gouvernement du Canada | Funded by the Government of Canada | Canada

SODEC Québec

Révision
Mathilde Godbout

Correction d'épreuves
Anne-Marie Bilodeau

Conception graphique
Julie Rivard

Mise en page
Alejandro Natan

Image de couverture
Charles-Antoine Giarrusso Bourguignon

Dépôt légal : 3e trimestre 2025

- Bibliothèque et Archives nationales du Québec
- Bibliothèque et Archives Canada

Imprimé au Canada
D6220-1 [01]

Remerciements

J'aimerais exprimer ma plus profonde reconnaissance aux personnes qui m'ont apporté leur aide dans la rédaction de ce manuel. L'esthétisme des figures a grandement bénéficié des talents de Charles-Antoine Giarrusso-Bourguignon. Je remercie mes deux relectrices, Astrid Tirel et Nathalie Giarrusso et mon relecteur, Michel Langlois, pour leurs nombreuses idées et pertinentes suggestions.

Merci à l'équipe des Presses de l'Université du Québec ayant travaillé sur le projet, en particulier Julie Simard et Gabrielle Dubé, ainsi qu'aux personnes responsables de la révision et de la correction, notamment Mathilde Godbout.

Enfin, je veux rendre hommage à toutes les personnes qui œuvrent dans le domaine du développement des affaires que j'ai croisées pendant toutes ces années et qui m'ont fortement inspiré, sans le savoir, tout au long de la réaction de ce texte.

Table des matières

Liste des encadrés, figures et tableaux

Encadrés

Figures

Tableaux

Liste des sigles et acronymes

BATNA	*Best Alternative to a Negotiated Agreement*
B2B	*Business-to-Business*
B2C	*Business-to-Consumer*
CRM	*Customer Relationship Management*
IA	Intelligence artificielle
RFI	*Request for information*
RFP	*Request for proposal*
RFQ	*Request for quote*
SCIAN	Système de classification des industries de l'Amérique du Nord
SCRM	*Social Customer Relationship Management*

Avant-propos

On peut dire que le développement des affaires est inscrit dans mon ADN. Mon arrière-grand-père a fondé une imprimerie en 1900, et de nombreux membres de la famille élargie ont contribué à la croissance de l'imprimerie Bourguignon tout au long du XXe siècle (Leroux, 2015). Mon père, qui a finalement acquis l'entreprise, a travaillé comme représentant auprès des industries pharmaceutique et cosmétique pendant 46 ans. Du côté maternel, mon grand-père et l'un de ses fils étaient vendeurs itinérants.

À mon père, je dois une leçon précieuse : le silence. Je me souviens encore de ma première visite de clientèle avec lui. Je ne comprenais pas pourquoi, tout au long d'une rencontre de 45 minutes avec une acheteuse, il avait à peine prononcé quelques mots. Je le trouvais bien discret. Après tout, une personne en vente ne doit-elle pas vendre ses produits et ses services ? Il m'a fallu quelques années d'expérience pour saisir l'importance de se taire plutôt que de parler. Le silence permet à la clientèle d'exprimer ses besoins, de dévoiler ses intentions lors de la négociation ou même de se convaincre par elle-même de la pertinence du produit ou du service. J'ai suivi les traces de mon père en faisant carrière dans le développement des affaires. J'ai d'abord été copropriétaire de mon propre commerce, avant d'occuper successivement les postes de représentant, de directeur des ventes et de vice-président des ventes dans le secteur de l'emballage imprimé, d'abord au sein d'une entreprise familiale, puis d'une multinationale. J'ai également développé des affaires en tant qu'entrepreneur à la tête de ma propre firme de consultation et de formation. Mon engagement académique est donc l'aboutissement de ce long parcours professionnel.

Avec l'avènement des communications numériques et le déferlement de l'intelligence artificielle, la vente et le développement des affaires ont connu une transformation radicale ces dernières années. Prendre le téléphone pour obtenir un rendez-vous n'est plus la seule option : une multitude de points de contact permettent désormais d'établir une relation.

Cet ouvrage traite principalement de la vente aux entreprises, par opposition à la vente aux consommateurs ou aux gouvernements. Son contenu s'adresse à trois types de lectorat. D'abord, il permettra aux propriétaires de petites et moyennes entreprises qui n'ont pas encore développé de compétences en vente de découvrir des stratégies et des tactiques qui les aideront à faire croître leur entreprise. Ensuite, il soutiendra l'apprentissage des étudiants et étudiantes en gestion qui envisagent une carrière en vente ou qui souhaitent éventuellement lancer leur propre entreprise. Enfin, les spécialistes évoluant en périphérie des ventes et amenés à collaborer occasionnellement avec la force de vente y trouveront également leur compte.

J'ai éprouvé énormément de plaisir à écrire ce manuel. J'espère que vous en aurez tout autant à le lire.

CHAPITRE 1

Une introduction au développement des affaires

« Rien n'arrive avant que quelqu'un ne vende quelque chose. »

Adage de vente

Le développement des affaires, qui permet de balancer l'offre et la demande, est une des activités de marketing les plus anciennes (Jones et Tadajewski, 2016). Autrefois, dès que les opérations agricoles généraient des surplus, les fermiers et fermières se rendaient au marché pour y écouler leurs stocks. Des voyageurs de commerce ou des marchands agissant comme intermédiaires ont rapidement proposé des produits non périssables à des marchés de plus en plus lointains.

Celui que l'on a surnommé le Père de la vente, John Henry Patterson, a été le premier à structurer la vente et le développement des affaires tels que nous les connaissons aujourd'hui. Après avoir acquis en 1884 une firme en difficulté financière, fabriquant les caisses enregistreuses NCR, il impose à son équipe des principes de la vente moderne, tels que le script de vente, la formation, les territoires de vente protégés et les quotas, encore utilisés aujourd'hui (Selling Power, 2010).

Les caisses enregistreuses étaient une véritable révolution technologique. Aujourd'hui encore, la technologie continue de bouleverser le développement des affaires. Les multiples plateformes du Web, où les responsables des achats décident quand et comment se procurer les biens ou les services dont ils ont besoin, ont remplacé le voyageur de commerce qui proposait ses produits à une clientèle ébahie.

Le développement des affaires désigne le processus d'identification et de poursuite des opportunités d'expansion et d'amélioration d'une entreprise. Il implique toute une série d'activités, dont l'étude de marché, la planification stratégique, le réseautage et la vente, qui visent toutes à reconnaître de nouvelles occasions de revenus et de rendement. Le développement des affaires est généralement motivé par le besoin de rentabiliser des investissements, de maximiser l'utilisation des ressources humaines et matérielles, de faire croître les ventes, d'augmenter les bénéfices ou d'obtenir un avantage concurrentiel.

Le développement des affaires se divise en deux volets : le développement des produits et des services, et le développement des marchés. Le premier englobe non seulement la création de nouveaux produits et services, mais aussi l'innovation sous toutes ses formes, notamment l'instauration de structures adaptables au modèle d'affaires. Il repose sur la recherche marketing, l'ingénierie et la créativité. Le développement des marchés, quant à lui, mobilise également la recherche marketing ainsi que le service des ventes. Ces deux volets doivent travailler en synergie pour assurer un développement des affaires efficace.

La figure 1.1 présente la matrice d'Ansoff (1957), qui propose une façon simple de comprendre les stratégies de développement des affaires.

FIGURE 1.1 / **Matrice d'Ansoff**

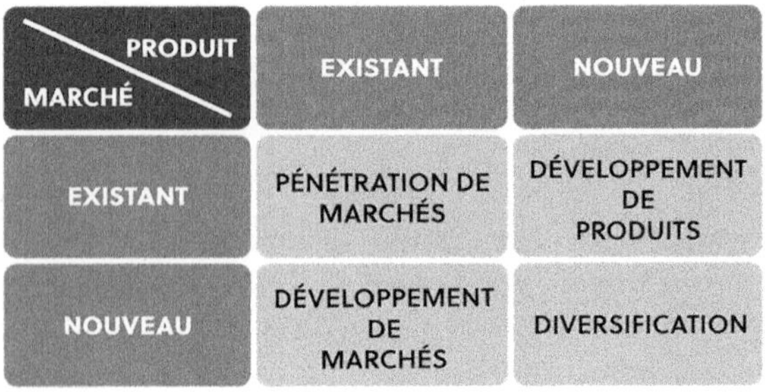

Source : Adapté de Ansoff (1957).

La pénétration de marché repose essentiellement sur la force de vente, car elle consiste à vendre davantage de produits sur les marchés actuels de la compagnie avec ses produits existants. En d'autres termes, il s'agit de convaincre la clientèle d'accroître ses achats ou encore la détourner de la concurrence au profit de sa propre société. La force de vente mettra en œuvre une stratégie de différenciation en collaboration avec l'équipe de marketing. L'entreprise peut aussi chercher à gagner des parts de marché en optimisant ses processus afin d'obtenir un avantage concurrentiel, notamment en réduisant ses coûts et ses prix. La pénétration de marché est l'une des stratégies les plus accessibles, car la force de vente maitrise déjà le marché et les produits. Toutefois, une approche de pénétration de marché trop agressive peut exacerber la concurrence, qui cherchera alors à défendre ses parts de marché.

Le développement de marchés consiste à offrir les produits existants dans de nouveaux marchés. Ces marchés peuvent être géographiques, par exemple lorsqu'une entreprise qui fait affaire au Québec cherche à prendre de l'expansion en Ontario. Il est aussi possible de découvrir des fonctions inédites aux produits actuels susceptibles d'intéresser différents segments du marché. La connaissance du produit demeure un atout majeur dans cette stratégie.

En revanche, la direction des ventes devra mettre en place une structure, embaucher du personnel, définir des modes de rémunération, identifier des circuits de distribution et recourir à différents médias de communication. L'équipe de vente doit comprendre les besoins de cette clientèle naissante et analyser les forces et les faiblesses de la nouvelle concurrence, le cas échéant.

Le développement de produits ou de services repose souvent sur les informations fournies par la force de vente concernant des besoins non satisfaits du marché actuel. Une fois les nouveaux produits ou services créés, les membres de l'équipe de vente doivent acquérir les connaissances essentielles pour communiquer à la clientèle les bénéfices de ces innovations. La direction des ventes peut adapter la structure de l'équipe afin de prendre en compte les différences techniques importantes entre les produits existants et les nouveautés. Lorsqu'une firme invente des produits ou des services sans prendre le pouls de la clientèle au moyen d'études de marché ou de retour d'expérience de sa force de vente, elle risque de souffrir de la *myopie du marketing*. Les professionnelles et professionnels de la vente disposent néanmoins de l'avantage de bien connaître le marché, même s'ils doivent relever le défi de présenter ces nouveautés.

Finalement, **la diversification** suppose que la compagnie développe de nouveaux produits pour de nouveaux marchés. La diversification repose principalement sur la haute direction, qui choisira souvent de procéder par voie d'acquisition, compte tenu de son manque de connaissances sur ces produits et ces marchés. Dans ce cas, l'entreprise bénéficie également d'une force de vente déjà implantée sur le territoire et parfaitement familiarisée avec son offre.

Les comptables savent que le développement des affaires a une incidence majeure sur les liquidités de l'entreprise puisqu'un investissement important sera indispensable avant d'enregistrer les premières ventes. Les aspects juridiques entrent également en ligne de compte, en particulier lors du développement des affaires dans d'autres villes, provinces ou pays. Ce manuel se concentre sur les facettes du développement des affaires qui relèvent essentiellement de la force de vente. Je laisserai à d'autres le soin de détailler les processus d'innovation et la structure de capital nécessaire à la diversification.

Le développement des affaires ne doit pas s'exercer de façon ponctuelle lorsque les ventes sont basses, par exemple, mais devrait être une préoccupation continue de la direction, parce que le marché évolue, la technologie progresse rapidement, la concurrence s'adapte, les besoins de la clientèle changent, ou encore des compagnies fusionnent, éliminent des gammes de produits ou disparaissent. Vos produits et vos services peuvent devenir caducs dans de telles circonstances.

Le développement des ventes consiste à exécuter la stratégie actuelle de l'entreprise auprès de la clientèle cible identifiée. Développer les ventes est une fonction tactique alors que développer les affaires est stratégique. Cela signifie-t-il que le département des ventes ne devrait pas se mêler de stratégie? Bien au contraire. Comme l'équipe de vente travaille beaucoup à l'extérieur de l'organisation, elle représente l'un des meilleurs outils de collecte de données et de veille de marché. Les membres de l'équipe doivent donc contribuer en identifiant des segments inédits, en proposant de nouveaux produits ou services, et peuvent même suggérer des opportunités d'alliances avec des prestataires, ainsi que des fusions et des acquisitions avec des entreprises concurrentes.

1/ Transaction ou relation

Avant de vous lancer dans une opération de développement des affaires, vous devrez déterminer si votre organisation, votre industrie et votre clientèle effectuent des transactions ou s'engagent dans des relations. On parle de vente transactionnelle lorsque l'échange est ponctuel et ne nécessite aucun suivi de part et d'autre. La plupart des achats de la population sont transactionnels, même si de nombreuses compagnies utilisent des programmes de fidélité ou des communications personnalisées pour tenter de créer des transactions répétées, ce qui pourrait s'apparenter à du marketing relationnel. Dans les faits, en raison de l'interdépendance des firmes dites B2B (*Business-to-Business*) lors de leurs échanges, les relations y sont plus critiques (Chisam et coll., 2022). Le marketing aux consommateurs ou aux consommatrices *Business-to-Consumer*, dit B2C, se distingue généralement du marketing aux entreprises B2B, par une plus grande clientèle qui effectue des transactions financièrement moins importantes. En vente transactionnelle, la valeur est intégrée dans le produit ou le service et elle est échangée contre une compensation financière alors qu'en vente relationnelle, une partie de la valeur est générée par les relations elles-mêmes.

La vente relationnelle, dont il est question dans ce manuel, découle de la volonté des deux parties de s'engager dans une véritable relation à long terme, formée de transactions répétées, plus fréquentes en B2B. Les pratiques de la vente relationnelle se sont considérablement transformées ces dernières années, entrainant une situation paradoxale. D'un côté, les échanges se font de plus en plus sur des plateformes de commerce en ligne, sans intervention humaine. De l'autre, la complexité des services et des solutions offerts exige des contacts fréquents et répétés entre l'équipe de vente et la clientèle (Arli et coll., 2018). Pour soutenir le choix des responsables des achats d'effectuer des transactions numériques en l'absence de spécialistes en vente, Arli et coll. (2018) recommandent notamment de donner plus de responsabilités à l'équipe de vente interne. Si les membres de cette équipe peuvent être efficaces pour la préservation des relations, les membres qui travaillent sur la route seront privilégiés lorsque la vente est délicate, ou encore pour bâtir des relations ou tenter de les réparer. La vente individuelle fait place à la vente en équipe lorsque l'ensemble du processus se complexifie, surtout dans les grandes organisations (Arli et coll., 2018).

En marketing relationnel, la valeur n'est pas seulement contenue dans le produit, comme c'est généralement le cas en B2C, mais également dans la relation, faite d'échanges d'informations, notamment sur les meilleures façons d'exploiter cette valeur grâce à l'utilisation du produit ou du service.

La clientèle acceptera de s'engager dans une relation si l'une ou l'autre des situations suivantes se présente :

- Complexité du produit ou du service : Un groupe d'achat confronté à la compréhension de caractéristiques techniques complexes ou à la comparaison de nombreuses solutions accessibles sur le marché voudra compter sur une personne en vente, d'autant plus si cette dernière dispose d'informations pertinentes difficiles à trouver sur le Web.
- Importance de la transaction : Cette dimension peut se mesurer par la valeur monétaire de la transaction, mais elle peut aussi s'évaluer en fonction de l'impact que cette acquisition peut avoir sur les opérations de l'organisation cliente. Cette valeur ou cet impact de la transaction déterminera la position hiérarchique des individus qui prendront part à la décision. Ces derniers voudront obtenir le plus d'informations possible sur le partenaire avec lequel ils veulent s'associer.
- Perception du risque et de l'incertitude : La complexité du produit ou du service et l'importance de la transaction peuvent créer un sentiment de risque et d'incertitude, ce qui influencera la fréquence des rencontres entre les deux parties.
- Expérience et connaissances des membres de la centrale d'achat : Elles permettent de réduire le risque ou l'incertitude.
- Pression temporelle : Lorsque la clientèle est pressée, elle pourra devoir se fier à des signaux de qualité, comme la marque, pour prendre une décision.

Une fois la transaction finalisée, deux facteurs déterminent le désir de maintenir des relations (Prior, 2021). D'abord, la durée de l'approvisionnement a une incidence sur la fréquence des interactions entre l'entreprise cliente et son fournisseur, selon les critères évoqués précédemment. Plus un achat est important et plus il se répète, plus l'entreprise cliente souhaitera poursuivre la relation et maintenir un canal de communication. Le second facteur est la nécessité de développer des relations auprès de personnes qui occupent différentes fonctions et qui œuvrent à divers niveaux hiérarchiques. Certaines situations, comme l'implantation d'un logiciel dans l'ensemble de la compagnie, faciliteront l'insertion de la personne en vente au sein de l'entreprise cliente. En revanche, dans d'autres cas, la personne responsable des achats cherchera à limiter la création de relations trop profondes avec ses collègues. Elle pourrait craindre qu'à long terme, ces relations l'empêchent d'obtenir le maximum de valeur de cette collaboration, parce que ses collègues pourraient vouloir favoriser un fournisseur ou une fournisseuse, par exemple, en raison de sa capacité à simplifier leurs tâches. L'aptitude de la personne en vente à contourner le ou la responsable des achats, sans entrer en conflit, pourra l'aider à maintenir la relation et à mettre en valeur son offre plutôt que celle de la concurrence.

2 / Une carrière en développement des affaires

La société ne valorise pas tellement les métiers de la vente ou du développement des affaires, peut-être en raison de pratiques douteuses de quelques individus comme ce couple de courtiers immobiliers condamnés en 2024 pour avoir créé de fausses offres d'achat pour encourager la surenchère (Jolicoeur, 2024). Il n'est donc pas étonnant que bon an mal an, selon un sondage de la firme de recherche marketing Léger, les personnes qui exercent dans le domaine de la vente soient celles en qui le public a le moins confiance (Archambault, 2024).

L'appellation des emplois en vente est variée. Selon une étude réalisée entre 2020 et 2022 sur les postes en ventes affichés sur LinkedIn et sur des sites de recrutement de 5 pays, une équipe de recherche a recensé 10 dénominations différentes, notamment celles de représentants des ventes, de développeurs d'affaires et de spécialiste des ventes qui totalisent 52 % des postes (Elhajjar et coll., 2023). Les compétences les plus recherchées, dans l'ordre, étaient : les habiletés en communication, la maitrise d'outils technologiques comme les logiciels de gestion de la relation client (*Customer Relationship Management* [CRM]), le leadership, les aptitudes en négociation, la gestion du temps et l'intelligence d'affaires. Toujours selon la même étude, voici quelques-unes des caractéristiques personnelles les plus désirées dans ces offres d'emploi, également classées par ordre d'importance : la motivation, le dévouement, la résilience, l'adaptabilité et l'énergie. Une autre étude révèle que les gestionnaires cherchent également à recruter des personnes possédant des compétences technologiques et analytiques (Peesker et coll., 2022). Enfin, en ce qui concerne les responsabilités, la priorité est donnée à la capacité de transformer une entreprise cliente potentielle en cliente réelle ainsi qu'au développement des affaires. Ces éléments figurent parmi les tâches principales, aux côtés de la gestion des relations et de la présentation du produit ou du service à la clientèle. On constate que, malgré l'omniprésence des technologies, les gestionnaires recherchent avant tout des spécialistes capables d'interagir directement avec la clientèle à différents niveaux hiérarchiques et qui occupent différentes fonctions.

Elhajjar et coll. (2023) ont aussi effectué une analyse pour identifier les activités réalisées par les personnes en vente. Cette équipe de recherche en a recensé 122, rassemblées en 10 catégories :

1. La vente relationnelle, dont les actions d'écouter, de poser des questions et de bâtir la confiance ;
2. La rédaction, qui se traduit par la création de contenus et l'écriture de commandes ;
3. Le divertissement de la clientèle ;
4. La prospection ;
5. Le travail à l'ordinateur, qui regroupe des activités telles que le CRM, les courriels, la recherche sur le Web et la surveillance des médias sociaux ;
6. La formation et le recrutement ;
7. Le support au produit, avec la gestion des plaintes ou la supervision de l'installation ;
8. Les activités éducatives, comme assister à des conférences ;
9. Le soutien aux canaux de vente pour aider la distribution ;
10. L'analyse, soit l'étude des données ou la lecture d'enquêtes sur la clientèle.

Une carrière en développement des affaires offre de nombreuses trajectoires. On trouve des personnes en vente dans les commerces de détail, qui comprennent les concessionnaires automobiles, les lunetteries, ou les boutiques de vêtements. La vente au grand public et aux entreprises inclut la vente de services, entre autres de télécommunications, de produits financiers, d'assurances et de services immobiliers. Dans les services aux entreprises, on ajoutera notamment l'organisation de conférences ou d'expositions, la vente d'espaces publicitaires dans les médias ainsi que la vente de services parfois externalisés, tels que les services juridiques, comptables, de nettoyage ou de sécurité.

Dans la vente de produits aux entreprises, on trouve ceux destinés à l'entreprise elle-même, comme des équipements, et ceux qui entrent dans la fabrication d'un bien, notamment des ordinateurs intégrés aux automobiles, que l'on qualifie de composants. Par ailleurs, des établissements achètent des marchandises réservées à la revente au grand public, par exemple les croustilles et les boissons gazeuses.

Généralement, une personne aura la possibilité d'œuvrer d'abord en tant que représentante au service à la clientèle ou encore au développement des ventes. Dans le premier cas, le travail consiste à soutenir les membres de l'équipe de vente qui sont sur la route dans l'exécution des commandes; dans le second, les tâches s'articulent autour de la recherche de firmes à prospecter afin qu'une autre personne prenne le relais pour présenter le produit ou le service. Une personne peut aussi commencer dans des postes de représentation auprès des commerces de détail. La vente consiste alors à faire une tournée des magasins pour offrir de la formation, s'assurer d'une mise en valeur des produits, etc., puisqu'une équipe spécialisée négocie les contrats à l'échelle nationale entre les compagnies de fabrication et les grandes bannières. Dans certaines industries offrant des produits complexes, une personne recrutée pour son expertise pointue arbore souvent le titre de représentante ou représentant technique.

Après quelques années d'expérience, une personne pourrait obtenir un poste en gestion des comptes majeurs ou à la direction des ventes. Dans la plupart des cas, ce genre de poste permet de percevoir un revenu de 100 000 dollars et plus. Cette rémunération attractive peut inciter la communauté étudiante à envisager une carrière en vente.

Alors, comment choisir ? Vous pouvez d'abord identifier les industries avec lesquelles vous avez des affinités. Ainsi, une personne peut faire carrière en développement des affaires dans le secteur informatique, si elle possède des connaissances dans ce milieu et si elle éprouve un intérêt.

La rémunération, qui sera abordée dans le prochain chapitre, varie de la commission pure au salaire fixe. La moyenne des émoluments en vente ne permet pas d'estimer précisément les gains envisageables, car plusieurs paramètres entrent en jeu. Ainsi, le revenu augmente généralement avec le degré de scolarité requis et les années d'expérience demandées. Les honoraires dépendent également de la valeur du produit ou du service, ainsi que de la position hiérarchique des interlocuteurs et interlocutrices. L'étendue et les spécificités de la zone géographique desservie influencent aussi le salaire. Par exemple, une personne chargée de visiter une clientèle au Canada et aux États-Unis recevra probablement une meilleure rémunération qu'une autre dont le territoire se limite à la ville de Québec. Le cycle de vie du produit ou du service peut également influencer la rémunération. En général, promouvoir un produit en phase de lancement génère des revenus plus élevés que la vente d'un produit arrivé à maturité.

La croyance populaire veut que les personnes travaillant en vente aient une personnalité flamboyante. En réalité, le travail de représentation exige avant tout une écoute attentive des besoins avant d'orienter progressivement la discussion vers l'achat à l'aide de questions. Ce travail suppose aussi d'être renseigné et de ne transmettre que les informations pertinentes à la prise de décision. Le développement des affaires repose également sur la création et l'entretien de relations. Souvent, vous devrez discuter avec des comptables pour justifier l'achat de votre produit ou de votre service. Si vous êtes à l'aise avec les chiffres, cela peut vous donner un avantage.

En conclusion, le travail en développement des affaires évoluera considérablement au cours des prochaines années afin d'anticiper et de suivre la révolution technologique en cours. Il n'en demeure pas moins qu'une carrière en développement des affaires reste l'une des plus stimulantes qui soient.

3 / Un nouveau processus de vente et d'achat

Depuis quelques années, la communauté académique a proposé de nombreux modèles de vente pour suivre l'évolution des préférences de la clientèle. L'une des premières propositions ayant marqué la mutation des pratiques est le modèle de vente adaptative (Weitz et coll., 1986), qui incite la personne en vente à ajuster sa présentation selon chaque destinataire et chaque contexte. Cette adaptation n'est réalisable que si la personne en développement des affaires construit son savoir en collectant des informations réutilisables dans différentes situations. (Spiro et Weitz, 1990). Quelques années plus tard, face aux bouleversements brusques dans l'environnement des affaires et à la diffusion du concept d'agilité en approvisionnement ou en production, Chonko et Jones (2005) ont suggéré que la force de vente devienne agile, c'est-à-dire rapide et flexible. Aujourd'hui encore, l'agilité en vente demeure essentielle dans un monde où les transformations de toute nature créent des turbulences majeures (Bourguignon et coll., 2021). Le service des achats se spécialise de plus en plus et vise à obtenir des produits et des services générant de la valeur pour son organisation. Les équipes commerciales ont ainsi structuré leur approche autour d'un concept de vente axé sur la valeur (Töytäri et Rajala, 2015). Or, la valeur perçue est subjective, contextuelle et variable dans le temps : un produit ou un service peut générer de la valeur pour une personne dans un contexte donné et à un moment précis, tout en étant inintéressant pour quelqu'un d'autre dans une autre situation ou à une autre période. Puisque les spécialistes de l'approvisionnement effectuent aujourd'hui une partie de leur parcours d'achat en ligne sur diverses plateformes numériques, les organisations de vente doivent une fois de plus faire preuve d'agilité en adoptant une approche hybride où les tâches sont réparties entre les représentantes et représentants itinérants, le personnel conseiller du service à la clientèle et une plateforme de vente transactionnelle (Thaichon et coll., 2018). Pour réussir aujourd'hui, une organisation de vente doit intégrer plusieurs de ces modèles. L'équipe doit faire preuve d'adaptabilité, adopter une approche agile, vendre en mettant l'accent sur la valeur et se déployer de façon hybride pour répondre aux attentes de la clientèle. Le développement des affaires continuera également d'évoluer sous l'influence des avancées technologiques, notamment l'intelligence artificielle, l'Internet des objets, les médias sociaux, les robots conversationnels, les applications mobiles, l'impression 3D, la réalité augmentée et les logiciels automatisés de relation client. Mais d'autres éléments demeurent inchangés, notamment le besoin pour la clientèle d'être accompagnée par une véritable personne pour certaines décisions.

Le processus de vente a longtemps été calqué sur la démarche d'achat. On peut retracer les origines de cette approche tactique aux années 1920 (Moncrief et Marshall, 2005). Ce processus différait légèrement selon qu'il s'agissait de celui du consommateur ou de la consommatrice ou de celui de la centrale d'achats en entreprise. Ces processus n'ont pas beaucoup changé.

Comme l'illustre la figure 1.2, la première étape du processus d'achat en B2C commence par la reconnaissance du besoin, par exemple lorsqu'une personne réalise qu'elle doit ou souhaite changer de véhicule. À l'étape de la recherche d'information, elle consulte les sites Web des concessionnaires, demande des avis à son entourage, en personne ou sur les réseaux sociaux, et fait des essais routiers. Elle hésite entre trois véhicules et doit maintenant peser le pour et le contre à l'étape de l'évaluation des choix. Une fois la décision prise, elle procède à l'achat du véhicule et commence à l'utiliser, ce qui correspond à l'étape de l'achat et de la consommation. Après quelques mois, elle partage son expérience avec son entourage et exprime sa satisfaction, ce qui représente le comportement post-achat.

FIGURE 1.2 / **Processus d'achat du consommateur ou de la consommatrice**

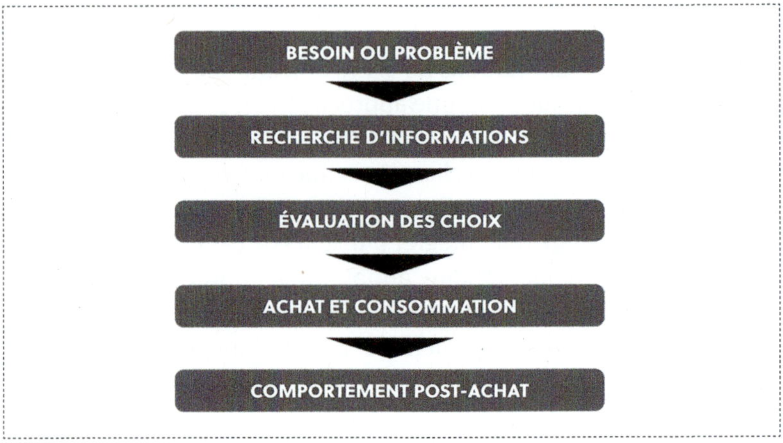

Source : Adapté de Engel et coll. (1995).

Pour le processus d'achat en entreprise (B2B), illustré dans la figure 1.3, après la reconnaissance d'un besoin ou la découverte d'un problème, une recherche d'information est effectuée si la société connaît peu le produit ou le service. Lorsque l'entreprise détient ou obtient l'information, elle établit des spécifications précises définissant les caractéristiques requises. La centrale d'achat publie alors un appel d'offres afin d'obtenir le meilleur prix pour la qualité voulue. Les responsables étudient les soumissions reçues et sélectionnent la proposition la plus avantageuse. Une fois la décision prise, l'entreprise émet un bon de commande et commence à utiliser le produit ou le service. Après une période qui peut varier de quelques mois à un an, la centrale d'achat procède à une évaluation formelle du produit, du service et même du prestataire.

FIGURE 1.3 / **Processus d'achat d'une entreprise**

Source : Adapté de Robinson et coll. (1967).

L'approche traditionnelle prévoit que le rôle de la personne en vente consiste à accompagner le ou la gestionnaire des achats lors de chacune des étapes de ce processus comme l'illustre la figure 1.4. D'abord, elle doit s'assurer que son entreprise soit considérée au moment de la reconnaissance du besoin, soit par sa notoriété dans le marché ou par les actions proactives de la personne en vente. Cela étant, les personnes en représentation commerciale doivent connaître les besoins de la clientèle et les bénéfices recherchés. Une fois que la personne en vente comprend bien comment résoudre le problème ou satisfaire le besoin, elle peut présenter une offre adaptée. C'est aussi à cette étape que la clientèle peut exprimer des objections, poser des questions ou faire part de ses préoccupations. Si la personne manifeste un intérêt, une phase de négociation s'engage, à moins que la personne en vente ne bénéficie d'un avantage concurrentiel décisif lui permettant d'imposer ses conditions, ou que le ou la responsable des achats dispose de multiples options dans l'environnement concurrentiel, ce qui lui permet d'obtenir les avantages souhaités, comme un prix plus bas. La négociation aboutit dans certains cas à une entente à la suite de laquelle la personne en vente propose un service après-vente pour maintenir la relation et, éventuellement, obtenir d'autres contrats.

FIGURE 1.4/ **PROCESSUS DE VENTE**

Source: Dubinsky (1980).

Ces processus ont perduré de nombreuses années dans la formation en vente et sur le terrain. Puis, on a commencé à les nuancer en expliquant que le ou la responsable des achats pouvait entrer et sortir de la démarche à différents moments. La numérisation des relations, des transactions, et des moyens d'information a profondément transformé ces processus, de sorte que l'on parle maintenant du parcours client. Les dernières recherches en développement des affaires insistent aussi sur l'importance de planifier soigneusement chacune des interactions avec la clientèle. De plus, les réponses aux objections étaient autrefois considérées comme une étape distincte jusqu'à ce que l'on constate que, dans les faits, la personne en vente doit réagir aux préoccupations tout au long du processus.

L'approche relationnelle en développement des affaires privilégie le renforcement stratégique des relations plutôt que leur gestion tactique. Ce nouveau processus combine des étapes tactiques de la vente, exposées dans la figure 1.4, et des phases de l'évolution de la relation décrites par Dwyer et coll. (1987), soit la découverte, l'exploration, l'expansion, l'engagement et la dissolution. Le processus présenté à la figure 1.5 est non linéaire. Il fusionne les étapes d'exploration et d'expansion, exclut la dissolution de la relation - jugée incompatible avec le développement des affaires - et ajoute la planification ainsi que la réponse aux préoccupations comme des éléments périphériques et permanents. Au cours de la phase d'exploration, l'une ou l'autre des parties pourrait décider de revenir à l'étape de la découverte. Le processus n'est pas délimité dans le temps puisque, par exemple, la découverte peut s'étendre sur plusieurs années en raison de changements internes : rotation des effectifs, acquisitions d'entreprises, lancement de nouveaux produits ou services. La figure 1.5 illustre ces nouvelles étapes et ce nouveau processus, qui constitueront la structure des chapitres suivants.

FIGURE 1.5 / **Nouveau processus de vente et d'achat**

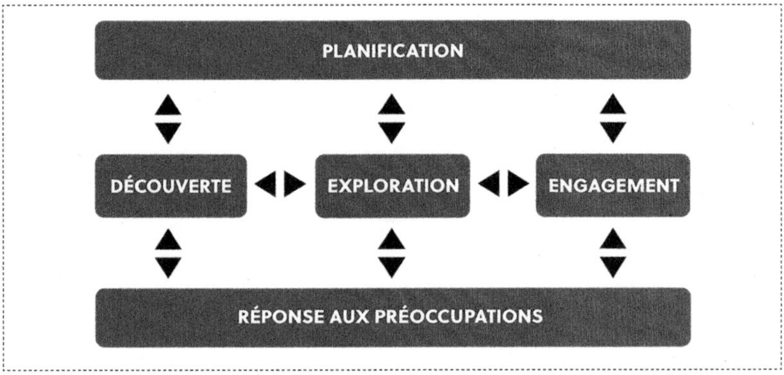

Les flèches bidirectionnelles horizontales expriment la possibilité que la relation évolue dans un sens comme dans l'autre. Même lorsqu'une entreprise s'engage avec un prestataire, elle peut décider de revenir à la phase d'exploration, par exemple si des éléments essentiels à l'engagement demeurent en suspens. Les flèches bidirectionnelles verticales symbolisent la circulation de l'information dans les deux directions, ce qui permet à la planification et à la réponse aux objections de se nourrir des interactions entre les parties ou de les influencer.

Le chapitre 2 portera sur la planification et la réponse aux préoccupations. S'il semble évident que la planification doit être réalisée en amont du processus, vous serez peut-être surpris d'y trouver également la réponse aux préoccupations ou aux objections. Encore une fois, il importe de distinguer l'approche stratégique de l'approche tactique. Dans les faits, la réponse aux objections a lieu durant les échanges avec la clientèle, mais l'équipe de vente connaît généralement à l'avance la plupart des objections qui seront soulevées, notamment la sempiternelle: «Vous êtes trop cher». C'est pourquoi ce manuel insiste sur l'importance de préparer les réponses aux objections en amont. Si la personne en vente n'est pas en mesure de défendre son prix plus élevé que la concurrence, elle ne devrait pas se présenter devant un responsable des achats à l'étape de la découverte. La planification et la réponse aux préoccupations interviendront également dans les deux étapes suivantes, celles de l'exploration et de l'engagement, cette fois non pas pour traiter de leurs aspects stratégiques, mais plutôt de leurs aspects tactiques.

Le chapitre 3 traitera de la découverte. À cette étape, les membres de la centrale d'achats chercheront à approfondir leur compréhension du produit ou du service, de l'industrie, de la concurrence, de la situation financière des éventuels fournisseurs et du macro-environnement. Ils souhaiteront en savoir plus sur les différentes personnes représentant les entreprises. De la même façon, les équipes de vente viseront en premier lieu à identifier les firmes à prospecter ainsi qu'à mieux connaître leur industrie, leurs fournisseurs et leur clientèle. Les membres de l'équipe de vente tenteront de rencontrer le plus grand nombre possible de personnes au sein de ces firmes afin de développer des relations.

À l'étape de l'exploration, présentée dans le chapitre 4, la personne en vente voudra d'abord se faire accepter par la clientèle, et surtout, faire entériner sa proposition. Cela se traduira par la mise en valeur des bénéfices de son offre par rapport à celle de sa concurrence. De son côté, la personne responsable des achats cherchera à déterminer si le fournisseur sera un partenaire fiable et s'il pourra répondre aux besoins actuels et futurs de son entreprise.

Enfin, le chapitre 5 traitera de l'engagement mutuel des deux parties. Entreprendre une négociation constitue déjà la première étape d'un rapprochement qui aboutira à la signature du contrat.

CHAPITRE 2 /

Planifier le développement des affaires

La planification est une composante fondamentale du succès en développement des affaires. Elle nécessite d'appréhender son environnement, de comprendre plus précisément sa clientèle, et enfin, d'organiser ses activités afin d'atteindre les objectifs stratégiques de l'entreprise.

1 / Comprendre son environnement d'affaires

Le développement des affaires implique une évaluation honnête des compétences et des ressources de son entreprise, une recension des opportunités et des menaces, ainsi qu'une appréciation de sa position par rapport à celle de sa concurrence. Il demande des efforts humains et financiers conséquents et, sans un soutien solide de toute l'organisation, la stratégie pourrait échouer.

1.1 / Se connaître d'abord

La première étape pour comprendre son environnement est de déceler les forces et les faiblesses de son organisation et de ses produits et services. Qu'il s'agisse d'un commerce de détail, d'une usine de pièces automobiles ou d'une firme comptable, vous possédez déjà des forces qui peuvent vous différencier de votre concurrence. Si ce n'est pas le cas, préparez-vous au pire. Soyez honnête et identifiez également les faiblesses susceptibles de freiner votre progression en développement des affaires.

Voici quelques questions à se poser avant d'entamer une telle démarche :

- L'entreprise dispose-t-elle des ressources matérielles, humaines et financières pour développer de nouveaux marchés ou pour concevoir de nouveaux produits et services ?
- L'équipe de vente peut-elle compter sur le soutien des autres départements de l'entreprise ?
- Les produits et services proposés suscitent-ils un réel engouement sur le marché ?
- Le niveau d'endettement permet-il une expansion et le recours à de nouveaux capitaux ?
- Les systèmes en place permettent-ils d'accumuler et d'analyser une grande quantité de données sur la clientèle ?

Le développement des affaires génère souvent des coûts sous-estimés. Une personne en vente peut engendrer des frais de rémunération dépassant les 100 000 dollars, ainsi que des dépenses annuelles de représentation atteignant 50 000 dollars, notamment si le poste implique de fréquents déplacements en avion.

1.2 / L'identification des opportunités et des menaces

Pour identifier ses opportunités et ses menaces, il est essentiel de comprendre deux types d'environnements : le micro-environnement et le macro-environnement.

Le micro-environnement regroupe toutes les parties prenantes qui gravitent autour de la compagnie comme les fournisseurs, la clientèle, la concurrence, les partenaires financiers externes, et même le public en général. Par exemple, un fournisseur pourrait avoir développé un nouveau produit s'intégrant parfaitement dans votre production, vous procurant ainsi un avantage concurrentiel. Aujourd'hui, les entreprises ne se font plus concurrence individuellement : ce sont des chaines logistiques toutes entières qui s'affrontent. Investir dans le développement des affaires est inefficace si le réseau de fournisseurs en amont ne peut répondre aux récentes exigences de la clientèle. Dans plusieurs industries, les fournisseurs présentent leurs produits et leurs services à la force de vente de l'entreprise permettant ainsi de former l'équipe commerciale pour qu'elle assimile les spécifications des différents éléments du produit. La compréhension des produits pour la force de vente ne se limite donc pas à la connaissance de ceux de sa propre entreprise, mais aussi à la connaissance des composantes principales qui émanent d'un réseau de fournisseurs. Prenons l'exemple d'une équipe de vente spécialisée dans les fermes de toit, une structure généralement en forme de A qui supporte la toiture d'une maison. Idéalement formée à l'usine qui fabrique ces produits, cette équipe devrait également maitriser les caractéristiques des différents matériaux achetés auprès des fournisseurs, comme le bois ou les ferrures, afin de mieux expliquer les spécificités du produit.

Les partenaires financiers externes peuvent être sollicités pour obtenir des capitaux additionnels en développement des affaires. Quant au public, comme le voisinage immédiat des installations manufacturières, il peut exercer une influence sur l'avenir de l'organisation en particulier si des préoccupations émergent quant à sa réputation en matière d'enjeux environnementaux ou sociaux. Enfin, les deux derniers éléments du micro-environnement, la concurrence et la clientèle, seront abordés plus en détail dans les deux prochaines sections.

Le macro-environnement regroupe les facteurs qui peuvent agir sur une grande partie des entreprises qui évoluent dans le même marché. Vous connaissez peut-être déjà l'acronyme PESTEL qui facilite la mémorisation des environnements politique, économique, socioculturel, technologique, écologique et légal. Tous ces facteurs ne génèrent pas systématiquement des opportunités et des menaces, mais les personnes responsables du développement des ventes doivent les surveiller pour anticiper les turbulences qui pourraient influencer les conditions du marché. Concevez seulement que le gouvernement bannisse l'utilisation d'un des composants qui entrent dans la fabrication de vos produits en raison des risques environnementaux, ou qu'un pays étranger impose des tarifs douaniers sur l'importation de vos produits. Ces menaces entraveraient sérieusement votre développement des affaires. À l'inverse, une opportunité pourrait résider dans votre expertise avancée dans l'utilisation d'une technologie. Imaginez que vous maitrisiez davantage l'intelligence artificielle que votre concurrence, par exemple, pour l'exécution de certains services juridiques. Cet avantage concurrentiel présenterait une occasion pour conquérir de nouveaux marchés.

1.3 / L'analyse de la concurrence

Si les firmes qui sont vos clientes et celles qui pourraient le devenir ne peuvent faire la différence entre vous et votre concurrence, votre produit ou votre service représentera une commodité à leurs yeux et vous n'aurez donc pas besoin d'une force de vente. Vous devrez trouver le moyen de vendre au plus bas prix possible. Toutefois, attention à ne pas tomber dans le piège de la banalisation où la clientèle finit par considérer tous les produits d'un marché comme des commodités, réduisant ainsi la concurrence à une simple guerre de prix (D'Aveni, 2010). Même un produit ou un service en apparence banal peut se différencier de la concurrence. Il y a plusieurs années, j'ai discuté avec le président d'une compagnie de minerai de fer. Je lui ai demandé si le prix était établi sur les grands marchés internationaux en tant que commodité. Il m'a rapidement détrompé: son minerai permettait des applications spécifiques qui leur donnaient l'opportunité de le vendre avec une prime.

Cela dit, le tableau 2.1 présente des exemples d'attributs qui pourraient faire une différence aux yeux de la clientèle. Libre à vous de l'adapter selon vos besoins. Voici comment l'utiliser:

1. La clientèle attribue une note de 1 à 5 à chacun de ces attributs, 5 représentant l'importance la plus élevée selon ses besoins.
2. Ensuite, votre clientèle ou vos prospects évaluent votre prestation ainsi que celle de chacune des entreprises qui vous font concurrence sur ces mêmes attributs, toujours sur une échelle de 1 à 5.
3. En multipliant l'importance de chaque attribut par la performance de l'entreprise, vous obtenez un score permettant de comparer les entreprises entre elles.

Ce processus vous aide ainsi à établir votre positionnement par rapport à la concurrence.

TABLEAU 2.1 / **Analyse de la concurrence**

		Ma compagnie		Concurrent A		Concurrent B	
	Importance sur 5	Note sur 5	Note x Importance	Note sur 5	Note x Importance	Note sur 5	Note x Importance
Attribut du produit							
Emballage primaire attrayant	1	5	5	4	4	3	3
Solidité de l'emballage secondaire	2	4	8	5	10	2	4
Fraîcheur du produit	3	3	9	1	3	1	3
Marque	4	3	12	2	8	5	20
Attribut de la compagnie							
Réputation	5	4	20	3	15	4	20
Installations	4	2	8	4	16	3	12
Facilité de commande	3	3	9	5	15	2	6
Assistance technique	2	4	8	3	6	1	2
Attribut de la personne en vente							
Rapidité à répondre	1	5	5	2	2	2	2
Orientée client	2	3	6	1	2	3	6
Honnêteté	3	1	3	4	12	4	12
Suivi	4	2	8	5	20	5	20
Qualité de présentation	5	4	20	3	15	1	5
Capacité à conseiller	1	3	3	2	2	2	2
Total			**124**		**130**		**117**

Source : Adapté du modèle multi-attribut de Fishbein et Ajzen (1975).

Dans ce tableau, on remarque que, pour la clientèle, la qualité de la présentation de la personne en vente et la réputation de l'entreprise sont les deux attributs les plus significatifs, chacun obtenant une note de 5. Votre entreprise obtient des résultats plutôt satisfaisants sur ces aspects, avec une note de 4. Cependant, pour trois autres attributs auxquels la clientèle accorde une importance de 4, les scores de l'entreprise sont plus faibles : 3 pour la marque, 2 pour les installations et 2 pour le suivi. Il serait donc essentiel pour l'entreprise de faire plus d'efforts sur sa marque et sur ses installations et, pour la personne en vente, d'améliorer le suivi auprès de la clientèle afin de mieux se démarquer de la concurrence.

2/ Comprendre les besoins de la clientèle

La planification doit intégrer de la recherche formelle et informelle, la création de relations avec la clientèle, la compréhension de la prise de décision chez les membres de la centrale d'achats, l'identification des multiples points de contact entre l'entreprise et sa clientèle, la détermination des éléments de valeur pour les responsables des approvisionnements, la préparation de réponses aux objections et l'importance de la marque et de la proposition unique de vente.

2.1/ La recherche accessible à toutes les organisations

La personne en vente doit posséder une compréhension approfondie des bénéfices que son produit et son service peuvent procurer à la clientèle, en mesurant l'adéquation entre, d'une part, les caractéristiques et les spécifications techniques du produit, et d'autre part, les besoins de la clientèle. Dans le tableau 2.1, c'est la recherche marketing qui permet au fournisseur de connaître l'importance de chaque attribut aux yeux de la clientèle.

La recherche marketing n'est pas exclusivement réservée aux grandes entreprises. Elle demeure accessible aux organisations de toutes tailles. Dans le milieu académique, la recherche scientifique suppose des procédures formelles et éprouvées, souvent adoptées par les compagnies disposant de ressources plus importantes. Cependant, les plus petites entreprises peuvent choisir certains éléments de la recherche de marchés sans pour autant épuiser leur budget, en mettant à contribution leurs équipes de vente et de développement des affaires.

La lecture des commentaires de la clientèle sur les réseaux sociaux ne demande pas de compétences statistiques poussées. Il en va de même pour les entretiens non structurés menés avec la clientèle par des personnes en vente ou par la direction de développement des affaires. La création d'un sondage de satisfaction est aujourd'hui simple et accessible et ses résultats peuvent constituer une véritable mine d'or pour établir le positionnement de l'entreprise, pour ajouter ou retrancher des produits et des services, ou pour ajuster le service client. Le secret est d'approcher la recherche comme un exercice permanent de l'équipe de développement des affaires, afin de saisir l'évolution des besoins et des activités de la concurrence.

Le développement des affaires nécessite une connaissance approfondie de l'environnement interne et externe de sa clientèle. Cela suppose que la force de vente ou le service du marketing comprenne la concurrence à laquelle fait face cette clientèle, identifie ses fournisseurs et sa propre clientèle, et recueille un maximum d'informations sur son environnement technologique et légal ainsi que sur ses ressources humaines et matérielles. Les entreprises qui vendent directement aux consommatrices et aux consommateurs chercheront à obtenir des informations sur les quartiers où ils résident ou sur les emplois qu'ils occupent. Le géomarketing permet souvent de cerner avec précision les habitudes de consommation de la population en fonction de son lieu de vie. Pour les commerces de détail, ces informations s'avèrent essentielles.

2.2 / Créer des relations dans toute l'organisation

Comme les approvisionnements sont généralement gérés par un comité plus ou moins formel, souvent sous la responsabilité d'une centrale d'achats (qui sera abordée dans la prochaine section), il est primordial d'établir un maximum de relations significatives avec le personnel des entreprises clientes. Cinq éléments résument la nature de ces relations (Chisam et coll., 2022) et sont illustrés à la figure 2.1.

FIGURE 2.1 / **Les relations entre les organisations**

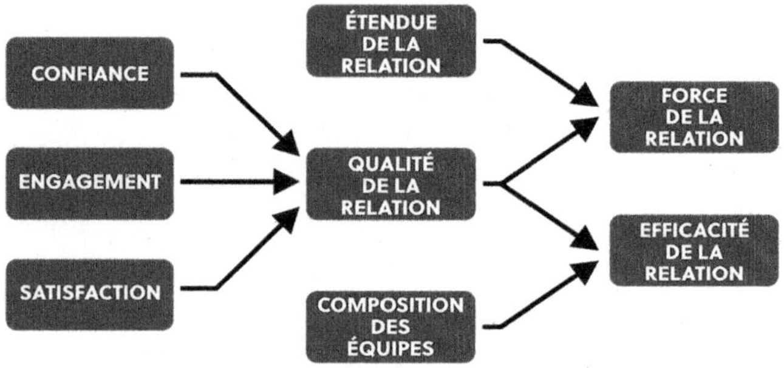

Source : Adapté de Chisam et coll. (2022).

Le premier élément est l'étendue de la relation, soit le nombre de personnes liées entre elles. Ainsi, si la ou le spécialiste des approvisionnements d'une entreprise quitte son poste, la personne en vente peut conserver le contact avec l'entreprise grâce à ses relations avec d'autres membres du personnel chez la firme cliente.

Le deuxième est la qualité de la relation, un concept multidimensionnel reposant sur la confiance entre les partenaires, l'engagement et la satisfaction à l'égard de la relation (Lasrado et coll., 2022). La confiance permet de réduire l'incertitude, de renforcer la coopération et facilite la résolution des conflits (Morgan et Hunt, 1994). Elle joue aujourd'hui un rôle accru, notamment en matière de sécurité des transactions numériques (Arli et coll., 2018). L'engagement correspond à la conviction que la relation est suffisamment importante pour que les deux parties s'y investissent pour la préserver (Morgan et Hunt, 1994). La satisfaction favorise le maintien de la relation : lorsque les interactions sont positives, les parties chercheront à les prolonger (Lasrado et coll., 2022).

Le troisième élément est la composition du groupe de personnes avec lequel le service des ventes interagit. Si ce groupe est diversifié, ses membres peuvent détenir un pouvoir décisionnel particulier et exercer une influence positive.

Le quatrième élément est la force de la relation, soit la capacité à surmonter les conflits potentiels. Elle résulte de l'interaction entre la qualité et l'étendue de la relation. Plus une personne de l'équipe de vente entretient des contacts à différents niveaux hiérarchiques au sein de l'entreprise cliente, et plus ces interactions sont fondées sur la confiance et l'engagement, plus la résolution des différends s'effectuera dans un climat de coopération.

Le cinquième élément, l'efficacité de la relation, résulte de l'interaction entre la composition de l'équipe et la qualité de la relation. La personne en vente se doit de maintenir des relations de confiance avec un groupe varié d'individus qui détiennent un certain niveau d'autorité pour prendre des décisions rapidement, afin que les échanges puissent permettre aux deux parties d'atteindre leurs objectifs.

2.3 / La centrale d'achats

Les approvisionnements ne relèvent pas uniquement de la personne désignée sous le nom d'acheteuse. Les décisions importantes sont généralement déléguées à plusieurs membres du personnel qui font partie d'une centrale d'achat de façon formelle ou informelle. Au sein de ce groupe, différents individus joueront des rôles spécifiques en fonction de leurs responsabilités. Par exemple, l'usagère ou l'usager utilise le produit ou le service, tandis que l'initiateur ou l'initiatrice est celui ou celle qui identifie un besoin au sein de l'organisation (Webster et Wind, 1972). Une personne en vente commettrait une erreur en concentrant tous ses efforts uniquement sur la personne responsable des approvisionnements. Le développement des relations implique donc de se lier avec des interlocuteurs et des interlocutrices à tous les niveaux hiérarchiques et dans divers départements. L'encadré 2.1 illustre les bénéfices qui peuvent découler de relations établies avec différents acteurs ou différentes actrices au sein de l'entreprise cliente.

ENCADRÉ 2.1 / **Pour que ça colle**

J'ai toujours aimé visiter les usines de ma clientèle et faire la connaissance des personnes qui utilisent nos produits. Je me souviendrai longtemps de cet entretien avec un opérateur sur une ligne d'embouteillage, qui m'a confié que j'étais le premier vendeur à m'intéresser à ses tâches et aux difficultés qu'il pouvait rencontrer. Son travail consistait à s'assurer que les étiquettes, pour lesquelles nous étions un des fournisseurs, adhéraient correctement aux bouteilles, tout en respectant des critères esthétiques.

Un jour, l'une de nos fabrications a enfreint les spécifications. L'acheteuse m'a contacté et, plutôt que de me faire des reproches, elle m'a simplement dit que l'opérateur voulait me parler. Ce dernier m'a montré comment notre erreur ralentissait sa production et m'a demandé de faire plus attention à l'avenir. Une bévue, qui aurait pu être coûteuse pour ma compagnie et dommageable pour la relation avec l'acheteuse, avait pu être atténuée par le développement d'interactions à tous les niveaux de l'organisation. Cet exemple illustre parfaitement l'importance de cultiver des relations à tous les échelons d'une entreprise cliente.

2.4 / Les points de contact et le parcours client

Les processus d'achat et de vente ne sont plus linéaires et les personnes en vente ne détiennent plus le monopole de l'information. La clientèle compte moins sur le membre de l'équipe de vente pour lui fournir les connaissances dont il a besoin puisque l'environnement numérique regorge de renseignements dont elle peut tirer profit avant de procéder à un appel d'offres ou même à une commande. La documentation des interactions entre la clientèle et le fournisseur, selon la perspective des points de contact, s'avère essentielle pour établir un parcours client et définir l'expérience client.

Les points de contact représentent les échanges directs et indirects entre la clientèle et un ou plusieurs fournisseurs. Ils englobent toutes les occasions où la clientèle pense au fournisseur ou effectue des actions relatives au fournisseur de façon consciente ou inconsciente, comme la visite de son site Web ou une discussion entre collègues à son sujet (Witell et coll., 2020). Ces points de contact peuvent se produire entre personnes occupant différents niveaux hiérarchiques ou fonctions, entre une personne et une machine, par exemple lorsqu'un employé de l'entreprise cliente consulte le site Web du fournisseur, ou encore entre machines, notamment dans le cadre de l'Internet des objets. Ils peuvent également survenir à différentes étapes du parcours client et impliquer des intermédiaires sur lesquels le fournisseur exerce peu d'influence, comme un commerce de détail. Avec l'essor du numérique, les lieux où se déroulent les points de contact se sont multipliés, qu'il s'agisse des répertoires d'entreprises, des influenceurs et influenceuses, des blogues, du clavardage ou des médias sociaux. L'équipe de vente doit donc être consciente de cette diversité de canaux et adapter ses stratégies en conséquence.

L'expérience client est la somme de toutes les interactions numériques, physiques ou sociales entre la clientèle et son fournisseur (Bolton et coll., 2018). Elle comprend des éléments cognitifs, sensoriels, affectifs, comportementaux et sociaux (Lemon et Verhoef, 2016). Les interactions numériques peuvent se produire pendant que la clientèle navigue sur le site Web de l'entreprise ou encore lors d'un échange avec un robot conversationnel. L'interaction physique intervient notamment lorsque la personne manipule un produit ou lorsqu'elle est immergée dans un environnement physique par le biais de la réalité virtuelle. L'interaction sociale concerne les échanges entre humains, qu'ils soient médiés ou non par la technologie, comme lorsque les deux parties participent à un tête-à-tête, en visioconférence. L'évolution de la technologie transforme donc l'expérience client et les personnes en vente doivent adapter leur travail en conséquence. L'équipe de vente doit profiter au maximum de ces interactions sociales pour optimiser l'expérience client.

Le parcours client, une notion qui émane du marketing numérique, demeure unique parce que chaque membre de la centrale d'achat vit une expérience subjective et individuelle avec tous les points de contacts, directs ou indirects, contrôlés ou non par le fournisseur (Purmonen et coll., 2023). Chaque point de contact contribuera à l'examen du fournisseur, que ce soit en phase de préachat, d'achat, d'usage ou de situation d'évaluation après l'achat (Halb et Seebacher, 2021). Les équipes de vente constatent aujourd'hui que le parcours de la clientèle avant l'achat se déroule souvent sans leur intervention. Le choix du fournisseur se fait parfois avant même qu'une personne en vente puisse intervenir, particulièrement lorsque le produit ou le service est simple et présente peu de risques, ou que les membres de la centrale d'achat sont très expérimentés.

2.5 / Les éléments de valeur pour la clientèle

Almquist et coll. (2018) ont identifié des éléments de valeur dans les échanges B2B, illustrés sous la forme d'une pyramide. À la base de la pyramide, la proposition doit répondre aux spécifications de la clientèle ainsi que respecter les lois et l'éthique à un prix acceptable. Il sera difficile, voire impossible de conclure une vente sans retrouver ces éléments de base. Au deuxième niveau, la centrale d'achat cherchera des éléments fonctionnels: l'offre permet-elle de réduire les coûts, d'augmenter les ventes ou d'améliorer le produit? Par le passé, et dans plusieurs industries, les fournisseurs se sont souvent limités à vendre ces éléments de valeur fonctionnelle. C'est à partir du troisième niveau que les entreprises peuvent véritablement se différencier de la concurrence. Ce palier de la pyramide inclut des éléments de valeur qui facilitent les échanges entre la centrale d'achat et le fournisseur, telles la réduction des efforts, une meilleure disponibilité, une gestion optimisée des risques, une flexibilité accrue, une intégration facilitée, plus de transparence et une expertise pointue. Au quatrième niveau, on introduit des valeurs personnelles, comme la réduction de l'anxiété ou l'extension du réseau de relations. Finalement, au sommet de la pyramide, les valeurs aspirationnelles permettent à la clientèle de développer sa responsabilité sociale, tandis que la vision du fournisseur peut l'aider à anticiper comment l'avenir transformera son modèle d'affaires.

2.6 / Les objections et les préoccupations

L'objection ou la préoccupation soulevée par la clientèle devient à la fois un obstacle à la conclusion d'une entente et une opportunité de clarifier votre offre, d'où l'importance de planifier vos réponses aux objections courantes. Vous devrez accorder une attention particulière aux préoccupations non verbalisées, car si une personne conserve ses doutes pour elle-même, vous n'aurez aucune occasion d'y répondre ni de l'accompagner dans son processus de décision. Une analyse approfondie vous permettra d'anticiper les objections qui pourraient survenir à n'importe quelle phase du processus d'achat. Mieux vous connaîtrez votre concurrence, plus vous pourrez mettre en avant les bénéfices distinctifs de votre offre pour répondre à une préoccupation. De la même façon, comprendre les défis auxquels fait face votre clientèle ou son industrie constitue un excellent moyen de se préparer aux questions susceptibles d'être soulevées lors de l'échange.

Parmi les étapes les plus critiques de la vente, la capacité à répondre aux préoccupations et aux objections s'avère essentielle. Vous ne pourrez pas conclure une transaction si le responsable des achats nourrit des doutes importants sur la décision qu'il s'apprête à prendre. Comme énoncé dans l'introduction, une personne en développement des affaires devrait avoir une assez bonne idée des préoccupations avant une rencontre et surtout des réponses à présenter lorsque des objections sont soulevées. Les préoccupations se regroupent en cinq grandes catégories (Ingram et coll., 2015).

La première catégorie est celle de la **reconnaissance d'un besoin ou d'un problème** par la clientèle. La phase de qualification, dont il sera question plus en détail au chapitre 3, doit vous permettre d'écarter les personnes ou les compagnies qui n'ont aucun besoin pour le type de produits que vous vendez. Cependant, la perception par rapport au besoin n'est pas toujours clairement définie. Vous pourriez savoir que le besoin existe dans des entreprises semblables, sans que la personne à qui vous vous adressez en ait pleinement conscience.

La deuxième catégorie de préoccupation concerne **votre offre**. Même si la personne ou l'entreprise reconnaît un problème à régler, elle n'est pas forcément convaincue que votre produit ou votre service constitue la meilleure solution. C'est une objection fréquente lorsque l'on propose une nouveauté sur le marché. La centrale d'achats n'aime pas prendre de risques : c'est pourquoi vous fournirez des preuves de l'efficacité de votre solution. La centrale hésite parfois à modifier les processus de l'entreprise ou les composantes de ses produits, soit par habitude, soit parce que cela entrainerait de nombreux autres changements. Promouvoir les bénéfices de votre offre, et surtout leur valeur monétaire en matière d'amélioration de la productivité ou de diminution des coûts, est une merveilleuse façon de faire face à une telle objection.

Remplacer un fournisseur établi représente un obstacle majeur et constitue la troisième catégorie de préoccupations, souvent exprimée ainsi : « Êtes-vous la bonne entreprise pour répondre à mon besoin ou pour résoudre mon problème ? » Cette objection s'avère d'autant plus difficile à surmonter dans le cas où l'entreprise entretient une excellente relation de longue date avec son fournisseur ou son représentant. Toutefois, si une ou un membre de la centrale d'achat accepte de vous rencontrer, cela signifie qu'il est prêt à explorer d'autres options. Grâce à l'analyse concurrentielle réalisée en amont, vous connaissez les forces et les faiblesses de votre offre ainsi que celles de votre concurrence. Vous pouvez alors poser la question suivante: «Quelles améliorations apporteriez-vous à la solution que vous employez à l'heure actuelle pour la rendre encore plus performante?» Si cette approche ne porte pas ses fruits, envisagez de solliciter un allié au sein de l'entreprise qui pourrait soutenir votre proposition. C'est pourquoi il est essentiel de développer un réseau de relations internes à différents niveaux hiérarchiques et au sein de plusieurs départements. Attention cependant: quelques opportunistes pourraient simplement utiliser les informations obtenues pour décrocher de meilleures conditions avec leur fournisseur actuel. Si, après plusieurs présentations, vous constatez que l'entreprise ne manifeste aucun intérêt réel, mieux vaut ne plus la solliciter afin d'éviter de gaspiller temps et énergie. Restez toutefois attentif aux changements susceptibles de créer des opportunités comme des difficultés logistiques ou financières rencontrées par un fournisseur actuel, ou encore une restructuration au sein de l'entreprise ciblée.

La quatrième catégorie est celle du **temps**. Par exemple, la personne aimerait du temps pour y penser ou encore elle voudrait reporter la décision à l'année prochaine. Deux cas de figure expliquent généralement cette position. D'abord, vos arguments ne l'ont pas convaincue que les bénéfices justifient l'investissement, que votre offre est la meilleure sur le marché ou que le changement en vaut réellement la peine. Si vous sentez que l'une de ces raisons est à l'origine de son hésitation, approfondissez la discussion afin de mieux comprendre et d'y répondre en exposant une fois de plus les bénéfices de votre offre. Ensuite, le processus organisationnel peut ralentir la prise de décision s'il implique plusieurs niveaux d'approbations ou différents départements. Vous devrez alors persévérer en vous assurant que le processus est toujours en mouvement. Encore une fois, si vous avez développé votre réseau de contacts, des personnes au sein de cette entreprise pourront faire avancer le dossier.

Enfin, la cinquième catégorie d'objection, sans doute la plus fréquente, est celle du **prix**. Si on trouve que votre prix est trop élevé, vous devez comprendre pourquoi. Vous pourriez avoir mal exposé les bénéfices par rapport à ceux de la concurrence, ou votre solution pourrait excéder les besoins réels de la clientèle. Dans certains cas, baissez vos prix ne garantira pas l'obtention de la commande. Voilà pourquoi il est crucial de clarifier ce que signifie réellement «C'est trop cher».

TABLEAU 2.2 / **Catégories d'objections et stratégies possibles pour y faire face**

Catégorie d'objections ou de préoccupations	Stratégie possible
Je n'en ai pas besoin ou je n'ai pas de problèmes	• Qualifier les prospects • Faire prendre conscience du besoin ou du problème
Ce n'est pas la bonne solution	• Promouvoir les bénéfices comme les économies potentielles de temps ou d'argent
Ce n'est pas la bonne entreprise	• Identifier les faiblesses du fournisseur actuel • Trouver un allié ou un promoteur interne
Ce n'est pas le bon moment	• Promouvoir les bénéfices comme les économies potentielles de temps ou d'argent • Persévérer
C'est trop cher	• Comprendre les raisons sous-jacentes • Présenter de nouveau les bénéfices de votre offre • Si possible, ajustez votre offre aux besoins exprimés

Source: Adapté de Ingram et coll. (2015).

Pourquoi ne pas réunir quelques membres de l'équipe pour dresser une liste des objections et des préoccupations les plus fréquentes lors de vos interactions avec votre clientèle? Ensuite, testez des explications possibles et choisissez celle qui vous apparaît la plus convaincante. Nous reviendrons plus loin sur les réponses aux objections d'un point de vue plus tactique.

2.7 / La marque et la proposition unique de vente

Chaque année, Interbrand publie son classement des meilleures entreprises en fonction de leur équité de marque (Interbrand, 2025). Parmi les 20 premières, on en retrouve 5 qui ne vendent exclusivement qu'aux entreprises comme Cisco, Adobe, Oracle, IBM et SAP. D'autres entreprises en tête de ce classement fournissent un service aux consommateurs et consommatrices, mais tirent la majeure partie de leurs revenus de la vente aux entreprises, comme Google ou Instagram. Pourtant, de nombreuses entreprises de *Business-to-Business* (B2B) accordent peu d'attention à leur marque, estimant que son développement représente un processus coûteux, et pertinent uniquement pour le marché grand public. Il est souhaitable que votre marque, souvent associée au nom de votre société, bénéficie d'une certaine notoriété, puisque vendre un produit ou un service d'une entreprise inconnue est un défi considérable. Même si votre marque profite d'une notoriété, vous voudrez qu'on lui associe certaines qualités plutôt que de laisser le marché les définir. Quand on pense à une entreprise, on se fait une idée de la qualité de ses produits, de leur rendement, de leur performance et de leur prix. Cette image de marque se crée au cours des nombreuses interactions entre la clientèle et son fournisseur, qu'il s'agisse de visites sur le site Web ou de rencontres avec une personne en vente. L'image de marque peut aussi se construire lors d'échanges entre la clientèle et votre concurrence, d'où l'importance de véhiculer un message cohérent et puissant.

La marque permet de vous différencier de votre concurrence, de développer une fidélité envers vos produits, d'augmenter vos ventes et souvent d'obtenir un prix plus élevé (Kotler et Pfoertsch, 2006). Vous pouvez décider de ne rien faire et laisser les autres définir l'image de votre entreprise, ou alors, travailler activement à façonner celle que vous souhaitez mettre en avant. Pour ce faire, il n'est pas nécessaire de dépenser des millions de dollars en publicité au Super Bowl. Pour une plus petite entreprise, une proposition de valeur distinctive peut suffire à créer un positionnement fort.

La proposition unique de vente, développée dans le monde de la publicité au début des années 1960, est encore fort utile aujourd'hui pour renforcer l'image de marque. Elle consiste à formuler, en une phrase concise, les bénéfices offerts et la manière dont votre entreprise se distingue de la concurrence pour répondre aux besoins de sa clientèle. L'encadré 2.2 illustre un exemple de proposition unique de vente.

ENCADRÉ 2.2 / **La construction d'une proposition unique de vente**

L'entreprise pour laquelle je travaillais tentait de bâtir sa notoriété dans le marché new-yorkais des emballages de cosmétiques et de pharmaceutiques. De nombreux fournisseurs américains servaient déjà ces industries, et nous devions trouver une façon efficace de nous démarquer, malgré notre statut de petit joueur situé hors des États-Unis. J'ai donc rencontré certaines personnes du nord-est américain pour comprendre pourquoi elles faisaient affaire avec nous. J'ai ainsi appris deux choses qui m'ont étonné. D'une part, notre taille plus modeste nous permettait d'être plus flexible face aux demandes spécifiques. D'autre part, bien que nos délais de livraison fussent parfois légèrement plus longs que ceux de notre concurrence, nous respections systématiquement nos échéances, contrairement à d'autres fournisseurs. À partir de ce constat, j'ai développé le message : « Flexible et fiable ». Pour appuyer ce positionnement, j'ai acheté les droits d'utilisation d'une photo d'une contorsionniste qui occupait presque tout le panneau de notre kiosque pendant un salon professionnel à New York. Les visiteurs et visiteuses nous demandaient le lien avec notre secteur d'activité, ce qui nous permettait d'expliquer que cette image symbolisait notre flexibilité et notre fiabilité. Ce message a rapidement été adopté par toute l'équipe de vente, qui l'intégrait systématiquement lors de ses rencontres avec la clientèle. Ce positionnement, qui a considérablement augmenté notre notoriété, établissait que nous fournissions des bénéfices différents de ceux de notre concurrence.

3 / Organiser ses efforts : structure, technologies et facteurs humains

Cette section ne prétend pas se substituer à un manuel spécialisé de gestion des ventes. Elle vise à donner un aperçu des principes fondamentaux de l'organisation des efforts en développement des affaires, qui inclut la structuration de l'équipe, l'implantation de technologies de vente et la prise en compte des facteurs humains.

Il se peut qu'en tant que personne entrepreneure, vous soyez seule à développer vos affaires. Toutefois, vous pourriez devoir organiser et superviser une force de vente plus ou moins importante, ce qui sera abordé dans les prochains paragraphes.

3.1 / La structure

Deux décisions importantes apparaissent rapidement en développement des affaires. La première consiste à choisir entre créer sa propre force de vente ou externaliser cette fonction. L'externalisation présente certains avantages, notamment en permettant à l'entreprise de se concentrer sur son exploitation et de bénéficier de l'expérience d'une firme spécialisée dans certains marchés. Cependant, cette option entraine une perte de contrôle sur les ventes et limite l'établissement de relations directes et privilégiées avec la clientèle, contrairement à une équipe interne dédiée au développement des affaires. La deuxième décision est de choisir entre une structure généraliste, dans laquelle chaque membre de l'équipe vend tous les produits de l'entreprise à toute la clientèle, ou une structure spécialisée, dans laquelle l'équipe est répartie selon les catégories de produits ou les segments de clientèle. Les firmes qui utilisent la structure généraliste offrent généralement une gamme de produits plus restreinte et moins complexe, tout en s'adressant à une clientèle limitée ou relativement homogène. À mesure que l'offre s'élargit ou que les produits requièrent des connaissances techniques plus spécialisées, la structure évolue vers une spécialisation par gamme de produits. La spécialisation par types de clientèle devient pertinente lorsque les segments affichent des besoins fortement différenciés. Il est même possible d'envisager une structure dans laquelle une ou un membre de l'équipe est dédié à une seule gamme de produits pour un segment précis de clientèle. Si quelques firmes représentent une grande partie de vos revenus, vous pourriez constituer une équipe de vente distincte pour servir cette clientèle clé. Ce point sera approfondi à la section 6 du chapitre 5.

Tôt ou tard, le marché québécois, voire canadien, pourrait vous sembler trop restreint pour assurer votre croissance. Dans ce cas, il y a fort à parier que vous envisagerez de vous tourner vers les États-Unis puisque, bon an mal an, notre voisin américain demeure la principale destination des exportations québécoises. Cependant, il est important de se prémunir contre les risques politiques inhérents à chaque marché d'exportation en équilibrant les revenus provenant de chacun. Si vous avez connu du succès jusqu'ici avec le développement de vos affaires, la vente à l'international représente un tout nouveau défi. La réglementation, le taux de change, les pratiques commerciales et la culture sont autant de facteurs pouvant déterminer la réussite ou l'échec de ce développement à l'étranger. Vous devrez choisir entre une force de vente qui se déplace entre le pays exportateur et le pays importateur, ou une équipe de vente établie directement sur place. Le niveau d'adaptation du produit, les communications destinées au marché cible et les circuits de distribution seront des éléments déterminants dans ce choix.

Une fois la structure en place, il est important de délimiter des territoires en fonction de la charge de travail. Cela implique d'évaluer le temps consacré aux tâches administratives, le nombre de clientèles potentielles, les déplacements entre celles-ci et le temps passé avec chaque personne sur une base hebdomadaire ou mensuelle. Cette analyse permet d'estimer le nombre de représentants et de représentantes nécessaires pour développer un marché donné. Par exemple, si les membres de votre équipe de vente travaillent 160 heures par mois, dont 40 heures dédiées aux déplacements et 40 heures aux tâches administratives, il leur restera 80 heures par mois à consacrer à la clientèle. Si vous identifiez 240 entreprises québécoises susceptibles d'être intéressées par votre produit ou par votre service et que vous souhaitez que chaque représentant ou représentante passe une heure par mois avec chacune d'elles, vous devrez alors compter sur une équipe de trois personnes (240 entreprises ÷ 80 heures disponibles par représentant).

Finalement, vous devrez définir des objectifs individuels et territoriaux. L'acronyme SMART (spécifique, mesurable, ambitieux, réaliste et temporel) s'avère très pratique pour se rappeler qu'augmenter les ventes n'est pas un objectif complet. Un objectif SMART permet, par exemple, de préciser qu'il s'agit d'une augmentation de 5% en Ontario d'ici 2 ans. Quant à savoir si cet objectif est à la fois ambitieux et réaliste, il reviendra à vous et à votre équipe d'en juger. Ces cibles pourront ensuite être converties en quotas individuels ou d'équipe. Les quotas représentent des objectifs auxquels on rattache souvent une forme de rémunération comme des bonis.

3.2 / Les technologies de développement des affaires

Les technologies numériques offrent une multitude d'opportunités pour le développement des affaires. Elles deviennent un avantage stratégique lorsqu'elles permettent des gains de productivité ou une meilleure adéquation entre les besoins de la clientèle et les services offerts par l'entreprise. Cependant, leur intégration représente des défis, notamment en ce qui concerne l'acquisition des compétences numériques au sein des équipes de développement des affaires (Fischer et coll., 2022), la difficulté d'équilibrer la vie personnelle et professionnelle ainsi que la création d'attentes non comblées par l'adoption d'une technologie.

Les technologies de vente regroupent l'ensemble des outils permettant aux personnes en vente d'atteindre leurs objectifs. Elles incluent notamment le Web, les logiciels de gestion de la relation client (*Customer Relationship Management* [CRM]), les médias sociaux, et l'intelligence artificielle. L'adoption de ces technologies dépend de facteurs situationnels et personnels (Agnihotri, 2021). Parmi les facteurs personnels, on retrouve principalement la facilité d'utilisation perçue et l'utilité perçue (Davis et coll., 1989). Les facteurs situationnels regroupent, entre autres, le soutien organisationnel, la formation et l'influence des pairs.

Vous devriez adopter deux stratégies complémentaires avec le Web : l'une axée sur la collecte active de renseignements et l'autre sur le marketing. D'abord, les moteurs de recherche octroient l'accès à une profusion d'informations utiles à la prospection, telles que des nouvelles sur des entreprises, des rapports financiers et des publications de médias sociaux. L'utilisation des données recueillies sur le Web peut aussi contribuer à affiner la stratégie de sollicitation de la clientèle. Ensuite, votre département de marketing doit s'assurer que le site Web de l'entreprise comporte non seulement les renseignements les plus récents, mais qu'il permet la meilleure expérience utilisateur possible. Par exemple, dans une perspective de marketing entrant, votre site devrait offrir une fonctionnalité permettant d'accéder à du contenu non commercial apportant une valeur ajoutée à votre clientèle actuelle potentielle. Malheureusement, les entreprises québécoises ne comprennent pas toujours l'importance de leur site

Web. Dans un sondage réalisé en 2023 auprès de 177 compagnies québécoises, on apprenait que le site Web de 60 % d'entre elles datait de plus de 3 ans (Fino et Bang Marketing, 2024).

Les **CRM** comptent parmi les plateformes technologiques les plus utilisées en développement des affaires. Très tôt dans l'histoire de la vente, il a fallu structurer les registres de la clientèle de façon à pouvoir la joindre facilement. Si la fonction comptable de l'entreprise avait déjà constitué ses propres répertoires pour gérer la facturation et la réception des paiements, chaque personne en vente s'organisait elle-même pour administrer sa liste de contacts. L'un des premiers systèmes de gestion est apparu dans les années 1950 sous la forme du Rolodex, qui répertoriait la clientèle par ordre alphabétique sur de petites fiches en carton ou permettait de classer les cartes de visite.

Le classeur Excel est devenu une des premières applications numériques utilisées à cette fin, bien qu'il n'ait pas été conçu pour cet usage. Il permet plusieurs manipulations, par exemple le classement de la clientèle par ordre alphabétique, selon l'importance de ses achats ou en fonction de son emplacement. Plusieurs compagnies utilisent encore Excel comme CRM, principalement parce qu'il ne génère pas de coûts additionnels pour les entreprises qui possèdent la suite Office. Cependant, cette méthode demeure limitée, car elle ne permet pas une collaboration aussi efficace que les véritables CRM.

Les CRM ont émergé avec le développement des bases de données marketing dans les années 1980. Par la suite, les logiciels de planification des ressources ont intégré un module CRM, puisqu'ils contenaient déjà une grande partie des informations nécessaires pour la force de vente.

Aujourd'hui, les CRM vont bien au-delà de simples fiches clients. Ils centralisent des données provenant des échanges de courriels, du Web et des médias sociaux (d'où l'appellation de CRM social, qui sera abordé plus loin). Plusieurs d'entre eux intègrent désormais l'intelligence artificielle afin de formuler des recommandations à la force de vente sur les meilleures firmes à prospecter ou encore pour détecter les firmes clientes les plus susceptibles de se désengager. Ces outils permettent d'étudier une grande quantité de données et de fournir des indications sur les comportements passés des entreprises clientes, afin d'anticiper les ventes.

Les systèmes de gestion de la relation client maximisent l'utilisation des ressources et améliorent la satisfaction de la clientèle en offrant un service personnalisé. Le CRM est l'un des outils privilégiés de la force de vente, du service à la clientèle et du marketing, puisque les informations peuvent être partagées entre les différents collaborateurs et collaboratrices en temps réel. Ce partage est d'autant plus essentiel que, pour servir les grandes entreprises clientes, plusieurs membres de l'équipe du fournisseur doivent établir des relations avec leurs homologues au sein de la firme cliente. Par exemple, une personne responsable des comptes débiteurs du fournisseur pourrait contacter la responsable des comptes créditeurs de l'entreprise cliente. De même, un membre du service des approvisionnements pourrait vouloir s'adresser directement au service à la clientèle ou au département de l'expédition pour obtenir des renseignements sur une livraison, sans passer par la personne en vente.

On parle aujourd'hui de CRM social (SCRM), car les logiciels intègrent des applications d'«écoute» des conversations sur les **médias sociaux** en vue de mieux comprendre les besoins de la clientèle. Les SCRM permettent aussi de promouvoir l'engagement de la clientèle pour les contenus générés par l'entreprise. En fait, une stratégie de contenus sur les médias sociaux sera incomplète si elle n'est pas combinée aux systèmes de CRM permettant d'exploiter ces données à des fins commerciales. Toutefois, même si les compagnies investissent des sommes importantes dans les SCRM, les résultats ne sont pas toujours au rendez-vous, en partie parce que leur déploiement est inadéquat (Perez-Vega et coll., 2022).

La démocratisation de l'**intelligence artificielle (IA)** ne peut être passée sous silence. L'IA se définit comme une technologie qui peut «prédire certains résultats, proposer des stratégies alternatives et procurer des recommandations» (Agnihotri et coll., 2023, p. 309). Son influence sur les membres de l'équipe de vente et leurs pratiques pourrait se manifester de diverses façons dans les prochaines années. D'abord, l'IA permet déjà d'obtenir une meilleure connaissance du marché et d'optimiser la conversion des firmes à prospecter en clientes (Paschen et coll., 2019). L'Internet des objets, qui permet aux équipements de la clientèle de communiquer directement avec leur fabricant, favorise l'acquisition de connaissances sur le comportement des usagers et usagères (Paschen et coll.,

2019). L'IA peut de la même façon aider les personnes en vente à en apprendre davantage sur la concurrence, entre autres en collectant des données sur les médias sociaux et en les classant sous forme d'informations pratiques pour la force de vente (Paschen et coll., 2019). L'IA peut aussi prédire certains comportements de la clientèle afin d'acheminer du matériel promotionnel personnalisé ou de créer du contenu comme des blogues ou des livres blancs (McClure et coll., 2024). Les équipes de vente pourraient se montrer réticentes à adopter certains systèmes basés sur l'IA. Il est donc essentiel de les former à ces nouveaux outils (Paschen et coll., 2020). Retenons que, si l'IA permet d'améliorer les efforts des membres des équipes de vente, elle ne les remplacera pas, particulièrement lorsque des compétences sociales ou émotionnelles sont requises (Paschen et coll., 2020).

Le développement de l'apprentissage machine et de l'IA permet d'optimiser les processus grâce à l'analyse prédictive des ventes. Cette approche consiste à extraire des perspectives inédites à partir de données existantes, par exemple en évaluant la probabilité de conversions ou le potentiel de ventes additionnelles (Habel et coll., 2023). Cependant, les cadres de vente et leurs équipes n'adoptent pas spontanément les recommandations formulées par l'IA pour améliorer leur performance en prospection ou en négociation. Les résultats doivent être pertinents pour démontrer le bien-fondé de l'analyse prédictive, voire prescriptive. De plus, certaines conditions peuvent maximiser les bénéfices de l'IA, comme un environnement caractérisé par une forte hétérogénéité de la concurrence et de la clientèle, par un marché sujet à des turbulences ou par l'autonomie limitée des personnes en vente dans leur processus décisionnel (Habel et coll., 2023).

Pourtant, Gartner (2023) prévoit que l'IA révolutionnera complètement les tâches et les compétences des équipes de vente d'ici quelques années, notamment en perfectionnant le contenu des messages de vente à valeur ajoutée et même en les automatisant. Le lectorat de ce manuel devra rester à l'affût de cette évolution, car les applications et les logiciels offriront de plus en plus d'options pour améliorer la performance à chaque étape du processus de vente. Malgré ses promesses, le contenu généré par l'IA, que ce soit pour la formation ou pour fournir une rétroaction aux membres de l'équipe de vente, pourrait recevoir un accueil tiède en raison d'un manque de contact humain (McClure et coll., 2024). Une dépendance excessive à l'égard de ces technologies pourrait également limiter l'interprétation de certains facteurs importants, en particulier dans l'analyse des données. En quelques années, l'IA est passée de l'analyse descriptive à l'analyse prédictive, puis, depuis peu, à une analyse prescriptive, qui détecte non seulement des tendances, mais émet aussi des recommandations d'interventions. Cette transition s'est opérée tant pour des données structurées, comme des revenus par firmes clientes, par région, ou par gamme de produits, que pour des données non structurées, comme les publications et commentaires rédigés par la clientèle sur les réseaux sociaux. Ces avancées contribuent aussi à une automatisation accélérée des tâches de l'équipe de vente, notamment avec les robots conversationnels et les assistants virtuels, ce qui, en principe, permet aux personnes en vente de se consacrer davantage à des activités à plus forte valeur ajoutée pour la clientèle. Adam et coll. (2023) ont comparé la réaction de la clientèle face à une interaction avec une agente ou un agent de vente humain et un agent de vente automatisé. Ils concluent qu'en B2B, la clientèle préfère s'intéresser à une offre proposée par un humain plutôt que par un robot. Paradoxalement, elle se montre moins encline à partager ses coordonnées avec un agent humain au-delà du premier contact par courriel, préférant souvent un échange numérique avant de planifier une rencontre. Bien que cette étude ait été menée dans un contexte spécifique qui limite sa généralisation, une automatisation trop rapide des tâches de vente pourrait nuire à la relation si l'expérience client n'est pas prise en compte.

Si les personnes en vente sont convaincues que ces technologies leur permettront d'améliorer leur efficacité, et si elles ont confiance en leurs propres compétences numériques, elles adopteront plus facilement ces outils et pourront les exploiter à leur plein potentiel (Mullins et Agnihotri, 2022).

3.3 / Les facteurs humains du développement des affaires

Pour réussir le développement des affaires, les responsables de l'équipe doivent embaucher les bonnes personnes, les rémunérer convenablement, veiller à leur bien-être physique et mental et s'assurer qu'elles adoptent un comportement éthique.

Pour **embaucher les bonnes personnes**, il faut établir un processus de sélection fiable et rigoureux afin de prédire la performance des individus. Comme les personnes en vente occupent souvent des postes bien rémunérés et entretiennent des relations privilégiées avec la clientèle, l'adéquation entre leur profil, la culture de l'entreprise et celle de la clientèle doit être soigneusement examinée. Ces évaluations passent par diverses méthodes, telles que les entrevues, les tests psychométriques et les simulations de situations de vente. Une fois que vous aurez identifié la bonne personne, vous lui soumettrez une offre écrite dans laquelle vous décrirez les tâches et préciserez les attentes de l'organisation. La candidate ou le candidat retenu devra signer une entente de confidentialité, afin de protéger les informations sensibles, ainsi qu'une entente de non-concurrence pour prévenir tout risque de départ avec la liste de la clientèle actuelle et potentielle.

La structure de **rémunération** représente un autre facteur clé de l'organisation de la force de vente. Cette structure n'est pas à prendre à la légère puisqu'elle peut avoir une incidence non négligeable sur la performance. Avant de choisir entre une rémunération fixe (souvent sous forme de salaire), une rémunération variable (généralement sous forme de commissions) ou une rémunération mixte (combinant un salaire de base et des commissions), vous devrez aligner votre choix avec les objectifs stratégiques de votre entreprise. Le type de rétribution variera selon que vous voudrez accroître les ventes, augmenter les profits ou hausser la satisfaction de la clientèle. Votre force de vente adoptera les comportements que vous récompenserez.

Vous devrez établir un plan de rémunération formel, simple et équitable, garantissant que les membres de l'équipe de vente soient rémunérés selon les mêmes critères. Une composante individuelle peut être envisagée, à condition qu'elle ne génère pas de sentiment d'injustice. Un programme de rémunération bien conçu peut contribuer à réduire le taux de roulement de la force de vente et à établir un budget prévisible pour les dépenses commerciales. Il doit s'inscrire dans la durée, car des modifications fréquentes risqueraient de susciter du mécontentement, quelques membres de l'équipe pouvant se sentir désavantagés par ces ajustements. Les changements indispensables devraient être annoncés suffisamment à l'avance et être discutés avec l'équipe de vente lors d'entretiens individuels ou d'équipe.

L'évolution rapide de la technologie, qui multiplie les rencontres virtuelles avec la clientèle et réduit potentiellement les déplacements, pourrait justifier une révision des modalités de rémunération.

Chaque mode de rémunération présente des avantages et des inconvénients, tant pour les membres de l'équipe de vente que pour l'entreprise. Pour une firme, une rémunération fixe permet de prévoir ses dépenses et de ne pas voir ses coûts augmenter en cas de hausse des ventes. Cependant, en période de baisse des ventes, la proportion de la rémunération par unité vendue augmente. La rémunération fixe permet aussi d'encourager les personnes à se concentrer sur des activités qui ne génèrent pas de revenu à court terme comme la formation ou le service à la clientèle. De plus, elle limite la tentation de privilégier les produits les plus faciles à vendre et incite plutôt à concentrer les efforts sur les produits plus rentables pour l'entreprise, même s'ils exigent davantage de travail. Enfin, une rémunération fixe favorise une meilleure couverture du territoire.

La rémunération fixe procure à la personne en vente une sécurité financière appréciable. Les recherches indiquent que les personnes salariées font preuve d'une plus grande loyauté envers leur employeur et se montrent généralement plus satisfaites de leur emploi (O'Donnell et Marsh, 2021). Les directions de vente estiment que les personnes rémunérées selon un revenu fixe travaillent mieux en équipe, se préoccupent plus de la clientèle et sont davantage motivées de manière intrinsèque. Ce mode de rémunération comporte aussi sa part d'inconvénients, notamment lorsqu'il garantit un revenu élevé à un personnel moins performant. La direction doit alors surveiller de près les activités des membres de l'équipe pour s'assurer que chacun et chacune déploie le maximum d'efforts. Cette forme de rémunération est appropriée lorsqu'il est difficile d'établir un lien entre les ventes et les efforts individuels. Par exemple, lorsque des dépenses très importantes en publicité encouragent l'achat de produits de grande consommation, la force de vente auprès des épiceries et des dépanneurs ne constitue alors qu'un des éléments contribuant à l'atteinte des objectifs.

La rémunération variable est avantageuse pour l'entreprise lorsqu'elle anticipe une baisse de ses ventes, car ses dépenses s'ajusteront à la diminution du chiffre d'affaires. En revanche, il peut être complexe de superviser les activités des membres de l'équipe qui voudront maximiser leurs revenus en augmentant rapidement leurs ventes. Si les produits ne requièrent pas de service après-vente, ou très peu, la rémunération variable peut s'avérer pertinente.

La rémunération mixte, qui combine les éléments fixes et variables, vise à atténuer les inconvénients de chacun des deux modes tout en optimisant leurs bénéfices. C'est d'ailleurs l'option privilégiée par la majorité des entreprises. Un bonus peut parfois faire partie de la rémunération mixte. Il consiste à verser un montant forfaitaire ou un pourcentage du revenu à l'atteinte de certaines cibles. Il peut être versé à différentes périodes, par exemple par trimestre ou par année. Toutefois, lorsque le versement du bonus est éloigné du résultat obtenu par une personne, il peut devenir contre-performant et perdre son effet incitatif. Si, après neuf mois, les personnes en vente constatent qu'elles ne pourront pas dépasser leurs objectifs annuels et qu'elles ne recevront aucun bonus, elles pourraient être tentées de ralentir certains processus d'achat afin de reporter des ventes à l'année suivante, pour maximiser leur prochaine bonification. Le bonus peut aussi récompenser la performance d'équipe.

Plusieurs facteurs organisationnels influencent la rémunération totale, notamment la taille de l'entreprise, sa réputation, le nombre de membres de l'équipe de vente, le type de produit, ou l'industrie dans laquelle elle évolue (Bowen et coll., 2023). Les caractéristiques de l'emploi affectent également la rémunération selon la fréquence ou le type de voyages, l'agilité requise ou le niveau d'instruction ou d'expérience nécessaire (Bowen et coll., 2023).

De nombreuses recherches ont étudié l'effet de la rémunération sur la performance organisationnelle et individuelle, ainsi que sur la motivation. La théorie de l'autodétermination, maintes fois démontrée, affirme que la motivation constitue l'un des facteurs les plus déterminants de la performance, la motivation intrinsèque ayant un effet encore plus marqué que la motivation extrinsèque. La motivation intrinsèque provient du plaisir que l'on éprouve à pratiquer une activité, tandis que la motivation extrinsèque est générée par des éléments comme une récompense. Cette théorie s'applique également au contexte professionnel en vente : si une personne dispose d'un certain contrôle sur ses tâches, bénéficie d'une autonomie, se sent compétente et perçoit un sentiment d'appartenance au sein de son équipe, sa motivation intrinsèque en sera renforcée

(Good et coll., 2022). Les récompenses, le salaire, les commissions, les bonis ou d'autres formes de rémunération en tant que motivations extrinsèques ont aussi un effet sur la performance, mais dans une moindre mesure. En fait, comme les personnes en vente ont des attentes spécifiques en matière de rémunération, décevoir ces attentes pourrait entrainer des conséquences notables sur la performance (Good et coll., 2022).

Les entreprises se préoccupent de plus en plus de la **santé physique et mentale** de leur équipe de développement des affaires. Une foule de facteurs contribuent à exposer les membres de l'équipe de vente à un stress personnel important qui peut éventuellement conduire à de l'insomnie, de l'irritabilité et même à de l'épuisement professionnel (Peasley et coll., 2020). D'abord, les personnes en vente, comme d'autres catégories de personnel, sont soumises à des facteurs de stress dans leur vie personnelle, qu'il s'agisse de problèmes de santé pour elles-mêmes ou pour des membres de leur famille, ou encore de difficultés relationnelles avec leur partenaire. Ensuite, des facteurs de stress professionnels propres au travail de représentation s'ajoutent. Par exemple, certaines personnes dans les firmes clientes s'attendent à pouvoir rejoindre leur représentant en tout temps, puisque les outils technologiques le permettent. Cet effacement de la frontière entre la vie professionnelle et la vie personnelle contribue à accroître le stress. De tout temps, les personnes en vente doivent concilier les attentes de leur clientèle et celles de leur employeur, ce qui les place parfois au centre d'intérêts divergents. Enfin, une partie, voire la totalité, de leur rémunération dépend souvent de leurs ventes. Les études ont démontré qu'une augmentation de la composante de rémunération variable sur le revenu total multiplie le stress et le nombre de journées de congé de maladie, ce qui entraine une réduction des commissions sur les ventes et une diminution de la performance (Habel et coll., 2021).

Des facteurs ponctuels peuvent aussi mettre à l'épreuve la gestion des émotions chez les personnes en vente. Lorsqu'un client insatisfait menace de rompre la relation d'affaires, la personne en vente peut être en colère, furieuse, insultée, irritée, dégoûtée, triste, ou inquiète (Bourguignon et coll., 2020). L'intelligence émotionnelle qui repose sur la reconnaissance et le contrôle des émotions peut aider à faire face à ces situations stressantes afin de préserver la relation, tout en maintenant un niveau de stress plus bas (Bourguignon et coll., 2023).

La personne en vente dispose de beaucoup d'autonomie, car sa performance est généralement évaluée sur des résultats très concrets comme le niveau de ses ventes. En conséquence, elle doit gérer son emploi du temps en fonction de ses objectifs et éviter les tâches chronophages qui ne contribuent pas directement à l'atteinte de ses buts. Cela fait appel à une organisation rigoureuse des activités et surtout à une constante remise en question de leur bien-fondé au regard des résultats escomptés. Les tâches administratives doivent être réduites à leur strict nécessaire et les déplacements doivent être soigneusement planifiés.

L'échec en vente, qui s'illustre par une incapacité à faire progresser le processus d'achat, est inhérent au métier et les personnes en vente l'admettent elles-mêmes (McGowan, 2020). Il peut se manifester à toutes les phases d'une carrière, ainsi qu'à toutes les étapes du processus de vente comme la prospection, la présentation ou la négociation. L'échec ne découle pas seulement d'un manque de compétences ou d'efforts; une mauvaise stratégie ou une piètre direction de l'équipe de vente peuvent aussi y contribuer (McGowan, 2020). Une anticipation des conséquences d'un échec peut également susciter une perte de confiance en soi ou provoquer des émotions semblables à celles ressenties lors d'un affrontement avec la clientèle comme mentionné précédemment.

Le dernier facteur humain dont il sera question est l'**éthique,** qui représente un aspect important du développement des affaires, ne serait-ce qu'en raison du risque réputationnel que les membres de votre équipe peuvent faire peser sur l'entreprise. Quatre éléments influencent les comportements éthiques des personnes en vente (Manning et coll., 2016). Le premier repose sur les normes formelles de la société et celles qui codifient les relations entre les compagnies, comme les lois, les règlements et les contrats. Ces normes ne peuvent cependant pas caractériser tous les cas d'espèce et c'est la présence d'autres facteurs qui permettront de créer un climat propice aux comportements éthiques. Le deuxième élément concerne les politiques de l'entreprise, qui peuvent, par exemple, encadrer l'usage d'Internet, le don de cadeaux à la clientèle, ou la divulgation d'informations confidentielles. Le troisième facteur est l'influence des membres de la direction et des pairs. Plus une personne sera entourée de collègues aux comportements éthiques, moins elle sera tentée de commettre des gestes contraires à l'éthique. Enfin, le quatrième élément repose sur les valeurs de chaque individu, soit ses croyances qui l'accompagnent tout au long de sa vie.

L'éthique varie selon les pays et les cultures, ce qui complique encore plus la gestion des comportements éthiques. La direction devrait encadrer de plus près les membres de son personnel qui vendent dans des régions du monde où les pratiques commerciales peuvent différer des usages domestiques. L'évolution de la technologie pose aussi davantage de risques éthiques, notamment en ce qui concerne la visioconférence ou l'IA (Hartmann et coll., 2023).

Les normes éthiques évoluent, et ce qui pouvait paraître raisonnable il y a 20 ou 30 ans ne l'est peut-être plus aujourd'hui. Par exemple, dans certaines industries, un acheteur ou une acheteuse pouvait accepter une paire de billets pour assister à un match des Canadiens de Montréal il y a quelques années. Mais ce cadeau pourrait désormais s'avérer non éthique, car perçu comme un pot-de-vin visant à obtenir des commandes futures, plutôt qu'un geste de remerciement pour des commandes passées. L'éthique suppose aussi l'adoption de pratiques de vente transparentes en dévoilant notamment des frais ultérieurs qui s'ajoutent au prix de base payé. Lorsque vous présentez votre offre, assurez-vous d'annoncer toutes les additions qui ne seraient pas incluses dans votre prix comme les frais de livraison ou d'installation.

3.4 / La gestion de la performance

La gestion de la performance de l'équipe en développement des affaires doit être alignée sur les objectifs stratégiques. Si l'entreprise vise à augmenter ses parts de marché de 3%, on mesurera la performance de l'équipe en fonction de l'atteinte de cet objectif. Les buts organisationnels doivent ensuite être déclinés sur le plan individuel pour que chaque personne apporte sa contribution. Pour encourager la force de vente à adopter les bons comportements, la firme devra ajuster ses récompenses individuelles en fonction de l'atteinte de ses objectifs stratégiques.

Autant le tableau de bord est essentiel pour un ou une pilote d'avion, autant son utilité est démontrée en développement des affaires. Il comprend des mesures rétrospectives et des mesures prospectives. Ainsi, en développement des affaires, on évaluera souvent la croissance des ventes, une mesure rétrospective. Cet indicateur fournit de l'information sur le passé, mais ne permet pas d'entrevoir l'avenir à court et moyen terme. Pour cela, il faudrait s'appuyer sur des mesures prospectives, comme le nombre de visites ou le nombre de soumissions. Ces deux mesures ne prédisent pas parfaitement une progression future des ventes, mais elles peuvent au moins annoncer que, si toutes choses demeurent égales par ailleurs, les ventes devraient être au rendez-vous dans les prochains mois.

En résumé

La planification du développement des affaires s'aligne sur la planification stratégique. L'analyse des forces et des faiblesses de l'offre et des capacités de l'entreprise, l'identification des opportunités et des menaces issues des changements dans le macro-environnement ou les actions de la concurrence, ainsi que la compréhension des besoins de la clientèle et de son processus de décision sont des éléments clés de la fondation du succès en développement des affaires. Enfin, une planification efficace repose sur une organisation stratégique de l'équipe de vente afin d'optimiser sa performance.

CHAPITRE 3 /

La découverte

En développement des affaires, vous recherchez des firmes qui pourraient devenir clientes, tout comme les spécialistes des approvisionnements sont en quête de nouveaux partenaires. En fait, toutes les entreprises demeurent à l'affût pour repérer des acteurs et actrices de la chaine d'approvisionnement susceptibles de favoriser la croissance. Les activités de découverte des équipes de vente incluent la prospection, le contact avec des entreprises, la qualification et la préparation à une première rencontre. Avant de se pencher sur la découverte du point de vue des ventes, explorons-la sous l'angle des achats à la recherche de fournisseurs, un sujet rarement abordé en développement des affaires.

1 / La recherche de fournisseurs

Les entreprises qui achètent ne se contentent pas d'attendre que les fournisseurs les contactent. Plusieurs d'entre elles mènent des veilles actives sur l'innovation afin de rester à la fine pointe des améliorations possibles en matière de produits, de services ou de processus. Pour en apprendre davantage à votre sujet, elles disposent de quatre types de demandes d'information. La première est l'invitation informelle. Vous recevez un courriel ou un appel d'une compagnie qui souhaite obtenir des prix pour votre produit ou votre service. Cette première prise de contact ne doit pas être négligée. Si votre offre repose sur autre chose que le prix, il est judicieux d'inciter votre interlocuteur à prendre quelques minutes pour discuter de ses besoins et lui présenter les bénéfices distinctifs de votre solution, que ce soit en personne ou par téléphone. Les trois autres types de demandes sont mieux connus sous leurs acronymes anglophones : RFI, pour *request for information* ou demande d'informations ; RFP, pour *request for proposal* ou demande de propositions et RFQ, pour *request for quote* ou demande de soumissions.

Le RFI n'est qu'un premier contact établi par la centrale d'achats pour identifier les fournisseurs susceptibles de combler ses besoins. À ce stade, on ne sollicitera pas de prix. Cette demande d'informations regroupe des questions générales comme les noms de la présidence et des membres de son comité exécutif, l'adresse du siège social, le chiffre d'affaires, les produits et les services offerts, ainsi que la liste des bureaux ou des usines avec leurs coordonnées respectives. Elle peut également inclure des questions sur l'expérience et l'expertise de l'entreprise en lien avec les besoins de la firme qui en fait la demande, quelques noms faisant partie de la clientèle, les procédures documentées de contrôle de la qualité et les modalités de paiement. Si vous avez des intentions notables en développement des affaires, vous devriez compiler ces renseignements dans une base de données afin de répondre plus rapidement aux RFI.

Vous pourriez être appelé à fournir des propositions si vous êtes sélectionné sur la base des informations détaillées dans le RFI. Le RFP, contrairement au RFI, peut mener directement à la conclusion d'un contrat. Il exige des compagnies sélectionnées qu'elles soumettent une proposition pour résoudre un problème ou combler un besoin. Il fournit juste assez d'indications pour permettre de formuler une offre, tout en laissant aux firmes candidates une marge de créativité. Par exemple, un RFP pourrait être utilisé pour la construction d'une usine, afin de bénéficier de l'expertise des soumissionnaires dans une industrie en particulier.

À l'inverse, la demande de soumissions (RFQ) limite les types de réponses possibles. Elle inclut des quantités et des spécifications techniques détaillées afin de comparer les fournisseurs sur une base équivalente.

Lorsqu'une entreprise recherche des fournisseurs, l'équipe de développement des affaires a trois missions principales. D'abord, il faut s'assurer que l'entreprise est prise en considération au moment opportun. Bien que cette tâche repose en grande partie sur les actions du département marketing, les membres de l'équipe de vente peuvent renforcer la notoriété de l'entreprise en pratiquant la vente sociale sur les réseaux sociaux, un concept que nous approfondirons plus loin dans ce chapitre.

Ensuite, la personne en vente responsable d'un compte ayant soumis une demande de RFI, de RFP ou de RFQ doit s'assurer que le processus se déroule rondement en établissant les liens nécessaires entre son entreprise et celle qui a formulé la demande. En effet, les documents contiennent parfois des questions ambiguës, mal rédigées ou portant sur des aspects confidentiels que l'entreprise préfère ne pas divulguer à ce stade, afin de protéger sa propriété intellectuelle ou des informations confidentielles sur sa profitabilité. La personne en vente doit aussi vérifier que les fichiers fournis soient à la hauteur de l'image que son organisation veut projeter. Ce point sera approfondi au chapitre de l'exploration.

Finalement, l'équipe de développement des affaires pourrait être appelée à négocier les termes d'une entente si son entreprise remporte la mise avec le RFP ou le RFQ. Nous discuterons de cet aspect au chapitre 5 à propos de l'engagement.

2/ Prospecter

Même si le processus d'achat n'est plus strictement linéaire, un fournisseur souhaitant développer ses affaires ne peut rester passif et attendre que la clientèle vienne à lui. La prospection représente une activité qui relève à la fois du marketing et des ventes. Le marketing entrant consiste à optimiser les moteurs de recherche, à créer du contenu pertinent et intéressant sur le site Web de la compagnie et à animer les réseaux sociaux de façon à collecter des données sur des firmes qui pourraient être intéressées par votre offre. De multiples plateformes de génération de listes de compagnies sont maintenant disponibles avec Sales Navigator de LinkedIn et Apollo, sans compter celles qui émergeront dans les années à venir. Bien que cette tâche, souvent confiée au marketing, soit d'une importance majeure, nous nous attarderons ici à la prospection individuelle, c'est-à-dire celle effectuée par les personnes en vente elles-mêmes.

La recherche de nouvelles entreprises à prospecter est un élément vital de la pérennité des firmes et constitue la fondation de la stratégie de développement des affaires. Pour illustrer la prospection, revenons à l'époque de la ruée vers l'or, alors que des prospecteurs et prospectrices passaient au tamis le lit des rivières afin de trouver des pépites d'or. Au fond du tamis de la personne en vente, on retrouve des entreprises «suspectes», c'est-à-dire des sociétés qui pourraient devenir clientes, mais qui n'ont pas encore été qualifiées. Une fois cette qualification effectuée, elles deviennent des entreprises à prospecter, car vous avez confirmé qu'elles pourraient utiliser votre produit, bien qu'elles ne le connaissent pas encore ou qu'elles achètent déjà un produit similaire auprès de la concurrence.

Comme tout prospecteur ou toute prospectrice d'or, vous devez maintenant identifier les meilleurs cours d'eau pour trouver des firmes à prospecter. Quatre grandes sources d'information sont disponibles: l'environnement numérique, les références, les activités de réseautage et la vente sociale.

2.1 / L'environnement numérique

Au-delà des réseaux sociaux, l'environnement numérique propose un accès à une multitude d'informations, qu'elles soient gratuites ou payantes. Plutôt que de chercher de façon aléatoire, demandez-vous avec quelles industries vous connaissez le plus de succès. Cette analyse vous permettra ensuite d'acheter des listes de compagnies qui représentent vos meilleures opportunités. Il est essentiel de comprendre le classement des industries pour tirer pleinement profit des répertoires. De nombreuses régions du monde, comme l'Amérique du Nord ou l'Europe, ont développé une arborescence permettant de catégoriser les entreprises de biens et de services en fonction de leur domaine d'activités. Au Canada, on utilise le Système de classification des industries de l'Amérique du Nord (SCIAN). Vous pouvez consulter ce système directement sur le site Web du gouvernement canadien[1]. Prenons un exemple concret: si votre entreprise fabrique des sacs de papier, dont une gamme est précisément conçue pour l'emballage d'aliments secs destinés aux chiens et aux chats, vous commencerez par identifier le code SCIAN correspondant aux

---

1 https://www23.statcan.gc.ca/imdb/p3VD_f.pl?Function=getVD&TVD=1181553

compagnies de fabrication de nourriture pour animaux de compagnie. Une fois ce code obtenu, vous pourrez l'utiliser pour rechercher et acquérir les coordonnées des entreprises pertinentes dans les répertoires spécialisés, optimisant ainsi votre prospection. Vous pouvez également identifier des firmes à prospecter en explorant les nombreuses publications d'industries désormais accessibles en ligne. Vous trouverez aussi des entreprises d'intérêt en consultant des parutions généralistes qui s'intéressent au domaine des affaires.

2.2 / Les références

Obtenir une référence d'un client ou d'une cliente est un peu comme un laissez-passer qui vous permet d'accéder plus rapidement à une personne ou à une firme qui vous intéresse. Cette approche consiste à demander l'autorisation à des contacts influents d'utiliser leur nom pour vous donner de la crédibilité lors d'un appel ou, encore mieux, à obtenir directement des coordonnées de prospects susceptibles d'être intéressés par vos produits ou services. Cette pratique est assez répandue dans le domaine de l'assurance vie ou de l'immobilier, mais reste sous-exploitée dans d'autres industries. Il arrive aussi que certaines entreprises préfèrent ne pas voir leur relation d'affaires mise en avant pour en solliciter d'autres. En dehors de votre clientèle, d'autres personnes peuvent exercer suffisamment d'influence pour vous aider à décrocher un rendez-vous. La direction d'une chambre de commerce ou d'une autre association professionnelle peut également vous conseiller sur les opportunités qui pourraient se présenter pour vous et votre compagnie.

Vous ne devriez pas sous-estimer votre entourage : amis, famille et collègues de travail. Vous pourriez être étonné d'apprendre que votre oncle connaît très bien la présidence de l'entreprise où vous essayez d'obtenir une rencontre d'introduction depuis des mois. Les collègues, quel que soit leur niveau hiérarchique dans votre firme, peuvent aussi vous donner de bons tuyaux. En fait, les réseaux intraorganisationnels des personnes en vente contribueraient à près de 27 % de leur performance objective (Bolander et coll., 2015).

2.3 / Les activités de réseautage

À l'ère du travail à distance, le contact humain revêt une importance encore plus grande dans toutes les étapes du processus de vente. Deux types d'événements favorisent la création de liens stratégiques: le réseautage général et le réseautage spécialisé.

Les conférences d'intérêt général, comme le discours prononcé par un politicien ou une politicienne devant un regroupement de gens d'affaires, relèvent du premier type. Vous y croisez habituellement des spécialistes issus de divers secteurs d'activité. Toutefois, toutes les activités de réseautage général ne se prêtent pas nécessairement au développement des affaires. Vous devez bien identifier les activités dans lesquelles les compagnies qui pourraient s'intéresser à votre offre se retrouvent. Les activités de réseautage spécialisé peuvent inclure des cinq à sept ou des conférences réunissant des acteurs et actrices clés d'une industrie. Alors que les personnes qui participent aux événements généralistes ne cherchent pas toutes à établir des contacts d'affaires, celles qui assistent aux événements spécialisés cherchent souvent à établir des contacts stratégiques avec des partenaires actuels ou potentiels.

Un autre type d'activités de réseautage spécialisé est la foire commerciale, illustré à la figure 3.1. Ces grandes expositions sont, avant tout, destinées à favoriser la mise en relation entre les responsables des achats et les personnes en vente. Ces derniers ne trouveront pas de meilleure occasion pour se rapprocher d'un rendez-vous d'affaires significatif, puisque les gens qui fréquentent ces salons cherchent activement à découvrir de nouveaux fournisseurs. On pourrait croire que la pandémie et le développement des canaux numériques auraient mis fin à cette pratique, mais, apparemment, les responsables des approvisionnements et les personnes en vente ressentent toujours le désir d'interactions en présentiel. Les spécialistes des achats qui se présentent à une foire commerciale sont motivés par la nécessité d'obtenir de l'information sur les fournisseurs et sur les solutions possibles à leurs problèmes ou à leurs besoins (Vitali et coll., 2022). Ils sélectionneront ensuite les kiosques à visiter et, compte tenu du temps à leur disposition, ils recueilleront des dépliants ou engageront de véritables conversations avec les personnes en vente.

FIGURE 3.1 / **Foire commerciale à Barcelone**

Source : Ian Hughes, (2019)[2].

Certaines entreprises investissent des milliers de dollars dans l'apparence de leur kiosque, une opération sous la responsabilité du marketing. Ces investissements colossaux auxquels s'ajoutent les frais de voyage et de séjour de l'équipe de vente pourraient être vains sans une formation adéquate. La première action des personnes en vente en prévision d'un salon d'affaires est de promouvoir sa présence. Le département de marketing rédigera probablement du contenu à ce sujet sur le site Web et les réseaux sociaux, mais les personnes en vente doivent aussi s'impliquer en envoyant, par exemple, des courriels personnalisés à leur clientèle. Lors de la foire commerciale, vous pouvez donner un séminaire sur un sujet qui met en valeur, de manière indirecte, votre produit ou votre service. Le comportement des membres de votre équipe dans votre kiosque constitue l'un des aspects les plus critiques pour rentabiliser cet important investissement (Siskind, 1997). Ces membres doivent prévoir des pauses pour ne pas laisser paraître leur fatigue et ne pas céder à la tentation de s'asseoir dans le kiosque. Ils doivent éviter de s'engager dans de grandes

2 MWC 2019, CC BY 2.0, https://commons.wikimedia.org/w/index.php?curid=103067221.

discussions avec les collègues afin de maintenir un contact visuel avec les visiteurs qui défilent dans les allées et adopter une posture professionnelle et accueillante. Lorsqu'un visiteur établit un contact visuel avec vous, invitez-le à s'approcher de votre kiosque et, une fois qu'il semble disposé à vous écouter, utilisez une accroche percutante en lien avec votre produit ou votre service. Vous aurez soigneusement préparé cette accroche avec d'autres membres de l'équipe. Par exemple, si votre service permet d'économiser du temps vous pourriez commencer avec: «Combien de temps votre personnel passe-t-il à réaliser ceci ou cela?» Vous suivrez ensuite un script plus ou moins structuré, mais flexible, non pas pour conclure une vente[3], mais pour obtenir un rendez-vous afin de poursuivre une conversation plus approfondie dans les bureaux de cette personne. Rappelez-vous qu'elle ne dispose généralement que de peu de temps pour parcourir les kiosques de l'exposition.

La tâche de l'équipe de développement des affaires ne se termine pas avec la fermeture de la foire commerciale. Vous devrez étiqueter tous les contacts obtenus selon leur probabilité de s'intéresser sérieusement à votre offre. À certains et à certaines, vous enverrez une simple lettre de remerciements et quelques informations sur vos produits et services. Pour les opportunités plus solides, vous prendrez contact par téléphone ou par courriel afin de fixer un rendez-vous.

2.4 / La vente sociale: une approche numérique de l'équipe de vente

La vente sociale consiste à utiliser les canaux numériques, dont les médias sociaux, pour créer des points de contact tout au long du parcours de la clientèle, assurant ainsi une présence stratégique au bon endroit et au bon moment (Ancillai et coll., 2019). Elle peut se manifester au niveau individuel, lorsque les personnes en vente recourent aux médias sociaux pour recueillir de l'information sur le marché, démontrer leur crédibilité en publiant du contenu pertinent et interagir avec les personnes intéressées. Elle s'applique également au niveau organisationnel, lorsque l'entreprise établit une véritable stratégie à long terme de présence sur les réseaux sociaux ou définit des lignes directrices sur leur usage par les personnes en vente. La firme peut également encourager l'utilisation des médias sociaux en

3 Il existe évidemment des exceptions à cette règle, surtout dans les foires commerciales destinées aux consommateurs.

complément du *Customer Relationship Management* (CRM) et favoriser la collaboration entre la force de vente et le service du marketing. La vente sociale peut se manifester à toutes les étapes du processus de vente, que ce soit pour la recherche de firmes à prospecter ou pour le développement et le maintien des relations (Schmitt et coll., 2021).

Une des questions soulevées par la vente sociale est celle de l'autonomie accordée à l'équipe de vente. Si la compagnie limite ses représentantes et représentants à ne publier que du matériel autorisé, ceux-ci se contenteront de partager les parutions de leur organisation (Schmitt et coll., 2021). À l'inverse, si l'entreprise les encourage à rédiger leurs propres messages, elle doit s'assurer de les former à créer du contenu approprié. Cette approche d'autonomisation doit aussi mettre en avant les bénéfices à long terme de la vente sociale pour la force de vente, afin qu'elle perçoive son utilité.

La clientèle a une opinion paradoxale de la vente sociale (Ancillai et coll., 2024). D'une part, les responsables des achats ne suivent pas officiellement les personnes en vente sur les médias sociaux et n'interagissent pas avec leurs publications pour des raisons de confidentialité. De plus, ils se plaignent d'être trop sollicités par le biais de messages non personnalisés. D'autre part, ces mêmes responsables semblent employer les médias sociaux pour rechercher et évaluer des fournisseurs potentiels, mais privilégiant les pages de médias sociaux des entreprises plutôt que celles des membres de l'équipe de vente.

En dépit de ces réactions assez tièdes de la clientèle à l'égard des médias sociaux, la vente sociale exerce une influence positive sur la performance des personnes en vente qui les utilisent pour recueillir des informations sur les firmes clientes ou celles qui pourraient le devenir, pour développer leur réseau de contacts et pour générer de l'engagement (Terho et coll., 2022). Cette utilisation augmenterait leur crédibilité en démontrant leur expertise et valoriserait leur image de marque personnelle (Barney-McNamara et coll., 2020).

3 / Contacter les firmes identifiées

Établir un premier contact peut être intimidant pour les personnes en vente, même les plus expérimentées. Différentes techniques existent pour faire connaissance en personne, par téléphone ou par courriel.

3.1 / Les approches en personne

Les rencontres en personne surviennent généralement lors d'activités de réseautage. L'approche auprès de personnes inconnues lors d'une foire commerciale est plus facile que dans d'autres circonstances, car, en principe, les gens qui visitent les foires sont ouverts à discuter avec vous. Dans d'autres contextes, l'établissement d'un premier contact peut s'avérer plus difficile d'où l'importance de bien s'y préparer. Cela consiste d'abord à découvrir qui participe à ces rencontres. Selon les politiques de l'organisation d'un événement, vous pourriez obtenir la liste des individus qui assistent à l'activité. Ou encore, vous pourriez joindre quelqu'un habitué à ces rendez-vous afin d'en apprendre davantage.

Vous devez également planifier votre approche pour aborder les gens. Une bonne tactique est de cibler les personnes seules, car elles sont inconfortables de n'avoir personne à qui parler. La personne que vous approcherez sera reconnaissante que vous mettiez fin à sa contemplation du plafond ou de ses chaussures. Vous pourriez choisir de participer à ces activités avec quelqu'un d'autre, mais cela comporte un risque : vous serez tenté de rester avec cette personne pendant toute la durée de l'événement, alors que votre objectif devrait être de rencontrer entre trois et cinq nouvelles personnes.

Pour engager la conversation, intéressez-vous sincèrement à votre interlocuteur en lui posant des questions ouvertes et en pratiquant les techniques d'écoute active qui seront abordées plus loin. Ensuite, cette personne devrait souhaiter en apprendre un peu plus sur vous et ce sera alors l'occasion de fournir plus de détails sur votre occupation. Ne donnez pas d'emblée cette information, mais trouvez plutôt une façon originale d'expliquer comment vous aidez votre clientèle à atteindre ses objectifs. Cette courte présentation se nomme **l'argumentaire éclair**, ou *elevator pitch* en anglais. Ce terme fait référence au fait que vous devriez être capable de résumer la valeur de votre travail en aussi peu de temps qu'il en faut pour monter en ascenseur jusqu'au quinzième étage d'un immeuble. La principale difficulté de l'argumentaire éclair est d'adopter un ton naturel et d'éviter de répéter des énoncés vides de sens. Le nombre de personnes en vente qui ont appris cette technique est tel que plusieurs s'attendent désormais à y être confrontés lors des cinq à sept.

La première étape de l'argumentaire éclair consiste évidemment à vous présenter. La poignée de main semble retrouver sa place, du moins en Amérique du Nord et en Europe, après la pandémie. Si vous entretenez des relations d'affaires avec des interlocuteurs issus d'autres cultures, prenez le temps de vous informer sur les usages en vigueur (encadré 3.1).

ENCADRÉ 3.1 / **La vente auprès d'autres cultures**

La plupart d'entre vous développeront des affaires au Canada, aux États-Unis et possiblement en Europe du Nord où, sans vouloir généraliser, les conventions culturelles s'apparentent aux nôtres. Toutefois aux États-Unis, on démontre plus de familiarité qu'au Québec, alors qu'en Europe du Nord, les contacts sont un peu plus formels que ce soit par les formules de politesse utilisées que par la tenue vestimentaire. Certaines règles peuvent changer en Amérique latine ou dans des pays d'Europe du Sud ou d'Afrique du Nord. Par exemple, la hiérarchie peut être plus manifeste et les subalternes ont tendance à émettre moins ouvertement leur opinion. Dans certaines cultures, les contacts physiques entre hommes et femmes sont proscrits : la poignée de main sera alors remplacée par un hochement de tête ou une légère inclinaison du corps. À l'inverse, d'autres pays privilégient l'accolade et la bise, ce qui peut surprendre des Nord-Américains peu habitués à ces pratiques. Ce manuel ne prétend pas brosser un portrait exhaustif des différences culturelles, mais vise plutôt à sensibiliser les membres de l'équipe de vente et les encourager à se renseigner sur les coutumes locales avant de commettre un impair qui pourrait compromettre toutes les tentatives futures de développement des affaires.

À la deuxième étape de l'argumentaire éclair, vous présenterez votre entreprise de façon à créer une solide impression à votre interlocuteur. Imaginez que vous vendiez des poutres d'acier[4]. Comment allez-vous capter l'attention de la personne responsable des achats ? « Nous sommes des spécialistes en structure. D'ailleurs, nous sommes l'un des deux partenaires pour la reconstruction du toit du stade olympique [c'est la phrase hameçon]. Saviez-vous que [ici vous pourriez inclure un bénéfice clé] ? » Raconter une courte histoire d'une

4 Exemple inspiré de la compagnie https://www.canam.com/

trentaine de secondes illustrant comment votre produit a procuré une réelle valeur à une entreprise cliente, sans avoir l'air de pousser la vente, pourrait contribuer à faire bonne impression. Dans le chapitre 4 sur l'exploration, je reviendrai plus en détail sur cette pratique du *storytelling*.

À la troisième étape, votre objectif est d'obtenir une carte professionnelle ou la permission de noter le nom et le numéro de téléphone de votre nouveau contact dans votre cellulaire. Encore mieux, posez cette question : « Me permettez-vous de vous ajouter à mes relations sur LinkedIn ? » Vous pourrez, par la suite, transférer ces renseignements dans votre CRM.

L'étape finale est essentielle, car elle repose sur un suivi téléphonique ou par courriel visant à fixer un rendez-vous. Commencez par établir un contact social en rappelant un point que vous avez particulièrement apprécié de la courte rencontre que vous avez eue. Je vous donnerai plus d'informations sur le contact social dans le chapitre 4.

3.2 / L'appel à froid

Même si, aujourd'hui, les responsables des achats effectuent une grande partie du processus par eux-mêmes, voire en approchant directement le fournisseur, l'appel à froid, ou *cold call*, demeure un outil utile pour promouvoir ses produits et ses services qui n'ont pas encore accumulé un capital de notoriété suffisant pour se hisser en tête des résultats des moteurs de recherche. L'appel à froid consiste à contacter des entreprises suspectes ou qualifiées qui ne vous connaissent pas, par opposition à un appel tiède où l'on communique avec quelqu'un qui vous connaît ou qui a entendu parler de votre entreprise, et à l'appel chaud, qui vise une personne déjà disposée à en apprendre davantage sur votre offre. En développement des affaires, les personnes en vente devraient concentrer leurs efforts sur les entreprises les plus susceptibles d'être intéressées par leur proposition. L'appel à froid se déroule en trois étapes : la planification, la présentation et la relance.

À l'étape de la **planification**, vous devrez clairement identifier votre objectif. Cherchez-vous à obtenir une commande immédiate ou à prendre un rendez-vous ? Aussi, vous devriez essayer d'en apprendre le plus possible au sujet de l'entreprise et les raisons pour lesquelles elle pourrait bénéficier de vos produits ou services. Recueillez également des renseignements sur la personne avec laquelle vous envisagez de vous entretenir, comme son parcours professionnel, son rôle dans l'entreprise et, bien sûr, la bonne prononciation de son nom. Ces informations vous seront utiles pour établir un contact lorsque vous réussirez à joindre la personne. Sachez aussi que quelques membres du personnel filtrent les appels, ce qui ajoute une barrière supplémentaire à votre tentative de joindre les responsables des approvisionnements. Prévoyez comment réagir à une personne qui vous demande d'envoyer votre documentation. Planifiez aussi le meilleur moment pour appeler. Chaque personne que vous tentez de joindre aura ses préférences. Vous augmenterez peut-être vos chances en tentant de les joindre à 7 h 30 ou à 17 h 30, hors des heures de bureau, mais ce n'est pas garanti. C'est pourquoi il est important de varier ses canaux de communication, en alternant, par exemple, l'appel téléphonique et l'envoi d'un courriel. Certaines personnes ont déjà répondu à l'un de mes courriels à 2 h du matin.

Vous devrez aussi rédiger des scripts d'appel (encadré 3.2). Trop de personnes improvisent lors d'un appel téléphonique, de la rédaction d'un courriel ou de l'envoi d'un message. Pourtant, cette première communication est déterminante : elle peut soit ouvrir la porte à une discussion, soit mener à un refus. Ici, rien ne peut être laissé au hasard. La personne en vente doit développer, tester et maitriser plusieurs scripts pour mettre toutes les chances de son côté, entre autres parce qu'on ne répondra pas à un premier message sur une boite vocale ou à un premier contact par les réseaux sociaux. Répéterez-vous un script identique lors du deuxième appel ? Cela démontrerait un manque de créativité ou même d'intérêt. Vous pourriez recourir ici, comme lors de la rencontre en personne, à la narration d'un récit véritable du succès d'une compagnie qui a utilisé vos produits ou vos services. Sauriez-vous relater cette histoire en quelques lignes ? Nous verrons plus loin comment le *storytelling* peut éveiller l'intérêt et créer une connexion avec votre interlocuteur.

ENCADRÉ 3.2 / **La rédaction du script d'appel**

Assurez-vous d'abord que votre script parle de l'entreprise de votre interlocuteur plutôt que de la vôtre. Pour vous assurer de respecter cette recommandation, comparez le nombre de mentions à la première personne (nous, nos, notre) à celles de la deuxième personne (vous, vos, votre). Si ces dernières sont plus nombreuses, votre message est centré sur le client ou la cliente, ce qui est idéal. Autrement dit, concentrez-vous sur les bénéfices que procurent votre produit ou votre service plutôt que sur ses caractéristiques. Par exemple, au lieu de simplement dire:«Notre produit offre une vitesse plus rapide que ceux de notre concurrence», poursuivez avec: «ce qui signifie que vous réduirez vos coûts de main-d'œuvre», ou «ce qui signifie que vous augmenterez votre productivité». Calculez le gain monétaire pour donner encore plus de poids à vos arguments: «ce qui signifie que vous réduirez vos coûts de main-d'œuvre de 2% dans le département de finition. Comme vos coûts de main-d'œuvre représentent 100000 dollars, l'économie annuelle de 2000 dollars vous permettra de rentabiliser l'écart entre le prix de notre produit et celui de la concurrence en seulement huit mois».

De plus, privilégiez une approche qui suscite la curiosité et l'intérêt immédiat de votre interlocuteur. Évitez les détours inutiles: le temps de la personne responsable des achats est précieux, et vous devez lui faire savoir que vous respectez cela. Soyez bref et convaincant.

Le script peut être divisé en quatre parties. La première consiste à se nommer et à mentionner clairement le nom de votre entreprise. Les gens aiment savoir à qui ils ont affaire.

La deuxième partie du script vise à établir un lien immédiat avec votre destinataire, de manière concise, car celui-ci dispose de peu de temps. Si vous devez laisser un message sur une boite vocale, vous pouvez mentionner une information qui démontre que vous avez effectué des recherches sur l'entreprise ou sur la personne à la phase de la planification. Si vous obtenez la communication avec la personne, vous utiliserez ces informations pour poser une question pertinente visant à établir un premier lien avec votre interlocuteur ou votre interlocutrice.

La personnalité de votre interlocuteur, dont nous parlons plus en détail plus loin, influencera le temps que vous consacrerez à cette étape. Vous pouvez tenter d'aborder un sujet personnel si vous identifiez un intérêt commun. Par exemple: « J'ai lu votre article de blogue pour la Société d'ornithologie. J'ai moi-même participé à une activité d'observation des hirondelles des granges l'an dernier, et j'ai trouvé cela fascinant. Quel est votre site d'observation préféré?» Si votre contact vous interroge sur la pertinence de cette question, vous pouvez sourire et reconnaître que vous vous êtes momentanément écarté du sujet. Comme il est difficile d'identifier le caractère de l'autre en quelques secondes, il est souvent plus prudent d'orienter la conversation vers son entreprise plutôt que vers des sujets personnels. Par exemple, « J'ai remarqué que votre société a récemment acquis ABC. Comment se passe l'intégration des nouveaux employés? » (si vous vendez un logiciel de gestion des ressources humaines).

À la troisième partie de votre script, vous établissez clairement la valeur de votre offre pour votre clientèle et ce qui vous distingue de votre concurrence. Par exemple : « J'aide les compagnies comme la vôtre à augmenter la rétention du personnel afin de réduire leurs coûts de recrutement. Comment la pénurie de main-d'œuvre actuelle affecte-t-elle la productivité de votre société? » Vous pouvez poser ce genre de question parce que vous avez effectué des recherches et que vous savez que l'entreprise affiche souvent des postes à pourvoir. En posant une question ouverte, vous invitez la personne à exposer son problème, auquel vous avez déjà une solution adaptée. En l'amenant à exprimer ses défis, vous créez une ouverture naturelle pour présenter votre offre. Tout au long de l'échange, appliquez les techniques d'écoute active afin de démontrer votre compréhension et d'adapter votre discours aux besoins exprimés. Évidemment, s'il s'agit d'un message téléphonique, vous ajusterez cette partie en conséquence.

La quatrième partie consiste à lancer un appel à l'action. Si vous êtes en conversation avec une personne, vous tenterez d'obtenir un rendez-vous pour faire la démonstration de votre produit ou de votre service, en proposant deux choix de dates et d'heure. Si les créneaux proposés ne lui conviennent pas, elle vous proposera d'autres disponibilités. Si vous laissez un message, vous donnerez vos coordonnées en invitant la personne à prendre contact avec vous. Assurez-vous d'énoncer votre numéro de téléphone distinctement et de le répéter deux fois, afin que la personne ne soit pas obligée de réécouter votre message pour noter vos coordonnées. Vous pouvez aussi mentionner que vous rappellerez une semaine ou deux plus tard si vous n'avez pas eu de nouvelles.

Il est recommandé de revoir vos scripts régulièrement. Sont-ils efficaces? Comment les réactions obtenues peuvent-elles vous aider à améliorer votre taux de conversion? Comment pouvez-vous partager les expériences des membres de l'équipe de vente avec les scripts pour les rendre encore plus performants?

La deuxième étape de l'appel à froid consiste à **présenter** votre script d'appel. Vous vous serez entrainé afin que votre message ait l'air le plus naturel possible. Personne n'aime avoir l'impression de parler à un robot. Le ton de votre voix doit être chaleureux et votre débit, dynamique et enthousiaste tout en demeurant crédible. Utilisez le nom de la personne à quelques reprises durant une conversation et assurez-vous qu'il soit mentionné au moins deux fois si vous laissez un message vocal d'une trentaine de secondes. Les personnes portent davantage attention à ce que vous dites lorsque vous vous adressez directement à elles.

La troisième étape de l'appel à froid est **la relance** ou le suivi après l'appel. Pour démontrer votre professionnalisme, il est recommandé d'envoyer un courriel qui résume la conversation, remercie votre interlocuteur pour le temps qu'il vous a accordé et qui confirme la prochaine étape des discussions. Si vous avez laissé un message, vous pouvez également relancer la personne en lui envoyant un courriel. Les gens ont des préférences différentes quant aux moyens de communication. Vous devrez recommencer sans cesse parce que les responsables sont souvent trop occupés pour retourner les appels ou les courriels, ou parce que votre script d'appel les a laissés indifférents. Vous pourriez devoir relancer la personne à qui vous souhaitez parler jusqu'à sept ou huit reprises avant de réussir à lui parler ou d'obtenir une quelconque réaction.

La persévérance est une qualité essentielle en développement des affaires. Combien de personnes en vente prennent la peine de faire un deuxième appel ou de rédiger un troisième message ? Très peu apparemment. Pourtant, c'est souvent ce qui permet de se démarquer. Imaginez-vous à la place d'un responsable des achats : choisiriez-vous un fournisseur qui abandonne au premier silence ? Pour éviter cette perception, il est essentiel de structurer ses relances et d'exploiter son CRM pour programmer des rappels stratégiques au bon moment.

4/ La qualification

Aucune compagnie ne peut vendre de tout à tout le monde. La force de vente doit donc impérativement connaître le marché de son entreprise, ses produits et ses services. En d'autres termes, vous cherchez à identifier les personnes ou les organisations prêtes à payer le prix que vous proposez pour ces produits ou ces services dans un territoire donné. Votre équipe de vente doit être en mesure de démontrer la valeur de votre offre auprès de ces personnes et de ces compagnies. Souvent, vos ventes passées peuvent vous renseigner sur la clientèle idéale pour votre firme. Cependant, il peut aussi être intéressant d'explorer des marchés connexes aux caractéristiques semblables.

La prochaine étape de la découverte est la qualification. Inutile d'investir des mois à tenter de convaincre une entreprise des avantages de votre offre si celle-ci ne correspond pas réellement à ses besoins. L'objectif n'est pas d'accumuler le plus grand nombre de prospects, mais de détenir un portefeuille d'opportunités solides et prometteuses. À quoi bon chercher de l'or dans une rivière stérile? De la même manière, pourquoi engager des discussions avec une entreprise qui n'a ni l'intérêt ni la capacité d'acheter votre solution?

Pour qualifier les compagnies, IBM a développé une méthode, dans les années 1950, résumée par l'acronyme BATA (BANT en anglais), soit le besoin, l'autorité, la temporalité et l'argent. La personne en vente doit apprendre rapidement si l'entreprise a besoin de son produit. Elle devra aussi identifier qui détient le pouvoir décisionnel pour conclure l'achat. Ensuite, l'entreprise prévoit-elle d'acheter à court terme ou repousse-t-elle sa décision à plus tard? Enfin, la personne en vente doit vérifier si l'organisation dispose du budget nécessaire pour investir dans le produit ou le service.

5 / Se préparer à une première rencontre

Dans la phase de découverte, une personne en développement des affaires doit savoir que ses vis-à-vis auront des personnalités différentes, que certains modes de communication sont plus efficaces que d'autres, que la maitrise de l'écoute active permet d'accélérer l'exploration et que démontrer son expertise renforce la confiance.

5.1 / Les personnalités des responsables des achats

L'adaptabilité est l'une des compétences clés en développement des affaires. Savoir s'ajuster à toutes les personnalités est l'une des aptitudes les plus recherchées en vente (Sujan et coll., 1988). Évidemment, les personnes en vente ne peuvent pas faire passer des tests psychométriques pour déterminer la meilleure approche. C'est pourquoi les psychologues industriels David Merrill et Roger Reid ont proposé, au début des années 1960, un modèle qui classe les personnalités en quatre types, répartis selon deux axes : l'assertivité et la sociabilité. Sur l'axe horizontal, une personne assertive aura tendance à exprimer clairement ses demandes, fera preuve d'une grande confiance en elle, prendra des initiatives, assumera des risques et affichera un fort esprit de compétition. Une personne moins assertive aura tendance à hésiter avant de formuler une demande, voire à s'en abstenir. Elle sera moins compétitive, prendra moins de risques, mais se montrera généralement plus collaborative. Sur l'axe vertical, une personne sociable, tournée vers les relations humaines, réagira plus vivement aux émotions des autres ainsi qu'aux siennes. Elle fera souvent preuve d'une plus grande empathie bien que cela puisse parfois nuire à son contrôle émotionnel. À l'inverse, une personne moins sociable, orientée vers la tâche, pourrait faire preuve de moins d'empathie, puisqu'elle est moins tournée vers les gens. À la figure 3.2, ces deux axes se croisent donc pour distinguer quatre types de personnalités : les aimables, les analytiques, les fonceurs et les expressifs.

Les personnes **aimables** démontrent une sociabilité élevée et s'affirment peu, ce qui les rend affables et fiables. Elles prennent des décisions lentement et privilégient la sécurité plutôt que le risque. Lorsque des spécialistes en développement des affaires discutent avec elles, il est essentiel de prêter attention à leurs émotions et d'exprimer également ses propres ressentis pour instaurer un climat de confiance.

Les personnes **analytiques** font preuve de persévérance, se soucient des détails et sont avides de données et de faits. Toutefois, leur faible assertivité peut les rendre plus réservées ou distantes. Cette personnalité cherche d'abord la sécurité en posant de nombreuses questions. Son processus décisionnel peut sembler lent, bien qu'elle respecte généralement les échéances. Lors d'un échange avec une personne analytique, soyez préparé, présentez des arguments clairs en détaillant les inconvénients et les bénéfices de chaque option. Tenez-vous-en aux faits.

FIGURE 3.2 / **Quatre types de personnalités**

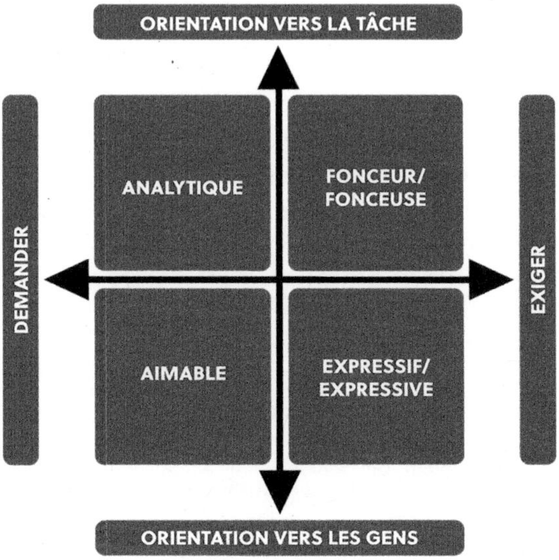

Source : Adapté des travaux de Merrill et Reid (1981).

Comme les analytiques, les personnes **fonceuses** sont orientées vers la tâche, mais elles se distinguent par leur assertivité marquée. Elles défendent fermement leur opinion et se concentrent sur les résultats, prêtant moins d'attention aux émotions des autres cherchant souvent à prendre le contrôle des discussions. Elles sont reconnues pour leur prise de décision rapide. Face à elles, les spécialistes en développement des affaires doivent être précis et bien organisés, au même degré qu'avec l'analytique, et ce, tout en faisant preuve d'efficacité. Un manque de concision ou de structure risque de provoquer de l'impatience.

Les personnes **expressives** sont aussi rapides que les fonceuses, mais elles s'intéressent davantage aux émotions qu'aux faits. Elles sont enthousiastes et sociables, aiment être complimentées et sont sensibles aux témoignages et aux expériences vécues, qui renforcent un argumentaire. Si vous trouvez un terrain d'entente avec une personne expressive, assurez-vous de consigner les détails par écrit afin de structurer vos échanges.

La communication la plus difficile entre différentes personnalités est celle qui implique les dyades fonceuse-aimable et expressive-analytique. Les membres de chaque paire ont du mal à se reconnaître tant au niveau de l'assertivité que de la sociabilité. Croyez-vous qu'une personne en vente soit favorisée par une personnalité plutôt qu'une autre? En fait, ce qui importe le plus, c'est votre capacité à reconnaître votre style, à identifier celui de votre vis-à-vis et, surtout, à vous adapter. Cette souplesse et cette capacité d'adaptation sont bien plus déterminantes que votre propre style de personnalité. Vous devrez donc faire preuve de polyvalence.

Le tableau 3.1 présente les comportements à privilégier en fonction de chaque type de personnalité.

TABLEAU 3.1 / **Les comportements à privilégier**

Type de personnalités	Comportement à privilégier
Analytique	• Soyez ponctuel et très organisé • Ne vous attardez pas à bâtir une relation • Démontrez clairement, par vos questions, que vous avez un but précis • Fournissez de la documentation • Ne soyez pas pressé de conclure
Fonceuse	• Allez droit au but dès le début de la rencontre • Cernez rapidement ses objectifs et axez votre présentation sur ceux-ci • Insistez sur les bénéfices et les résultats en présentant des faits et des statistiques
Aimable	• Prenez le temps de socialiser en abordant des sujets plus personnels • Adoptez un ton professionnel et amical • Donnez-lui le temps de bien comprendre votre proposition • Montrez-lui pourquoi votre solution n'est pas risquée
Expressive	• Donnez du rythme à votre présentation en adoptant un ton enthousiaste • N'insistez pas trop sur les faits et les détails • Demandez ses opinions et ses idées en étant prêt à recadrer la discussion • Montrez-lui comment votre solution lui permettra de se faire remarquer favorablement auprès de ses collègues

Source: Adapté de Merrill et Reid (1981).

5.2 / Trois modes de communication

Certains pourraient croire que, dans un monde numérique, la transmission par texto ou par courriel est la plus efficace. Ce n'est pourtant pas le cas, d'où l'importance de maitriser les trois modes de communication, connus sous le nom des 3 V. Ces modes regroupent la communication verbale, qui concerne les mots utilisés, la communication vocale, qui inclut le ton de la voix, et la communication visuelle, qui englobe la posture, les expressions faciales et le langage corporel.

Lors d'une communication en face-à-face, on emploie à la fois les mots, le ton de la voix et le langage non-verbal. Dans un tel contexte, c'est la somme pondérée des effets indépendants des trois modes qui composent l'ensemble de la communication. Deux chercheurs ont déterminé que les mots, soit la composante verbale, ne comptent que pour 7 % de la communication, alors que le ton de la voix représente 38 % et que le langage non-verbal en constitue 55 % (Mehrabian et Ferris, 1967). Même si les chercheurs ont mis en garde contre une extrapolation des résultats de leurs recherches à d'autres contextes que celui qu'ils ont étudié, on doit reconnaître qu'une communication qui utilise les trois modes est plus efficace que si elle n'en utilise qu'un seul.

Ainsi, répondre à une plainte par courriel limite la capacité d'exprimer son empathie, puisque ni le ton de la voix ni le langage corporel ne sont perceptibles. L'ubiquité des moyens de communication numériques ne rend pas caduque la nécessité de voir les personnes. La visioconférence reste une solution de rechange à une rencontre en personne.

Si la communication non verbale joue un rôle aussi crucial, cela signifie que vous devrez d'abord porter attention à l'image que vous projetez, par exemple en souriant et en adoptant une posture ouverte. De plus, être attentif aux expressions du visage et au langage corporel de l'interlocuteur ou l'interlocutrice permet de mieux percevoir son niveau d'engagement, son impatience ou son intérêt pour la discussion.

5.3 / L'écoute active

Confucius a écrit : « Si l'être humain a deux oreilles et une bouche, c'est pour écouter deux fois plus qu'il ne parle. » Cette maxime s'applique particulièrement au développement des affaires. Écouter permet de mieux comprendre les besoins de la clientèle et d'adapter son offre en conséquence. Cependant, écouter passivement ne suffit pas : il faut aussi encourager le ou la destinataire à se livrer et à partager des informations essentielles.

Le psychologue Carl Rogers a développé une technique appelée l'écoute active. Adaptée au contexte du développement des affaires, cette technique repose sur plusieurs principes :

- Regarder son interlocuteur pour démontrer son intérêt et décoder son langage non-verbal ;
- Éviter d'interrompre et laisser des silences pour lui permettre d'ajouter certains points ;
- Écouter véritablement, sans préparer sa réponse pendant que l'autre s'exprime ;
- Encourager la personne à développer sa pensée et à exprimer ses sentiments par des gestes comme hocher la tête ou des mots d'acquiescement ;
- Poser des questions de clarification ;
- Répéter les idées de l'autre en reformulant certaines affirmations pour vous assurer d'avoir bien compris ;
- Utiliser l'écoute active à toutes les étapes de la vente, y compris lors de la réponse aux objections.

5.4 / Démontrer son expertise

Le service des achats peut trouver une grande quantité de renseignements en ligne, au risque d'être enseveli par les données. Une personne en vente peut faire une différence en organisant ces données pour simplifier la prise de décision. C'est pourquoi elle doit d'abord se tenir informée des caractéristiques techniques et des bénéfices de son produit ou de son service. Par exemple, si vous représentez une compagnie qui vend du papier destiné à l'usage de bureaux, notamment pour les imprimantes et les photocopieurs, chaque produit possède des propriétés spécifiques, que ce soit son poids, sa brillance ou encore l'origine de la matière première, certifiée ou non en vertu de certaines normes environnementales. La personne en vente doit être en mesure de mettre en avant les bénéfices de chacune de ces caractéristiques en fonction de l'ensemble des produits disponibles, mais aussi, parfois, en les comparant à ceux offerts par la concurrence.

De plus, la personne en vente peut être amenée à communiquer des informations sur sa compagnie, sur ses engagements environnementaux, sur ses états financiers ou sur ses plans d'expansion qui pourraient avoir un impact sur sa clientèle. Elle évolue également au sein d'une industrie soumise à diverses contraintes du macro-environnement, qu'elles soient légales, politiques, économiques ou sociales. Connaître son industrie, c'est non seulement avoir une bonne compréhension de sa concurrence, mais aussi de ses fournisseurs. Par exemple, la personne qui achète du papier peut très bien maitriser l'offre sur le marché, mais il se peut qu'elle en sache relativement peu sur les sources d'approvisionnement. Dans ce cas, la personne en vente doit être en mesure d'expliquer les différences entre les labels de certification environnementale des forêts ou les procédés chimiques utilisés dans la fabrication des papiers, notamment pour le blanchir.

Bien sûr, chaque membre de l'équipe de vente détient des informations de base sur sa clientèle qu'il peut récupérer sur le Web. Cependant, il est encore plus utile d'observer directement l'entreprise cliente ou d'analyser sa propre clientèle lorsque cela est possible. Si vous vendez de la bière artisanale, vous gagnerez à observer les rayons d'épiceries pour mieux comprendre les défis liés à la mise en valeur des nombreuses marques proposées. De même, vous devriez également comprendre le secteur dans lequel votre clientèle évolue. Je me souviens d'une formation donnée à des personnes en vente qui commercialisaient leurs marchandises dans des dépanneurs. Un des volets de cette formation visait à les renseigner sur les tendances à venir comme le développement d'une section de prêt-à-manger. Même si les produits offerts par les personnes en formation n'étaient pas associés avec le prêt-à-manger, ces nouvelles connaissances leur permettaient de démontrer leur expertise auprès des propriétaires de dépanneurs. Selon votre secteur d'activité et votre offre, vous devrez peut-être devoir investir entre 2 et 10 heures de travail par semaine, seulement pour rester à jour avec ces nouvelles connaissances.

En résumé

La phase de découverte vous a permis d'identifier des entreprises potentiellement intéressées par votre offre et de les contacter. Bien que le numérique occupe aujourd'hui une place centrale, la prise de contact en personne reste un atout majeur, car elle permet de créer une relation plus authentique et durable. Vous avez aussi constaté la nécessité de se préparer adéquatement à une première rencontre qui constitue la prochaine étape du développement des affaires : l'exploration. Dans le chapitre 4, nous verrons comment le ou la responsable des achats et la personne en vente s'évaluent mutuellement afin de déterminer s'ils souhaitent s'engager dans une collaboration à plus long terme.

CHAPITRE 4 /

L'exploration

L'exploration peut se dérouler en une seule rencontre lorsque les approvisionnements sont moins complexes, mais dans la plupart des cas, elle nécessitera plusieurs réunions. Dans les sections suivantes, nous examinerons la planification de l'exploration, l'exploration préliminaire, les méthodes de vente éprouvées à utiliser à ce stade, les tactiques pour répondre aux préoccupations, la présentation de l'offre commerciale, ainsi que le bilan du rendez-vous.

1/ La planification de l'exploration

Bien que nous ayons déjà abordé la planification sous un angle stratégique au début de ce manuel, elle joue aussi un rôle essentiel sur le plan tactique à chaque phase du processus. Il peut être utile de créer un modèle standardisé pour structurer les étapes clés de cette première phase d'exploration, comme illustrée dans le tableau 4.1. Ce formulaire devrait être utilisé avant chaque rencontre, et pas seulement lors de la première visite.

Vous y inscrirez notamment le nom des personnes participantes à la rencontre, qu'elles fassent partie de l'entreprise cliente ou de votre propre compagnie. De nombreuses réunions en développement des affaires impliquent plusieurs intervenants de chaque côté de la table.

Ensuite, il est recommandé de répondre à une série de questions pour clarifier votre objectif SMART (spécifique, mesurable, ambitieux, réaliste et temporel) pour la visite et identifier les ressources nécessaires, comme des diapositives, des brochures ou des échantillons.

Prendre quelques minutes pour remplir ce formulaire vous permettra de maximiser vos chances de succès lors de la réunion et de faire une excellente première impression.

TABLEAU 4.1 / **Planification de la rencontre**

Planification avant la rencontre			
Nom de l'entreprise cliente			
Date de la rencontre		**Heure de la rencontre**	
Personnes de l'entreprise cliente		**Personnes de mon entreprise**	
Nom	**Titre**	**Nom**	**Titre**
Quel est l'objectif général de la rencontre?			
Comment commencerez-vous la réunion?			
Comment créerez-vous des liens avec chacune des personnes présentes?			
Que voulez-vous apprendre durant ce rendez-vous?			
Comment ferez-vous ressortir les bénéfices et la valeur pour l'entreprise cliente?			
Quelles seront ses objections ou ses préoccupations?			
De quoi aurez-vous besoin pour la réunion?			
Comment définirez-vous le succès de cette rencontre?			
Comment entrevoyez-vous la prochaine étape?			

Source : Adapté de Kane et Zappulla (2016).

2 / L'exploration préliminaire

Réussir une bonne entrée en matière après avoir créé un lien relève de l'exploration préliminaire. Il s'agit d'établir un contact social efficace et de structurer l'entretien pour en assurer le bon déroulement.

2.1 / Socialiser

La première rencontre est une occasion de jeter les bases de la relation entre vous et la personne qui représente la compagnie que vous sollicitez. Vous n'avez qu'une seule chance de faire une première bonne impression. Assurez-vous d'avoir l'air confiant, mais pas arrogant. Portez une tenue vestimentaire adaptée à l'industrie de votre clientèle. La poignée de main, jadis incontournable, est devenue aujourd'hui un geste délicat, encore plus depuis la pandémie. Autant elle annonce le désir d'établir un véritable contact humain, autant elle peut créer un malaise en saison de transmission de virus ou même en raison de disparités culturelles. Observez le langage non-verbal de votre vis-à-vis. Si la personne hésite, vous pourriez adopter un signe de salutations différent comme un hochement de tête. Si elle vous tend la main, vous pourrez la présenter à votre tour avec une poigne à la fois ferme et délicate. Vous ne devez pas écraser les jointures de l'autre ou lui tendre une main molle. Une personne extrêmement chaleureuse qui commence à vous connaître pourrait même vous faire une accolade. Sauriez-vous comment vous comporter ? Pourquoi ne pas pratiquer ces gestes avec quelqu'un de votre entourage pour vous assurer que vous ne manquiez pas cette première amorce d'une conversation ? Quant au contact visuel, plusieurs ouvrages sur la vente recommandent de regarder la personne dans les yeux lors de la poignée de main. Or, dans certaines cultures orientales, un contact visuel, particulièrement s'il est soutenu, peut être perçu comme une forme d'agressivité. Encore une fois, faites preuve d'un sens aiguisé d'observation pour comprendre les attentes de l'autre personne.

Vous devez commencer la conversation en créant un lien social. Évitez les banalités sur la pluie et le beau temps. Ne commentez pas non plus des éléments du décor du bureau, car d'autres avant vous auront certainement fait des remarques sur la photo de famille ou le trophée de golf. De plus, assurez-vous de ne poser aucune question inappropriée, intrusive ou ambiguë. En revanche, une question pertinente sur la personne ou sur son organisation suscitera l'intérêt de votre vis-à-vis, ce qui démontrera que vous avez pris la peine d'effectuer des recherches. Mieux encore, trouvez une approche inattendue, un sujet que les trois personnes en vente qui vous ont précédé dans le bureau du service des achats n'ont pas utilisé. Il reste donc deux grandes catégories d'éléments de conversation : les plus récentes informations récoltées sur la firme ou les connaissances ou les activités que vous et votre contact partagez.

Par exemple, si l'entreprise vient d'annoncer des investissements importants, peut-être pourriez-vous chercher à savoir comment cela influencera le travail de votre vis-à-vis. Si vous connaissez la société où cette personne a travaillé auparavant, vous pourriez discuter de son expérience. Ces informations, obtenues grâce à votre recherche sur le site Web de l'entreprise ou la page LinkedIn de votre interlocuteur, vous distingueront de la majorité des autres vendeurs ou vendeuses. Écoutez attentivement la réponse à votre question et préparez-vous à relancer la conversation avec une autre interrogation pour démontrer votre intérêt. Restez concentré sur les besoins de l'autre personne et évitez, si possible, de recentrer l'échange sur vous ou votre employeur. Selon le style de personnalité de votre contact, cette question sera soit la dernière avant d'entrer dans le vif du sujet, soit l'occasion d'un échange plus long de 5 ou 10 minutes. N'oubliez pas votre objectif : obtenir une première commande ou un premier contrat. Il faudra donc repérer le bon moment pour transitionner naturellement vers la discussion d'affaires, à moins que vous ne passiez du temps avec la personne sur un terrain de golf ou au restaurant. Dans ces situations, la socialisation se prolongera certainement au-delà des 10 minutes.

Même si cette ouverture est particulièrement importante lors de la première réunion, elle demeure utile si vous ne rencontrez cette personne qu'une fois par mois. En vente relationnelle, la discussion sociale permet de tisser et de maintenir des liens durables.

2.2 / Structurer

Avant de plonger dans le vif du sujet, prévoyez une phase de transition entre la socialisation et l'exploration formelle en fonction des types de personnalités vus précédemment. Structurez la suite de la rencontre en exposant clairement les points à aborder et obtenez l'assentiment de votre interlocuteur. Une bonne entrée en la matière consiste à soumettre verbalement, ou même par écrit, un bref ordre du jour. Une fois que vous vous serez entendus sur le déroulement de la rencontre, vous pourrez aller de l'avant avec l'une des deux méthodes les plus reconnues dans le domaine de la vente, soit la méthode SPIN et la méthode Challenger que je présenterai brièvement dans les sections suivantes.

3 / Deux méthodes de vente éprouvées

Des dizaines de livres sur la vente paraissent chaque année, promettant tous un succès assuré grâce à une approche innovatrice. Pourtant, peu de ces formules résistent à l'épreuve du temps. En revanche, deux méthodes qui s'appuient sur des années de recherche auprès de milliers de personnes en vente devraient retenir votre attention, soit les méthodes SPIN et Challenger.

Ces deux approches insistent, comme mentionné précédemment, sur l'importance de promouvoir les bénéfices concrets qu'un client ou une cliente peut retirer de votre offre. Attention à ne pas confondre bénéfices, caractéristiques et avantages. Les caractéristiques désignent les composantes techniques de votre offre, par exemple, la vitesse d'un équipement, ou l'expérience de vos consultants et consultantes. Elles sont utiles pour de petites ventes, mais restent insuffisantes pour conclure des contrats d'envergure. Les avantages, eux, montrent comment une caractéristique peut résoudre un problème ou combler un besoin. Par exemple: « Notre appareil est plus rapide que ceux de la concurrence, ce qui signifie que vous pourrez accroître votre productivité. » L'expression « ce qui signifie que » permet d'établir le lien entre la caractéristique et l'avantage. Bien que l'avantage soit plus convaincant que la caractéristique, il reste générique et applicable à plusieurs clientèles. En revanche, si lors de vos discussions votre interlocuteur exprime un besoin précis, par exemple la nécessité de réduire ses coûts de main-d'œuvre, vous pouvez adapter votre discours ainsi: « Notre équipement est plus rapide que ceux de notre concurrence, ce qui signifie que vous pourrez réduire vos coûts de main-d'œuvre ». En formulant ainsi votre argumentaire, vous soumettez un bénéfice précis, ce qui augmentera encore plus vos chances de succès.

La méthode SPIN vise à découvrir les véritables besoins de la clientèle et à démontrer comment votre offre permettra de résoudre les problèmes en présentant un retour sur l'investissement. La méthode Challenger, quant à elle, remet en question les croyances du client et présente l'offre de façon assertive. La méthode SPIN est accessible à tous les vendeurs et vendeuses en raison de sa simplicité. En revanche, la méthode Challenger requiert davantage d'expérience, car elle demande une assertivité maitrisée pour être efficace sans brusquer l'interlocuteur. Voyons maintenant ces méthodes en détail.

3.1 / La méthode SPIN

La méthode SPIN (Rackham, 1988) repose sur une étude, menée auprès de 35 000 personnes travaillant en vente. Elle permet non seulement d'identifier les bonnes questions à poser, mais également de voir à quel moment les formuler. L'acronyme SPIN représente 4 types de questions stratégiques : Situation, Problème, Implication et Nécessité de rendement. Voyons maintenant ce que signifie chacune des étapes et le type de questions qui permettent d'obtenir l'assentiment.

Les questions de **situation** permettent de contextualiser les besoins, d'en apprendre davantage sur l'entreprise, son contexte concurrentiel, ses processus, etc. Elles sont utiles pour les membres de l'équipe de vente, mais n'apportent aucune valeur immédiate à la clientèle. Si une information pouvait être obtenue autrement, évitez de poser une question qui ferait perdre du temps à votre interlocuteur. Malgré tout, au début d'une relation, certaines questions restent incontournables pour proposer une solution taillée sur mesure. Vous pourriez, par exemple, chercher à savoir quelle quantité l'entreprise utilise par mois ou à quelle fréquence elle doit réaliser une tâche précise. Il peut aussi être pertinent de comprendre le processus décisionnel et d'identifier les membres de l'équipe d'achats et leur degré d'influence. Si vous pouvez obtenir le nom des fournisseurs actuels, cela vous aidera à ajuster votre approche et votre proposition. En général, vous éviterez de passer trop de temps dans cette phase de questionnement.

La deuxième est celle de la découverte du **problème**. Puisque vous avez obtenu une rencontre, c'est qu'il existe un problème ou un besoin. À vous maintenant de le mettre au jour et d'encourager la personne à en parler. Ce n'est pas très agréable de dévoiler ses difficultés, surtout en affaires, car cela démontre une faiblesse. Pourtant, cette étape est cruciale afin de faire prendre conscience que le problème cause des ennuis et que vous êtes en mesure d'y remédier. Vous pourriez demander: « Si vous possédiez une baguette magique, quelle serait la première chose que vous aimeriez améliorer dans votre situation actuelle? », ou encore: « Est-ce que cette situation nuit à vos opérations? » Comme le fait d'exposer ses problématiques peut s'avérer délicat, il est recommandé de ne pas trop s'attarder à cette étape. Passez à l'étape suivante, l'**implication**, une fois que vous avez bien compris le problème.

Vous explorerez ici les conséquences que ce problème entraine pour son organisation. En accompagnant la personne dans sa réflexion sur les risques associés aux difficultés, vous l'aidez à prendre conscience des coûts et des inconvénients. Le changement de produit, de service ou de fournisseur apparaîtra alors comme une option plus attrayante. Vous bâtissez ainsi une argumentation solide qui justifiera le prix demandé pour votre solution. À cette étape, vous pourriez poser des questions telles que: « Que vous causent ces difficultés que vous venez d'exprimer? » ou encore: « Quel est l'impact d'une plus grande maintenance de vos équipements sur votre compétitivité? » Ces questions peuvent également créer un malaise, mais elles sont essentielles pour amorcer la prochaine étape: la **nécessité de rendement**.

Les réponses obtenues précédemment vous aideront à construire la valeur de votre offre. Vous permettrez à votre clientèle d'envisager un meilleur avenir en lui faisant entrevoir les bénéfices qu'elle obtiendra en acquérant votre solution. Elle pourra maintenant se concentrer sur la solution et non plus sur le problème. Vous pourriez demander: « Quelles sont les économies que vous pourriez réaliser si votre personnel n'avait plus à travailler avec trois logiciels comptables différents? » Ou encore: « Combien de ventes additionnelles pourriez-vous conclure en réduisant de 50 % les temps de maintenance? » ou «De quelle façon croyez-vous qu'une nouvelle machine à café améliorerait la rentabilité de votre restaurant? »

Les quatre étapes de la méthode SPIN et des questions possibles pour chacune des étapes sont proposées au tableau 4.2.

TABLEAU 4.2 / **Étapes de la méthode SPIN**

Étape	Exemple de question
Situation	• Quelle quantité utilisez-vous par mois? • À quelle fréquence devez-vous...? • Comment prendrez-vous votre décision? • Qui est votre fournisseur actuel?
Problème	• Quelle serait la première chose que vous aimeriez améliorer? • Est-ce que cette situation nuit à vos opérations?
Implication	• Que vous causent ces difficultés que vous venez d'exprimer? • Quel est l'impact d'une plus grande maintenance de vos équipements sur votre compétitivité?
Nécessité de rendement	• Quelles sont les économies que vous pourriez réaliser? • Combien de ventes additionnelles pourriez-vous conclure? • De quelle façon croyez-vous qu'une nouvelle machine améliorerait la rentabilité?

Quand vous quitterez la rencontre, vous aurez fourni les arguments nécessaires pour que la personne puisse «vendre» elle-même votre solution aux autres membres de la centrale d'achat. Plus la vente est importante, plus les discussions internes prennent de l'importance en comparaison avec votre propre intervention, d'où la nécessité de munir votre vis-à-vis des arguments en faveur de votre solution.

La méthode SPIN permet de réduire les objections, car votre vis-à-vis devient un expert de la solution à son problème. Vous ne souhaitez pas que le responsable des approvisionnements connaisse toutes les caractéristiques de votre produit. Vous voulez que cette personne retienne que votre solution peut résoudre son problème ou satisfaire son besoin. La planification est tout aussi essentielle pour appliquer la méthode SPIN. Plus vous en saurez sur la firme cliente avant votre rencontre, moins vous devrez poser de questions de situation. Plus rapidement, vous identifiez 3 difficultés auxquelles la clientèle pourrait être confrontée et pour lesquelles vous avez la solution, plus efficace vous serez pour les mettre au jour, voire pour en révéler d'autres. Si vous prenez le temps de comprendre les implications générales de ce type de problèmes pour les entreprises, vous serez mieux préparé à aider votre clientèle à en mesurer l'impact, même si ces répercussions diffèrent de celles que vous avez prévues. Enfin vous devez pouvoir chiffrer la valeur de votre proposition pour conduire la clientèle vers les bénéfices du changement.

3.2 / La méthode Challenger

L'autre approche, également fondée sur une étude menée auprès de milliers de spécialistes de la vente, est la méthode Challenger (Dixon et Adamson, 2011). Elle repose sur trois actions clés: enseigner, adapter et prendre le contrôle de la conversation.

La première action consiste donc à **enseigner**. La clientèle apprécie les personnes en vente qui les amènent à réfléchir différemment en leur soumettant des idées innovantes pour optimiser leur croissance ou réduire leurs coûts. Ne vous contentez pas de présenter des idées génériques qui pourraient également servir à votre concurrence. Mettez plutôt en avant celles qui démontrent clairement pourquoi votre offre est la plus adaptée pour concrétiser ces nouvelles opportunités. Ces connaissances doivent également être applicables à un segment précis de votre marché. Pour réussir à vous démarquer grâce à l'enseignement, les auteurs recommandent de renforcer votre crédibilité en dévoilant, par exemple, les défis courants auxquels votre clientèle fait face et qui résonnent particulièrement avec votre interlocuteur. Ensuite, orientez la discussion vers des éléments qui valorisent votre approche, sans dévoiler immédiatement votre recommandation. Ce n'est qu'une fois que vous aurez amené votre clientèle à voir la situation sous un nouvel angle que vous exposerez les bénéfices de votre offre.

On reconnaît depuis longtemps l'importance d'**adapter** son message à l'industrie de la clientèle et à la réalité spécifique de chaque entreprise. Toutefois, la deuxième action de l'approche Challenger pousse cette adaptation encore plus loin en ajustant non seulement le discours aux besoins de l'organisation, mais aussi aux personnalités et aux rôles des différents décideurs et décideuses impliqués. L'objectif est donc d'identifier les priorités et les motivations de chaque membre de la centrale d'achats afin d'adapter l'argumentaire en conséquence. Par exemple, la direction des services financiers n'aura pas les mêmes préoccupations que les graphistes.

La troisième action, **prendre le contrôle de la conversation**, requiert un équilibre subtil entre la passivité et l'agressivité. Lorsqu'une personne résiste fortement à une proposition ou pendant une négociation, une personne en vente pourrait choisir de céder trop rapidement avec une attitude plus passive, ou, à l'inverse, adopter une posture trop offensive qui risquerait de nuire à la relation. En réalité, la majorité des personnes en vente ont tendance à battre en retraite plutôt qu'à confronter, en partie parce qu'elles perçoivent que le pouvoir repose entre les mains de la clientèle et que l'on n'a jamais cessé de leur dire que la clientèle passe en premier et se trouve au centre du succès de l'organisation. La méthode Challenger recommande au contraire une approche assertive qui consiste à exercer une pression calculée et observer la réaction de la clientèle face à cette posture plus affirmée. Toutefois, cette approche ne peut fonctionner que si l'équipe de vente maitrise parfaitement la valeur réelle de son offre et a la confiance et l'expérience nécessaires pour la défendre.

4 / Répondre aux préoccupations ou aux objections

Le chapitre 2 a déjà souligné l'importance d'anticiper les objections. Certains affirment que si l'on met suffisamment en avant les bénéfices d'une offre, on peut prévenir bon nombre d'objections (Rackham, 1988). Lorsqu'un client ou une cliente exprime une préoccupation, cela peut signifier que la présentation n'a pas suffisamment mis en lumière les bénéfices, ou que ceux-ci ne sont pas perçus comme assez convaincants. Donc, à moins que votre produit ou votre service ne s'impose de lui-même dans le marché grâce à des bénéfices incontournables, ou que vous maitrisez parfaitement les techniques SPIN ou Challenger, il y a fort à parier que vous devrez utiliser le processus de réponse aux objections qui suit. Ignorer une objection n'est pas une bonne stratégie, car cela réduit considérablement vos chances d'obtenir un engagement. La réponse aux objections se déroule en quatre étapes. La première consiste à écouter activement en reformulant certaines affirmations pour s'assurer d'avoir bien compris la véritable préoccupation de la clientèle. À la deuxième étape, vous voudrez découvrir les motivations sous-jacentes à l'objection en posant des questions.

Ce n'est qu'à la troisième étape que vous pourrez réellement réagir à l'objection, car vous disposerez alors de toutes les informations nécessaires. Plusieurs types de réponses sont envisageables. L'une d'elles consiste à contredire la personne quand l'objection se base sur des informations erronées, par exemple, si elle soutient que la concurrence offre une garantie de cinq ans sur son produit, alors que vous savez très bien que c'est faux. Si elle remet en doute les performances que vous annoncez, vous pourriez effectuer une démonstration ou lui proposer d'essayer le produit ou le service pendant un certain temps. Une autre approche consiste à présenter des témoignages ou des preuves visuelles issues d'expériences clients antérieures pour renforcer votre crédibilité. Enfin, votre produit ou service doit avant tout être perçu comme bénéfique pour le client ou la cliente. Une présentation soigneusement préparée mettra systématiquement de l'avant les bénéfices concrets, plutôt que de s'attarder uniquement sur les caractéristiques ou les avantages techniques.

Après avoir correctement répondu à l'objection, la quatrième étape consiste à vous assurer que vous avez effectivement répondu à cette préoccupation et que vous pouvez poursuivre votre présentation. Vous pouvez y aller de façon directe et demander: « Ai-je bien répondu à votre question?» Si la personne répond oui, vous pouvez enchainer de cette façon: « Puis-je continuer ma présentation?» Assurez-vous que toutes les objections ont été traitées, car il est difficile sinon impossible de conclure une vente si des doutes persistent.

5 / La proposition de vente

À un certain moment au cours des premières rencontres, vous devrez soumettre votre offre plus formellement, généralement sous forme écrite. Dans certains cas, cela se traduira par une simple soumission, mais dans d'autres, il sera nécessaire de présenter un document plus substantiel. Les sections suivantes aborderont plusieurs aspects essentiels de cette phase: l'utilisation des supports visuels, les techniques de présentation efficaces, les principes de persuasion, le rôle du *storytelling* (narration) et la notion de vente de valeur.

5.1 / Les supports visuels

La présentation visuelle en image ou en texte peut prendre plusieurs formes : un support papier, une vidéo, une présentation en réalité virtuelle, un diaporama de type PowerPoint, ou des images numériques. Si votre entreprise dispose d'un département de marketing, vous aurez probablement accès à de la documentation standardisée sur vos produits, sur vos services et sur l'organisation elle-même, souvent accompagnée d'une liste d'arguments pour illustrer la valeur de votre offre. Vous savez maintenant que cette documentation devra être adaptée à chaque entreprise et à chaque personne.

Une proposition écrite doit être soignée. Elle doit inclure une page couverture et une table des matières si le document dépasse 10 pages. Dès les premières pages, vous résumerez les attentes de la clientèle auxquelles vous proposez une solution. Ensuite, vous détaillerez votre solution dans les sections suivantes en insistant toujours sur les bénéfices. Puis, vous mettrez en avant des preuves tangibles et des témoignages crédibles afin de renforcer la valeur de votre offre, avant de dévoiler votre prix. Vous conclurez en présentant votre entreprise ainsi que les membres clés de votre équipe, soulignant ainsi votre expertise et votre capacité à répondre aux besoins de la personne.

Les preuves consistent à quantifier les bénéfices. Par exemple, si des tests démontrent que votre équipement accroît la cadence de 3 %, sans augmenter le personnel requis, et que cette cadence permet à la clientèle d'engranger 10 000 dollars de plus par année, la différence de prix entre votre produit et celui, moins coûteux, de la concurrence devient alors plus facile à justifier. Ces preuves sont élaborées en fonction des bénéfices qui comptent le plus. Il est essentiel de chiffrer les bénéfices autant que possible en fonction du prix. Ainsi, si vous êtes en mesure de prouver que l'emploi de votre produit entraînera des économies sur 18 mois, qui paieront d'elle-même votre offre, ou qui combleront l'écart entre votre prix et celui de la concurrence, il sera difficile de résister à un tel argument. Vous pouvez également utiliser les chiffres pour exposer les résultats de sondage de satisfaction de la clientèle ou démontrer la performance de votre produit comparée à celle des offres concurrentes.

Les compagnies qui vous connaissent peu seront d'autant plus rassurées lorsque vous présenterez des références d'entreprises qui font affaire avec vous. Pour renforcer leur confiance, les témoignages de clientèles satisfaites sont également des outils convaincants. La raison est simple : l'entreprise parlera toujours en bien de ses produits ou de ses services, mais rien ne vaut un avis externe. Observez comment vous-même tenez compte des opinions d'autrui pour choisir un hôtel, un restaurant ou pour évaluer des produits concurrents avant de faire un achat. Il peut être plus aisé d'obtenir des témoignages dans certaines industries que dans d'autres. Vous pouvez donc suggérer à la personne que vous sollicitez pour un témoignage de publier le texte de façon anonyme. Certaines personnes prétexteront ne pas avoir le temps de rédiger un témoignage, dans ce cas, proposez-leur de rédiger un brouillon basé sur votre connaissance de leur expérience qu'elles pourront lire et approuver ultérieurement. Vous pouvez également vous inspirer des nombreux exemples facilement accessibles sur des sites Web d'entreprises.

La section suivante de la proposition de vente aborde le prix. Selon les domaines, plusieurs façons de le soumettre sont possibles : un prix tout inclus, un prix décomposé, ou encore des options de prix. Le prix tout inclus permet de mettre en valeur tous les services compris, comme la garantie ou le service après-vente. Il est recommandé de quantifier la valeur d'un service inclus. Par exemple, plutôt que d'indiquer simplement que l'installation est offerte, mieux vaut préciser : « L'installation, évaluée à 1250 dollars, est incluse sans frais. » Cela permet de mettre en évidence les économies réalisées.

On peut décomposer un prix de différentes manières : par personne, à l'heure avec une estimation du temps requis, ou encore en fonction des options et accessoires d'un équipement. Les options de prix peuvent également inclure différents modèles ou niveaux de service, à l'image d'un menu de restaurant.

La dernière section de votre proposition de vente doit mettre en avant les membres de votre équipe. La clientèle sera rassurée de savoir qu'elle bénéficiera d'un soutien structuré, plutôt que de devoir compter sur une seule personne. Évidemment, si vous travaillez seul, il peut être plus compliqué de présenter une équipe, mais il est fréquent que des entrepreneurs s'associent de manière informelle avec d'autres spécialistes offrant des services complémentaires.

Mentionner ces collaborations peut renforcer la crédibilité de votre offre. De façon générale, vous devrez porter une grande attention aux choix des mots et des images que vous sélectionnerez dans votre proposition de vente. Par exemple, assurez-vous d'utiliser le pronom «vous» au lieu de «nous». Cela signifie que vous exposerez davantage les bénéfices que des caractéristiques qui peuvent vous paraître attrayantes, mais qui n'ont que peu d'intérêt pour la clientèle.

On dit qu'une image vaut mille mots. Même si vous vendez des services, n'hésitez pas à les illustrer visuellement dans votre proposition de vente. Préférez des images où l'on voit votre solution en action: plutôt que de présenter les composants d'un système d'arrosage de jardin, montrez-le une fois installé et en fonctionnement. Encore mieux, intégrez des personnes souriantes qui l'utilisent, renforçant ainsi l'impact émotionnel.

Outre la proposition écrite sur un support papier, vous pourriez aussi faire appel aux vidéos, à la réalité virtuelle ou aux diaporamas de type PowerPoint.

Les courtes **vidéos** connaissent un fort engouement sur les médias sociaux, et de nombreuses entreprises commencent à peine à exploiter ce format. Aujourd'hui, un simple téléphone cellulaire suffit pour produire un clip de quelques secondes avec une qualité suffisante.

Pour les compagnies disposant de budgets plus conséquents, la **réalité virtuelle** devient une option de plus en plus accessible pour présenter leurs produits ou leur entreprise. Il ne s'agit que d'une question de temps avant que la démocratisation de cette technologie, grâce à la baisse des coûts de production et l'allègement des casques, en fasse un outil courant. Certaines entreprises utilisent déjà la réalité virtuelle pour démontrer le fonctionnement de machines imposantes, mesurant plusieurs dizaines de mètres. Le développement de la réalité augmentée et de la réalité virtuelle offre également des opportunités pour les exposants dans les foires commerciales d'améliorer l'expérience client (Wieland et coll., 2024). Imaginez que vous vouliez démontrer l'efficacité de votre chaine de fabrication, trop volumineuse pour être déplacée sur un salon, la réalité virtuelle peut dorénavant permettre aux visiteurs d'explorer votre usine comme s'ils y étaient.

Parmi les supports visuels les plus couramment utilisés pour les présentations de vente, le **diaporama** de type PowerPoint reste incontournable. Toutefois, il peut devenir une béquille lorsqu'il est mal utilisé. Certaines normes s'appliquent à l'utilisation de support visuel, qui peut également être imprimé. Trois principes clés peuvent considérablement améliorer vos présentations. Premièrement, limitez la quantité de texte par diapositive. La plupart des spécialistes recommandent une taille minimale de police de 24 points afin d'assurer une lecture fluide et rapide. Deuxièmement, restreignez le nombre de points à six par diapo. Cela vous obligera à structurer vos idées de manière concise et vous évitera de lire le texte à l'écran au lieu d'interagir avec votre auditoire. Troisièmement, prévoyez environ une minute par diapositive. Ainsi, pour une présentation de 15 minutes, inutile de préparer 40 diapositives : 15 à 20 suffisent pour maintenir l'attention du public tout en restant efficace.

5.2 / La présentation de la proposition de vente

Si vous présentez votre proposition à une seule personne, vous vous appuierez peut-être sur votre expérience ou sur votre éloquence naturelle. Cependant, une trop grande confiance en vous peut nuire à votre présentation. Si vous vous êtes donné la peine de préparer une proposition, vous devez prendre le temps de bien la présenter. Vous pourriez même aller jusqu'à l'écrire mot à mot pour vous assurer de ne rien oublier. L'objectif n'est pas de réciter, mais d'avoir un repère en cas de trou de mémoire. La répétition vous permettra également de repérer d'éventuelles lacunes dans votre argumentation.

Si vous présentez devant plusieurs personnes, cette préparation devient encore plus nécessaire, car vous pourriez éprouver une certaine nervosité. Deux actions peuvent vous aider à maitriser ce trac : respirer consciemment et marcher en parlant, puisque le mouvement atténue les manifestations du trac. Assurez-vous de ne pas tourner le dos à votre auditoire pour regarder vos diapositives, ou pire, pour les lire. En prévoyant un écran de portable devant vous, vous pourrez suivre le déroulement en gardant un contact visuel avec votre auditoire.

Vous voudrez peut-être en profiter pour mettre en valeur quelques membres de votre équipe. Cependant, si ces personnes sont moins habituées que vous à parler en public, une répétition préalable sera essentielle. Cela leur permettra de mieux maitriser leur rôle et d'assurer une présentation fluide et convaincante.

La présentation est aussi l'occasion d'offrir une démonstration, l'un des moyens les plus efficaces pour mettre en valeur les bénéfices de votre offre. Que ce soit pour illustrer l'efficacité d'un logiciel, d'un produit portatif ou, avec le soutien de la réalité virtuelle, d'un produit ou d'un service non transportable, une démonstration rendra votre proposition plus concrète et convaincante. Lorsque la démonstration physique est impossible, les images vidéo peuvent prendre le relais. Essayez de solliciter les cinq sens de votre auditoire. Si vous possédez des échantillons d'un produit qui dégage une odeur agréable, invitez les gens à le sentir. Vous pourriez aussi faire entendre le signal sonore de votre produit, comme un avertisseur en santé-sécurité au travail, ou faire goûter un plat si vous œuvrez dans l'hôtellerie et la restauration. Si votre clientèle potentielle accepte de se déplacer, une démonstration inclura la visite d'une usine ou d'un hôtel pour l'organisation d'un congrès.

Après avoir consacré un temps considérable à préparer votre présentation, il peut être tentant de faire défiler les diapositives les unes après les autres, sans réelle interaction. Avez-vous pensé à des moyens pour encourager la participation? Poserez-vous des questions pour susciter l'engagement? Une présentation dynamique et participative est bien plus efficace qu'un monologue récité, comme le démontre l'exemple présenté à l'encadré 4.1.

ENCADRÉ 4.1 / **Une présentation trop bien préparée**

Un client nous avait demandé de présenter notre politique sur le développement durable. Pensant bien faire, notre équipe marketing avait préparé un document qui contenait plus de cent diapositives. Le représentant que j'accompagnais à titre de vice-président des ventes s'est dès lors lancé avec confiance puisqu'il avait bien répété. Au bout de quelques minutes, je sentais que le client s'ennuyait. Je lui ai demandé si nous répondions à ses attentes et il m'a expliqué qu'il souhaitait non pas une présentation détaillée, mais plutôt un résumé concis de notre position. Heureusement, nous avons rapidement corrigé le tir en réorientant la discussion. Cette expérience nous a appris une précieuse leçon : une présentation ne doit pas être un monologue figé, mais un échange adapté aux besoins de l'auditoire. Depuis cet incident, j'ai exigé de mon équipe de vente qu'elle adapte toujours le matériel fourni par le marketing avant toute rencontre avec une clientèle.

5.3 / La persuasion

Les techniques de vente s'intéressent depuis longtemps à la persuasion. Robert Cialdini (1990) a identifié six principes fondamentaux qui facilitent la prise de décision et influencent les comportements : la **réciprocité**, selon laquelle une personne se sent souvent encline à donner quelque chose en retour lorsqu'on lui offre une information précieuse, un essai gratuit ou un échantillon ; la **rareté** qui suscite le désir en rendant un produit ou un service exclusif ; l'**autorité** qui repose sur la tendance des individus à suivre les recommandations d'un expert ou d'une experte ; la **cohérence**, qui pousse une personne à maintenir une ligne de conduite en accord avec ses engagements et opinions passés ; l'**appréciation**, qui rend les propositions plus convaincantes lorsqu'elles proviennent d'une personne avec qui l'on ressent une affinité ; et finalement, la **preuve sociale**, qui influence fortement les choix en raison de la tendance à adopter un comportement déjà affiché par un nombre significatif de personnes. Selon l'auteur, si vous appliquez un ou plusieurs de ces principes dans votre approche commerciale, vous maximisez vos chances de persuasion, que ce soit en vente ou dans d'autres contextes professionnels.

Un autre auteur, Jay Conger (1998), propose un cadre en quatre étapes pour convaincre plus aisément son entourage ou sa clientèle. La première consiste à établir sa crédibilité en démontrant son expertise et en cultivant des relations solides. J'ai déjà insisté sur l'influence de l'expertise en développement des affaires, qui ne peut se réaliser qu'en maintenant ses connaissances à jour. J'ai également souligné l'importance d'étendre ses relations au sein de l'entreprise cliente. La crédibilité se gagne à force d'efforts constants et sans elle, la persuasion sera difficile. La deuxième étape propose de signaler les bénéfices pour les personnes que vous voulez convaincre. La troisième étape consiste à présenter des preuves pour soutenir votre argumentation. Enfin, Conger suggère d'adapter l'expression de ses émotions à son audience.

On voit bien que les techniques de persuasion ne sont pas si éloignées des principes évoqués jusqu'ici dans ce manuel, que ce soit la nécessité d'établir sa crédibilité ou de créer des relations pour que les personnes nous apprécient et achètent notre produit ou notre service. En revanche, les gens n'aiment pas qu'une personne en vente use de stratagèmes pour les convaincre. En fait, ils connaissent les tactiques et savent comment s'en prémunir (Friestad et Wright, 1994). La persuasion peut devenir un élément important en développement des affaires dans la mesure où la personne en vente croit que l'autre connaît ses intentions et agit en conséquence.

5.4 / La narration

La narration, connue dans le monde anglo-saxon comme le *storytelling*, est une technique efficace dans divers domaines, dont la vente et le développement des affaires. Elle constitue l'une des tactiques les plus puissantes dont dispose la force de vente pour influencer, éduquer et inspirer. Qu'est-ce qui la rend si utile? Tout d'abord, les récits établissent des liens entre les individus, les personnes et les idées. Ils servent de véhicule pour transmettre la culture, l'histoire et les valeurs qui unissent les gens. Ils vont au-delà de la simple création d'une émotion de connexion. Ils instaurent un sentiment de familiarité et de confiance, permettant à la personne responsable des achats de s'immerger dans l'histoire depuis sa propre perspective, ce qui favorise une réceptivité accrue. Un récit bien construit permet ainsi de rendre des idées complexes plus accessibles, tout en favorisant l'engagement et le bouche-à-oreille (Gilliam et Flaherty, 2015). Contrairement à une simple énumération de données ou à une discussion sur des concepts abstraits, les histoires se révèlent bien plus captivantes, mais aussi plus faciles à retenir.

Le récit de vente ne relève pas de la fiction, mais plutôt d'un documentaire structuré qui emprunte les règles de la narration. Le premier élément de cette narration est de choisir un ou des personnages réels auquel votre vis-à-vis pourra s'identifier. Un ou une propriétaire d'une entreprise en démarrage qui ne compte que quelques membres du personnel ne se reconnaîtra pas dans l'histoire d'une multinationale ; il est donc essentiel d'adapter l'exemple à la réalité de votre interlocuteur. Ensuite, vous devrez construire l'histoire en trois actes. Pour le premier, vous devez présenter vos personnages et le contexte dans lequel ils évoluent. Le deuxième acte fait survenir un problème, un conflit, une difficulté à surmonter. Enfin, le troisième acte consiste à résoudre ce conflit grâce à votre produit ou votre service. Un exemple de récit de vente est reproduit à l'encadré 4.2.

ENCADRÉ 4.2 / **Un exemple d'histoire de vente**

Le contexte et le premier acte : William (le personnage) a démarré une société de consultation en informatique qui a connu une croissance étonnante au cours des trois dernières années. Récemment, son équipe est passée de 6 à 12 employés et employées.

Le conflit et le deuxième acte : William a engagé quelqu'un pour s'occuper de la comptabilité, mais, bien vite, le logiciel qui lui donnait l'opportunité de suivre ses revenus et ses dépenses au lancement de son entreprise s'est révélé inefficace pour la facturation ou la paie du personnel. Les solutions offertes sur le marché étaient soit trop compliquées pour une PME, soit incompatibles avec son *Customer Relationship Management* (CRM).

La résolution du conflit et le troisième acte : Nous avons alors proposé notre logiciel Dcompte (nom fictif) qui s'avérait plus simple d'utilisation et permettait une intégration avec le CRM de la compagnie. William et son équipe recourent à Dcompte depuis quelques mois et peuvent maintenant se concentrer sur leur activité principale : la consultation en informatique.

Ces histoires, vous les racontez peut-être déjà entre les membres de votre équipe. Vous devez maintenant les cataloguer, dans un CRM ou ailleurs, pour être en mesure de les utiliser au moment opportun.

5.5 / La vente de valeur

La vente de la valeur consiste à traduire des bénéfices en termes monétaires, en vue de démontrer concrètement comment une offre permet de contribuer à la profitabilité de la firme cliente (Terho et coll., 2012). Bien que de nombreuses définitions de la valeur aient été proposées au fil des ans, il existe un consensus sur certains principes fondamentaux: la valeur est liée à l'usage d'un produit ou d'un service, elle est perçue par la clientèle et non dictée par le fournisseur, et elle suppose une contrepartie financière (Woodruff, 1997). Une autre approche de la valeur suggère qu'elle existe aussi au sein même de la relation parce qu'elle constitue la somme des interactions et des expériences passées ainsi que les anticipations au sujet de l'avenir de la relation (Minerbo et coll., 2021).

La vente de valeur repose sur trois étapes clés: la planification, l'implantation et l'exploitation (Töytäri et Rajala, 2015). Lors de la phase de planification, les entreprises doivent identifier les éléments qui génèrent de la valeur pour leur clientèle, puis élaborer une proposition de valeur accompagnée d'outils de vente adaptés. L'implantation consiste ensuite à sélectionner des entreprises auprès desquelles appliquer cette approche et à quantifier la valeur offerte. Finalement, à la phase d'exploitation, on devra confirmer la concrétisation de la promesse de valeur et rédiger des études de cas afin de démontrer, preuves à l'appui, comment l'offre a généré des bénéfices tangibles pour la clientèle. Un exemple réel de vente de valeur est présenté dans l'encadré 4.3.

ENCADRÉ 4.3 / **SKF: La vente de valeur de roulements à billes**

La compagnie suédoise de roulements à billes SKF utilise la vente de valeur pour commercialiser ses produits à un prix 50 % plus élevé que ceux de la concurrence, tout en obtenant des parts de marché plus importantes (Hinterhuber et coll., 2021). L'entreprise réussit ce tour de force en attribuant une valeur financière garantie, et non juste promise, à ses produits, en démontrant leur fiabilité supérieure, en prouvant une réduction du temps d'installation et en mettant en évidence un besoin moindre de lubrification. Même la rémunération des équipes de vente est partiellement reliée à l'augmentation de la profitabilité de la clientèle ou à sa satisfaction.

La récupération de la valeur par l'équipe de vente ou le service des achats constitue un enjeu majeur. Elle peut s'opérer de trois façons: par la négociation de prix, qui sera abordée au chapitre 5 et qui implique qu'une des parties réalise des gains au détriment de l'autre; par les économies de volume, qui permettent de réduire les coûts grâce à des volumes importants; ou encore par une approche collaborative, où les deux parties cocréent de la valeur, par exemple, en développant conjointement de nouveaux produits ou services (Minerbo et coll., 2021).

6 / Le bilan de la rencontre

Tout comme la planification d'une rencontre nécessite une préparation minutieuse, un bilan s'avère essentiel pour assurer un suivi efficace et planifier les prochaines étapes. Pour ce faire, vous pourriez utiliser une feuille telle que celle présentée au tableau 4.3. En fait, vous pourriez imprimer le formulaire de planification du tableau 4.1 au recto et le bilan du tableau 4.3 au verso.

TABLEAU 4.3 / **Bilan de la rencontre**

Bilan de la rencontre	
Qu'avez-vous accompli durant cette rencontre?	
Quels sont les éléments qui se sont bien déroulés pendant la rencontre?	
Quels sont les éléments qui se sont moins bien déroulés pendant la rencontre?	
Qu'avez-vous appris?	
Quelles sont les trois prochaines étapes?	
Quel est l'échéancier prévu pour leur mise en œuvre?	
Quelles ressources humaines et matérielles seront nécessaires pour les réaliser?	

Source: Adapté de Kane et Zappulla (2016).

Vous devriez remplir ce formulaire dès la fin du rendez-vous, pendant que les informations sont encore fraîches dans votre esprit. Vous ne pourrez pas planifier efficacement une prochaine rencontre sans d'abord réaliser un bilan de celle qui vient de se terminer, puisque bilan et planification sont intimement liés. Je vous propose quelques questions, mais n'hésitez pas à les adapter ou à en ajouter selon votre réalité. Le bilan permet d'instruire les membres de votre équipe de vos progrès ou des obstacles qui surviennent, mais aussi de mieux structurer la suite des échanges en tenant compte de la rétroaction obtenue.

En résumé

Au même degré que pour la phase de découverte, ce chapitre a mis en évidence l'importance de la planification. Maitriser différentes techniques pour établir un premier contact, identifier les besoins et mettre en valeur les bénéfices de son offre est essentiel pour progresser dans le processus de vente et d'achat. Une proposition de vente et une présentation bien préparée, intégrant des éléments essentiels comme des preuves, des témoignages, des démonstrations, des réponses convaincantes aux préoccupations et une narration engageante, facilitent la transition vers l'étape suivante: l'engagement.

CHAPITRE 5 /

L'engagement

Une fois que les deux parties auront exploré leur intérêt à poursuivre la relation, elles devront déterminer les termes de leur engagement par le biais de la négociation. Celle-ci devra d'abord être soigneusement préparée. Ensuite, la personne en développement des affaires sera encouragée à observer des étapes bien précises pour éviter de consentir trop vite aux demandes de l'autre partie. La négociation du prix fera l'objet d'une section particulière en raison de la fréquence à laquelle ce facteur surviendra durant les pourparlers. Ce chapitre abordera également la question de la négociation par courriel et par d'autres moyens numériques. L'engagement sera officialisé lorsque l'autre partie donnera son accord, et la personne en vente pourra user de quelques tactiques pour faciliter cette décision. Enfin, le chapitre se conclura avec des éléments qui permettront de pérenniser la relation, notamment par le service à la clientèle et le développement de partenariats avec la clientèle clé.

1/ La planification de la négociation

Une négociation réussie repose sur une préparation minutieuse, où rien n'est laissé au hasard. Cette planification implique la maitrise de quatre éléments : l'objet de la négociation, la compréhension des forces en présence et de leur pouvoir, l'identification des concessions possibles, ainsi que l'établissement d'un ordre du jour clair et d'une hiérarchisation des enjeux.

1.1 / L'objet de la négociation

La négociation portera souvent sur les enjeux suivants:

- Techniques: caractéristiques du produit ou spécifications requises par la firme cliente.
- Logistiques: échéancier, moyen de transport, emballage, service à la clientèle.
- Juridiques: confidentialité, propriété intellectuelle, garantie, résolution des conflits.
- Financiers: conditions de paiement, gestion des inventaires, prix.

Chacun de ces enjeux possède une valeur différente pour l'une ou l'autre des parties. Ainsi, la firme cliente peut préférer le produit du fournisseur à celui de toutes les possibilités offertes sur le marché, parce qu'il lui procure des bénéfices supérieurs à ceux des choix possibles sur le marché. Toutefois, cette préférence ne se concrétisera que si les efforts logistiques ou financiers exigés restent raisonnables. En d'autres mots, le coût total ne doit pas excéder la valeur des bénéfices obtenus. C'est pourquoi la personne en développement des affaires doit identifier avec précision les éléments de son offre pour lesquels la clientèle accorde de la valeur, soit les attributs du produit ou du service qui lui apporteront un bénéfice. La promotion de ces bénéfices est nécessaire lors des discussions.

Dans certains cas, l'objet de la négociation est clairement défini. Par exemple, si vous proposez trois modèles de foreuses avec des caractéristiques différentes, une fois le modèle choisi, la négociation portera sur des éléments précis tels que les délais de livraison, le service à la clientèle, les modalités de paiement et, bien sûr, le prix. Cependant, pour d'autres industries, la négociation peut s'avérer bien plus complexe. Prenons le cas d'un entrepreneur négociant avec une entreprise pour la construction d'une nouvelle usine. Un tel projet implique un devis pouvant s'étendre sur des centaines de pages et nécessitant des pourparlers sur des dizaines d'aspects techniques, logistiques et juridiques. Face à cette diversité, il est essentiel de bien cerner la valeur de votre offre et de la situer par rapport aux options offertes sur le marché, notamment celles proposées par la concurrence. Cette compréhension vous permettra de mieux justifier votre prix et de défendre votre position avec assurance.

Par ailleurs, avant d'entamer la négociation, il est primordial de définir vos objectifs. Bien entendu, vous devez connaître le prix minimum que vous êtes prêt à accepter, mais vous devrez encore considérer de nombreux autres facteurs. Souhaitez-vous renforcer la relation d'affaires en vue d'obtenir une part plus importante des achats futurs de cette entreprise? Avez-vous identifié des opportunités de collaboration à long terme? De plus, vous devriez identifier les buts de l'autre partie en recueillant un maximum d'informations sur le Web ou durant des discussions préliminaires. Par exemple, si vous savez qui décidera ou comment le choix se fera, vous pourrez jouer ces cartes lors de la négociation en présentant les bénéfices de votre offre qui résonneront le mieux auprès des différents membres de la centrale d'achats.

1.2 / Examiner les forces en présence et leur pouvoir

Il est essentiel pour la personne en vente de savoir qui participera à la négociation, car elle se déroule souvent en présence de plusieurs individus. Dans ce cas, elle devra déterminer s'il est pertinent de constituer sa propre équipe afin d'équilibrer les forces en présence. Elle doit également connaître le statut et les fonctions des personnes présentes au sein de l'entreprise cliente. En règle générale, il est recommandé d'inclure des personnes occupant des fonctions similaires à celles des membres de l'équipe de votre vis-à-vis, avec un niveau hiérarchique équivalent. Par exemple, si l'autre partie a intégré la direction technique dans son équipe de négociation, vous devriez envisager d'en faire de même. Le style social ou la personnalité des membres de l'équipe cliente doivent être pareillement considérés. Une personne en vente attentive à ces éléments pourra ajuster sa présentation en conséquence et préparer son équipe à interagir de manière plus efficace, en tenant compte des sensibilités et des dynamiques propres à ses destinataires.

Le pouvoir et la dépendance sont des facteurs sur lesquels réfléchir en prévision d'une négociation. Ces concepts de la théorie de l'échange social, initialement explorés par les sociologues à la fin des années 1950, ont été mis de l'avant notamment par Richard M. Emerson (1962), qui affirmait que «le pouvoir d'une personne équivaut à la dépendance d'une autre». En d'autres termes, dans une relation commerciale, le pouvoir exercé par une firme cliente sur un fournisseur découle directement de la dépendance de ce dernier à son égard. Le pouvoir et la dépendance découlent des choix possibles. Ainsi, plus une partie bénéficie de solutions de rechange et plus elle détient du pouvoir. À l'inverse, elle deviendra dépendante si elle dispose de peu de choix. Le pouvoir peut aussi résulter de la capacité à récompenser, par exemple, en achetant un volume important (le pouvoir d'achat), ou, au contraire, à sanctionner, en retirant des commandes au fournisseur. Si le pouvoir est disproportionné, alors la négociation a peu de chances de se produire puisque la partie la plus puissante pourra forcer l'autre à accepter ses conditions, que ce soit de façon diplomatique ou en usant de coercition. Un exemple frappant est celui d'Airbus en 2018, qui a incité les fournisseurs de l'ancienne C Series de Bombardier à réduire leurs prix de 20% afin de continuer d'approvisionner l'entreprise en pièces aéronautiques (La Presse, 2018).

Généralement, les acheteurs et acheteuses détiennent un plus grand pouvoir dans un marché arrivé à maturité, puisqu'ils disposent de nombreuses possibilités. Cependant, un fournisseur peut renverser cette dynamique s'il propose une solution distinctive ou si son industrie s'est consolidée, réduisant ainsi les options disponibles pour la clientèle. Le pouvoir dans une négociation repose sur plusieurs éléments clés:

- **La valeur que procure le partenaire**: Lorsqu'une entreprise représente une part significative des ventes ou des profits d'un fournisseur, elle exerce un pouvoir considérable sur lui. Toutefois, si cette clientèle est peu rentable, le fournisseur pourrait choisir de s'en détacher. À l'inverse, si son produit ou service est essentiel et difficilement remplaçable à court terme, la firme cliente se retrouve avec peu d'options, ce qui réduit son influence dans la négociation.
- **Les options**: Un fournisseur qui évolue au sein d'un oligopole sera moins exposé aux pressions puisque les options de la clientèle sont limitées. À l'inverse, dans un environnement concurrentiel où les possibilités sont nombreuses, le pouvoir sera entre les mains de la clientèle.
- **L'expertise**: Les plus grandes entreprises spécialisent souvent leurs responsables des achats en fonction d'industries précises dans lesquelles évoluent leurs fournisseurs, afin qu'ils deviennent des experts de ces domaines. Cette stratégie leur permet de mieux négocier en compensant l'avantage informationnel des équipes de vente. À l'inverse, les petites entreprises n'ont pas toujours cette capacité et il peut être complexe pour

elles de développer une compétence interne qui fait contrepoids à celle de la personne en vente. Les nouveaux produits et services soulèvent aussi la question de l'expertise. Ainsi, la compréhension des avantages et des inconvénients de l'intelligence artificielle constitue, à l'heure actuelle, un véritable enjeu pour les organisations, quelle que soit leur taille.

- **Le contexte** : Le contexte joue également un rôle clé dans la répartition du pouvoir et la dépendance entre les parties. Une entreprise qui vient de perdre un fournisseur clé en raison d'une faillite devra rapidement en trouver un nouveau. Dans une telle situation, son pouvoir de négociation sera alors considérablement réduit.

Une véritable négociation ne peut exister que lorsque les rapports de force sont relativement équilibrés et que chaque partie souhaite réellement conclure une entente. En d'autres termes, une forme d'interdépendance doit exister entre elles.

1.3 / Identifier les concessions possibles et les exigences

Êtes-vous prêt à faire des concessions? C'est évidemment une drôle de question à poser en vente, mais elle s'avère essentielle. Vous devez déterminer le seuil au-delà duquel vous serez prêt à vous retirer de la négociation. Cela ne sera possible que si vous avez identifié votre meilleure option à une entente négociée connue sous l'acronyme BATNA (*Best Alternative to a Negotiated Agreement*), un concept développé par Roger Fischer et William Ury. Si vous n'avez pas prévu de porte de sortie, vous pourriez devenir anxieux pendant la négociation, une émotion que votre interlocuteur pourrait exploiter à son avantage. Une solution de rechange viable ne s'improvise pas: elle repose sur un travail à long terme qui consiste à maintenir un portefeuille d'entreprises intéressées par votre offre. Ce travail vous permet non seulement de sécuriser des options de remplacement en cas d'échec de la négociation, mais aussi d'avoir confiance en la valeur de votre offre et en votre capacité à conclure des contrats rentables. Gardez en tête que votre vis-à-vis dispose fort probablement d'au moins une alternative à votre proposition. Une recherche approfondie avant la négociation vous aidera à déceler les différentes options dont il pourrait disposer. Vous serez ainsi mieux préparé à démontrer la valeur supérieure de votre offre.

Vous devez ensuite reconnaître les composantes de votre offre qui ont moins de valeur à vos yeux, mais qui pourraient être déterminantes pour l'autre partie. De la même façon, il est important de repérer ce qui compte réellement pour vous, mais qui, à l'inverse, est moins prioritaire pour votre vis-à-vis. Vous pourrez alors concéder des éléments considérés importants pour lui, mais qui comptent moins pour vous.

Pour mieux comprendre tous ces facteurs, consultez le tableau 5.1 qui présente un formulaire de planification de la négociation. Les deux premières composantes ont été complétées à titre d'exemple afin d'illustrer les options à prévoir avant d'entamer une négociation.

On remarque dans ce tableau que le prix revêt une importance capitale pour les deux parties, recevant une note de 5 sur 5. On obtient une importance commune de 25 lorsque l'on multiplie les deux scores. Cela signifie que le prix pourrait constituer l'élément le plus difficile à négocier surtout si les deux parties sont très éloignées. C'est pourquoi il est souvent préférable de commencer par s'entendre sur les points les plus faciles à régler, afin de démontrer qu'une entente est possible. Par ailleurs, le prix est généralement abordé en toute fin de négociation, car il peut dépendre d'autres facteurs. Par exemple, si la personne en vente accepte d'entreposer des produits destinés à la firme cliente six mois plutôt que la durée standard de trois mois, cela aura une incidence sur ses coûts et, par conséquent, sur ses prix. La colonne des commentaires permet d'inscrire des repères stratégiques, notamment le seuil minimum au-delà duquel le fournisseur ne peut pas descendre (10 dollars) et l'objectif souhaité, qui est souvent déjà connu de l'acheteur (10,75 dollars). Par ailleurs, on remarque une structuration progressive des concessions sur les délais de paiement : plutôt que d'accorder immédiatement 90 jours, l'offre passe de 30 à 60 jours, puis à 75 jours, réduisant ainsi l'ampleur de chaque concession. Cette approche donne l'impression que chaque nouveau compromis devient plus difficile à obtenir, ce qui incite l'acheteur à mieux mesurer ses demandes. La zone d'entente possible se situe entre le seuil minimum acceptable pour le fournisseur et le seuil maximum que la firme cliente est prête à céder.

TABLEAU 5.1 / **La quantification des éléments de la négociation**

Élément de négociation	Importance pour les ventes	Importance pour les achats	Importance commune	Commentaire
Prix	5	5	25	Prix souhaité et annoncé : 10,75 $ Palier 1 : 10,50 $ Palier 2 : 10,30 $ Palier 3 : 10,15 $ Palier 4 : 10,05 $ Prix minimum par unité : 10 $
Délai de paiement	5	3	15	Délai minimum (votre objectif) 30 jours Palier 1 : 60 jours Palier 2 : 75 jours Palier 3 : 85 jours Délai maximum (avant BATNA) 90 jours
Délai de livraison	4	4	16	
Durée du contrat	4	2	8	
Garantie	3	5	15	
Entreposage	3	4	12	
Service après-vente	2	4	8	
Emballage	2	3	6	
Fréquence des livraisons	1	5	5	

La figure 5.1 illustre cette dynamique en représentant le BATNA (meilleure solution de rechange à une entente négociée) et les limites de cette zone d'accord.

FIGURE 5.1 **BATNA et zone d'entente possible**

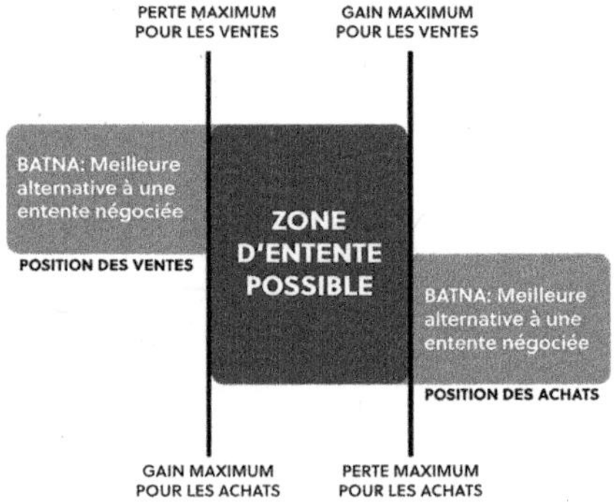

Source: Adapté de Fisher et coll. (2011).

1.4 / L'ordre du jour et la hiérarchisation des enjeux

Reprenons quelques éléments du tableau 5.1. La durée du contrat revêt une grande importance pour la vente avec une note de 4 alors qu'elle semble relativement secondaire pour l'acheteur avec une note de 2. Il s'agit probablement d'un des premiers points que la personne en vente voudra négocier, car il lui est plus favorable. Elle pourrait aussi chercher à l'échanger contre un élément de moindre importance pour elle, mais crucial pour l'acheteur, comme le service après-vente avec une note de 2 pour la vente et une note de 4 pour l'achat. Quant à la fréquence des livraisons, bien qu'elle semble être le point le plus facile à régler, il serait plus stratégique pour une personne en vente de garder cette concession en réserve. Pourquoi? Parce que cet élément représente une priorité élevée pour l'acheteur et qu'il exige peu d'effort du fournisseur. Il devient ainsi un levier intéressant à utiliser en fin de négociation pour obtenir des contreparties plus avantageuses.

1.5 / La gestion des émotions

Plusieurs émotions peuvent survenir avant, pendant et après une négociation. Vous devez donc vous préparer à y faire face afin d'anticiper comment elles pourraient nuire à la conclusion d'une entente favorable pour vous ou pour les deux parties. Brooks (2015) a identifié quatre grandes catégories d'émotions qui peuvent affecter une négociation : l'anxiété, la colère, la déception et le regret, ainsi que la joie et l'enthousiasme.

L'anxiété nous encourage à fuir une situation au plus vite. Ressentir de l'anxiété pendant une négociation pourrait vous inciter à faire une première offre désavantageuse pour vous, à répondre plus vite aux demandes de votre vis-à-vis et à finaliser l'entente trop rapidement. De plus, si vous exprimez cette anxiété, une personne négociatrice rusée pourra s'en servir contre vous pour vous soutirer des concessions. Pour réduire votre anxiété, vous devrez vous préparer et pratiquer des situations tendues comme lorsqu'une personne vous menace si vous n'acceptez pas ses demandes de concession. Plus vous vous familiariserez avec ces situations et moins vous éprouverez de l'anxiété.

Alors que l'anxiété est dirigée vers soi-même, la colère, elle, est dirigée vers autrui. Certaines personnes croient que la colère donne un avantage en négociation. Cela peut être vrai dans des situations où la relation à long terme n'a pas d'importance, mais en développement des affaires, c'est une stratégie risquée. La colère peut ébranler la confiance de l'autre partie, compromettre la négociation et, dans le pire des cas, mettre fin à la relation commerciale. Si vous perdez la négociation, vous pourriez ressentir de la déception ou du regret.

Manifester de la déception peut vous aider à gérer votre colère. La déception peut aussi advenir si la négociation se conclut trop vite. On peut ainsi avoir l'impression d'avoir abdiqué trop facilement, et cette déception peut se transformer en regret. Ce dernier survient plus souvent lorsqu'on réalise, après coup, qu'on a oublié de soulever un point clé ou de saisir une opportunité. Pour éviter cette sensation, je vous recommande de poser un maximum de questions pendant les pourparlers afin de découvrir toutes les occasions et de débusquer tous les éléments susceptibles d'avoir un impact sur l'avenir de la relation.

À l'inverse, si vous obtenez gain de cause, l'enthousiasme pourrait vous inciter à exprimer ouvertement votre satisfaction. Extérioriser de la joie ou de l'enthousiasme à la fin de tractations risque de faire sentir à l'autre personne qu'elle a cédé trop facilement et trop tôt, ce qui pourrait en outre conduire à une rupture de l'entente qui vient d'être conclue.

L'intelligence émotionnelle peut aider à gérer les émotions pendant une négociation. Elle repose d'abord sur la capacité à reconnaître et à comprendre les émotions de l'autre. Exprimer à voix haute votre perception de ses sentiments peut parfois désamorcer des tensions et favoriser un dialogue plus constructif. Si votre vis-à-vis manifeste de la colère, essayez de l'apaiser, et ce, même si vous croyez qu'il n'a aucune raison de le faire. Vous pourriez encore réclamer une pause de quelques minutes ou de quelques jours afin de réduire la tension. Si vous sentez la colère monter en vous, prenez un instant pour réfléchir avant de l'extérioriser. Une approche efficace consiste à transformer cette colère en tristesse. Par exemple, face à une demande inacceptable, plutôt que de vous emporter, vous pourriez exprimer votre regret en disant: « Je suis déçu de voir que cette demande pourrait mettre en péril notre collaboration. » Une telle formulation pourrait inciter l'autre partie à revoir sa position et à envisager une concession plus raisonnable.

2/ Le déroulement de la négociation

Vous venez de présenter votre offre et, sans attendre, votre vis-à-vis passe aussitôt à l'offensive: « Si vous ne réduisez pas immédiatement votre prix de 10 %, nous devrons envisager une autre option. » Voilà une des nombreuses tactiques employées par les responsables des achats les plus déterminés à remporter la mise coûte que coûte, ou encore par ceux qui disposent déjà d'une solution de rechange. Maitriser les tactiques de négociation de base permet aux personnes en vente de ne pas sacrifier des points de profitabilité sous prétexte d'obtenir le contrat. Cela est d'autant plus vrai que les spécialistes en approvisionnements sont souvent formés aux techniques de négociation avancées. L'objectif n'est pas de tromper, mais de se sentir parfaitement en contrôle du processus et de parvenir à un accord équilibré qui profite aux deux parties.

Deux types de négociation sont possibles. Ces dernières années, la **négociation intégrative**, ou gagnant-gagnant du *Harvard Negotiation Project* (Fisher et coll., 2011) a été très publicisée. Cette approche cherche à maximiser la répartition des ressources disponibles entre les parties pour augmenter leur satisfaction et pour préserver la relation. Elle est particulièrement recommandée lorsque les forces en présence sont équilibrées, que les enjeux sont complexes et multiples et que les parties disposent de suffisamment de temps pour explorer leurs intérêts respectifs plutôt que de se braquer sur leurs positions (Fleming et Hawes, 2017). À l'opposé, la **négociation distributive** repose sur un modèle plus compétitif, où l'un gagne et l'autre perd. Celle-ci survient au moment où l'une des parties détient plus de pouvoir, que la relation à long terme n'est pas une priorité et que la négociation repose sur un seul critère, souvent le prix. Cette section évoquera les deux types de négociation, car ils sont tous deux présents avant la conclusion du contrat.

2.1 / Connaître les tactiques de la clientèle et y répondre

Dans cette section, je présenterai cinq tactiques fréquemment utilisées en négociation par les responsables des achats : l'ultimatum, l'encouragement, la délégation de la décision à une autre personne ou à un comité, l'absence ou l'insuffisance de budget, et la promesse. Pour chacune, je proposerai une façon d'y répondre.

L'**ultimatum** peut prendre plusieurs formes et peut survenir à tout moment dans une négociation. L'un des scénarios les plus courants est le suivant : on vous met la pression en affirmant que si vous n'acceptez pas la requête présentée, un autre fournisseur, offrant des conditions plus favorables, obtiendra le contrat. Face à cette tactique, une première réponse consiste à demander l'identité de ce fournisseur ainsi que des précisions sur les conditions prétendument plus favorables. Cette approche permet d'éviter de négocier contre la concurrence imaginaire. Une préparation rigoureuse vous aura déjà permis d'anticiper ce type de comparaison et de structurer une réponse convaincante. À l'instant où on vous présente un ultimatum en début de négociation, vous devez d'abord tenter d'ignorer cette position et ramener la conversation sur les autres aspects de votre offre afin de vous assurer que vous avez bien expliqué sa valeur. Lorsque l'ultimatum survient plus loin dans la négociation, vous devrez vous appuyer sur votre plan où se trouvent les concessions que vous êtes prêt à accorder et aux arguments que vous avez préparés pour défendre votre position. L'ultimatum inclut une dimension temporelle en plus d'une composante de résultat, ce qui justifie de réclamer un sursis pour en discuter avec votre hiérarchie, car la décision dépasse votre marge de manœuvre. Ce procédé est un couteau à deux tranchants, puisqu'il vous permet d'une part de ralentir le rythme de la négociation et de faire comprendre à l'autre partie que vous cèderez difficilement, mais d'autre part, il envoie le signal que le ou la responsable des achats aurait intérêt à parler directement à votre direction la prochaine fois.

Des ultimatums qui n'en ont pas l'air se présentent sous différentes formes comme l'**encouragement**, la deuxième tactique utilisée par une personne responsable des approvisionnements. Elle ressemble souvent à une phrase de ce genre : « Allez ! Vous pouvez certainement faire mieux. » L'autre partie veut décrocher une concession, mais essaie de l'obtenir avec un ton amical. Ne vous détrompez pas. Si cela ne fonctionne pas, attendez-vous à une attaque frontale. Face à cette situation, vous devez ralentir le processus en réclamant du temps pour y penser, en proposant une contre-offre minime ou en faisant appel à votre hiérarchie.

Une autre tactique courante est la **délégation** de la décision à une tierce personne ou à un comité. Vous aviez pourtant vérifié que votre interlocuteur avait l'autorité nécessaire pour négocier, mais voilà qu'il annonce devoir consulter sa direction ou d'autres membres de son équipe avant de prendre une décision définitive. La première façon de répondre serait de demander quelle sera sa recommandation au comité. Vous devez l'inciter à se compromettre, car vous avez sans doute déjà octroyé quelques concessions et il est maintenant trop tard pour reculer, et toute nouvelle intervention risque d'entrainer des demandes supplémentaires. Cette tactique n'est pas nécessairement à votre désavantage. Vous pourriez à votre tour inviter votre direction à participer et en profiter pour changer le niveau de négociation. Plutôt que de discuter d'une entente pour un seul produit, pourquoi ne pas explorer la possibilité de proposer une solution plus globale ? Au lieu d'un contrat d'une durée d'un an, vous pourriez orienter les discussions vers un engagement de trois ans.

Si votre interlocuteur affirme que votre offre dépasse le **budget** alloué par son entreprise, il pourrait s'agir d'une autre tactique fréquemment employée en négociation. Face à cette situation, plusieurs options s'offrent à vous. Vous pouvez proposer un produit qui fournit de moins grandes performances ou un service qui compte moins de caractéristiques à un prix ajusté. Une autre approche consiste à assouplir les conditions de paiement, dans la mesure où cela demeure viable pour votre entreprise. L'objectif est de répondre aux contraintes budgétaires de la clientèle sans pour autant sacrifier inutilement votre marge de profit.

Enfin, la cinquième tactique repose sur la **promesse**. Votre vis-à-vis vous assure qu'une commande plus importante suivra, à condition que vous acceptiez de faire une concession immédiate. À moins d'un engagement écrit ou de l'existence d'une longue relation basée sur la confiance, une telle promesse n'a que peu de valeur en affaires. Vous devriez donc la traiter comme une simple tentative de concession et négocier en conséquence. Le tableau 5.2 propose un récapitulatif de ces réponses.

TABLEAU 5.2 / **Réponses possibles aux tactiques**

Tactiques	Réponses possibles
L'ultimatum	• Demander d'en savoir davantage sur les autres offres • Ralentir le déroulement de la négociation en faisant appel à votre hiérarchie
L'encouragement	• Ralentir le déroulement de la négociation en faisant appel à votre hiérarchie • Faire une contre-offre minime
La délégation	• Demander quelle sera la recommandation de votre interlocuteur ou votre interlocutrice • Faire aussi appel à votre hiérarchie pour étendre la portée d'une éventuelle entente
Le budget	• Réduire ou éliminer des composantes de votre offre • Proposer des conditions de paiement avantageuses
La promesse	• Ne pas en tenir compte à moins d'un engagement écrit ou de l'existence d'une longue relation basée sur la confiance

Maintenant que nous avons pris connaissance des tactiques pour obtenir des concessions, considérons à présent le processus de négociation comme tel.

2.2 / Le processus de la négociation

Le processus de négociation démarre avec une demande de concession présentée par le client ou la cliente. Avant de céder, vous devrez suivre un déroulement qui devrait, en principe, vous conduire à consentir le moins possible tout en réussissant à conclure un accord. Rappelez-vous que ces quatre étapes, soit la présentation d'une explication, la demande d'une contrepartie, la concession que vous acceptez de faire et la conclusion d'une entente, doivent être employées pour chaque demande de concession. Et chaque fois, vous devrez garder votre calme, une posture plus facile à maintenir lorsque l'on est bien préparé.

Lorsque l'on vous présente une requête, vous devez d'abord vous assurer que votre vis-à-vis a bien compris la valeur de votre proposition. C'est l'étape de l'**explication**. Vous devez bien appréhender les raisons qui motivent cette demande en invitant votre interlocuteur à s'expliquer. Vous devrez également récapituler votre proposition en tenant compte des précisions obtenues. Cette tactique vous permet de renforcer votre crédibilité et d'affirmer la valeur de votre offre. Évidemment, vous ne devez pas vous entêter à rester à cette étape, sinon l'autre partie pourrait être tentée de se retirer des discussions.

Une fois que vous avez bien compris les motivations derrière la requête et que vous avez eu l'occasion de réitérer les bénéfices de votre offre, vous entamez la deuxième démarche du processus: la **contrepartie**, qui constitue la base même de la négociation. L'objectif est simple: recevoir quelque chose en échange. Cette étape comporte trois sous-étapes: la question préalable, l'ouverture et la contre-proposition. Pour illustrer ces sous-étapes, imaginons que vous êtes prêt à négocier une réduction de prix de 10%, à condition que la marchandise soit livrée en deux envois plutôt qu'en quatre. La question préalable consiste à inciter l'autre partie à se rapprocher de votre demande de contrepartie sans lui demander de front. Vous pourriez, par exemple, l'interroger de cette façon: «Croyez-vous que vous pourriez aider notre département de logistique?» Une réaction positive n'engage pas encore votre interlocuteur, mais montre son ouverture au dialogue. Si la réponse est négative, vous pouvez quand même enchainer avec la prochaine sous-étape au cours de laquelle vous ferez preuve d'ouverture en énonçant plus clairement vos intentions sous-entendues dans la question préalable. Vous pourriez ainsi dire: «Si vous acceptiez de recevoir nos produits en deux livraisons plutôt qu'en quatre, je pourrais probablement fournir un effort additionnel. Croyez-vous que ce serait possible?» Une réponse positive vous conduira à la troisième sous-étape: la contre-proposition. «Puisque vous convenez de recevoir la marchandise de cette manière, je pourrais réduire votre prix de 4 %. Non seulement vous profitez d'un rabais très intéressant, mais en plus, votre service de réception devra remplir deux fois moins de formulaires. Qu'en pensez-vous ? » L'ampleur de votre contre-proposition dépendra évidemment du rapport de force entre les parties. Si vous êtes en position de pouvoir, vous pourriez accorder un rabais moindre, par exemple 2 %. En revanche, si vous faites face à une concurrence agressive, si vous tentez de gagner une nouvelle clientèle ou si votre produit offre peu de différenciation par rapport aux autres possibilités, il pourrait être nécessaire d'accorder un ajustement plus substantiel, jusqu'à 6 %.

Vient un moment où vous pourriez être forcé de faire une **concession** sans rien obtenir en retour, ce qui représente la troisième étape du processus. Par exemple, après avoir consenti 4 %, l'autre partie pourra vous signifier que ce n'est pas assez et que, si vous n'êtes pas disposé à faire mieux, elle devra accepter la proposition de votre concurrence. Voici quelques réflexions avant d'aller plus loin :

- Votre concession ne doit pas être si avantageuse par rapport à votre offre initiale sinon l'autre personne pourrait penser que votre prix ou vos conditions étaient artificiellement gonflés.
- Souvenez-vous que les concessions peuvent vous causer du tort à moyen et long terme, car la personne aux achats pourrait un jour changer d'emploi pour occuper un poste majeur dans une autre compagnie à qui vous n'aviez pas cédé autant. Ce genre d'information peut circuler assez rapidement au sein d'une industrie même si, en principe, les contrats imposent souvent des clauses de confidentialité.
- Rappelez-vous aussi que chaque fois que vous faites une concession, vous perdez de l'argent.

Voici maintenant des tactiques à utiliser lorsque vous faites face à une demande de concession :

- Vous pouvez changer l'objet de la concession, par exemple en déclarant ceci : « Je ne peux pas réduire votre prix davantage, mais que diriez-vous d'allonger les délais de paiement de 30 jours à 45 jours ? »

- Il peut être tentant d'adopter la tactique qui consiste à couper la poire en deux. L'autre partie exige 10 % de réduction, tandis que vous proposez 4 %. Il serait si simple de proposer : « Allons à mi-chemin à 7 %, d'accord ? » Pourtant, cette approche peut jouer en votre défaveur. Votre vis-à-vis aurait peut-être accepté 6 %, ce qui signifie que vous venez de céder un point de pourcentage supplémentaire sans raison. Pire encore, une personne plus aguerrie à la négociation, voyant que vous avez immédiatement consenti à 7 %, pourrait renchérir en demandant 9 %. Vous n'obtiendrez toujours pas d'entente et vous devrez encore sacrifier un autre point de pourcentage, à 8 % pour couper la poire en deux. Si le jeu en vaut la chandelle, l'autre partie acceptera peut-être de baisser sa demande à 8,75 %, etc.
- La concession suivante devrait systématiquement être inférieure à la précédente. Vous avez accordé 4 % au départ ; la prochaine concession se limitera à seulement un 2 % additionnel pour un total de 6 %. Votre vis-à-vis aura ainsi l'impression de devoir fournir un effort supplémentaire pour obtenir une nouvelle concession.
- Le temps est un autre facteur important à considérer. Vous devez argumenter longtemps à chaque demande de concession et prendre de plus en plus de temps pour céder. Votre première concession a demandé trois minutes de discussion, alors la suivante devrait en requérir cinq. Une tactique différente pour étirer le temps consiste à quitter la pièce pour passer un appel à votre direction, ou de retourner au bureau afin de vérifier si vous pouvez encore diminuer le prix.
- Soyez précis dans votre réduction de prix. Ne dites pas : « Je pourrais baisser le prix d'environ 5 %. » Cela laisse entendre que vous pourriez vous satisfaire d'une diminution de 6 %, et pourquoi pas de 7 %, ou de 7,5 %.

La quatrième étape vise à conclure une **entente** sur ce qui vient d'être négocié. Avant de consentir un élément qui semble majeur pour votre entreprise, vous devriez utiliser deux questions afin de vous assurer de ne pas recevoir de requêtes de dernière minute qui pourraient gruger encore davantage votre marge de profit. La première est la question de confirmation, qui pourrait ressembler à ceci : « À part ce facteur (le prix, par exemple), sommes-nous en accord sur les autres ? » Cela vous garantit que si vous effectuez une concession importante, vous ne ferez pas face à une multitude de demandes. Si l'autre partie veut en effet négocier de nouveaux aspects de votre offre, vous pourrez alors en tenir compte dans votre concession. La question suivante est celle d'engagement : « Si nous nous entendons sur cet élément, est-ce que vous pourrez émettre le bon de commande aujourd'hui ? » Vous vous prémunissez ainsi contre une éventuelle nouvelle phase de négociation avec un autre membre de la centrale d'achat. Ces deux questions vous permettent d'éviter de vous retrouver dans une spirale de concessions sans contrepartie. Tant que votre vis-à-vis ne s'engage pas officiellement, il sera toujours tenté de vous soutirer une nouvelle concession, voire de contacter une entreprise concurrente pour tenter d'obtenir un meilleur prix à votre détriment. La figure 5.2 schématise les étapes du processus présenté précédemment.

FIGURE 5.2 / **Processus de négociation**

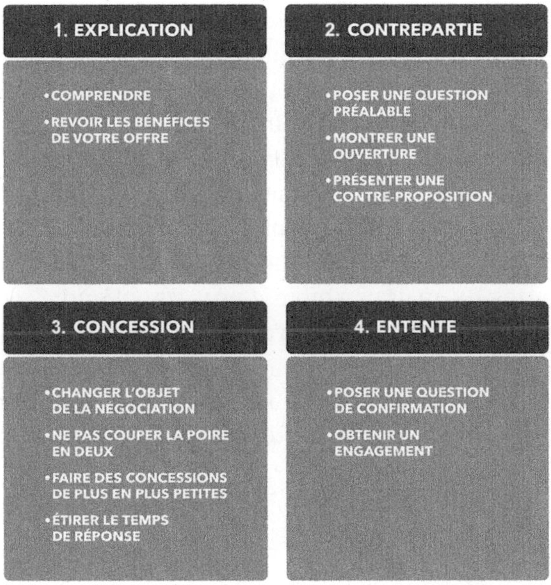

3/ Le prix, toujours le prix

Un des sujets de négociation les plus fréquents est le prix. La firme cliente peut croire que le prix annoncé n'existe que pour être négocié ou peut supposer que vous accorderez une réduction par crainte de perdre la vente. Si vous ne rencontrez jamais d'objection sur le prix, méfiez-vous! Cela signifie qu'il n'est peut-être pas assez élevé à moins que vous soyez devenu un expert pour démontrer la valeur de votre offre. Les pourparlers sur le prix devraient toujours être reportés le plus loin possible dans la négociation pour vous donner la chance de présenter les bénéfices avant de discuter du prix. Rappelez-vous que chaque fois que vous abaissez votre prix c'est une partie de votre profit que vous abandonnez. Chaque réduction peut même nuire à l'image de votre produit ou de votre société, car vous validez implicitement l'idée que votre offre ne vaut pas son prix initial. Si votre offre présente une qualité supérieure sur certaines caractéristiques, démontrez une relation claire entre le prix et la qualité. De même, si votre entreprise bénéficie d'une meilleure réputation sur le marché, soulignez l'impact du prix sur la sécurité et la fiabilité.

Réduire votre prix trop rapidement pourrait semer le doute chez votre vis-à-vis quant à la véritable valeur de votre offre. Cela peut également nuire à la confiance qu'il vous accorde et l'amener à se demander si d'autres entreprises, y compris ses propres concurrents et concurrentes, ont réussi à obtenir une réduction encore plus importante en négociant plus fermement. Si vous éprouvez des difficultés à défendre votre prix, c'est que vous êtes peut-être simplement moins bien préparé. Pourtant, vous savez que des objections sur les prix seront souvent présentes, que vous perdrez des commandes à cause de votre prix, mais non moins que vous obtiendrez des contrats malgré cette contrainte. Le prix n'est pas toujours le véritable point de blocage. Parfois, la compagnie ne dispose tout simplement pas du budget nécessaire, ou encore elle ne perçoit pas clairement la différence entre les bénéfices de votre offre et ceux de votre concurrence.

Vous pourriez utiliser différentes tactiques pour soumettre votre prix. Si votre produit ou service le permet, vous pouvez décomposer les caractéristiques ou les options. Par exemple, si votre produit inclut la maintenance d'un équipement pendant un an, vous pouvez évaluer la valeur de ce service et l'exclure de la proposition pour ajuster le prix. En consultation, on présente fréquemment plusieurs choix avec des prix différents.

Une deuxième tactique consiste à diviser le prix en fonction de l'utilisation. Par exemple, vous pourriez expliquer que votre logiciel ne coûte que quatre dollars par jour, par employé. Vous pouvez aller plus loin en comparant ce prix divisé avec le prix d'un autre produit de cette façon: «Pour à peu près l'équivalent d'un café quotidien, vous pourriez offrir l'assurance dentaire à tous vos employés.»

Vous pouvez aussi utiliser l'effet d'ancrage, selon lequel les gens ont tendance à évaluer la valeur en fonction d'un prix initial, si vous évoluez dans une industrie où la négociation de prix fait partie de la culture. En proposant un prix élevé au début de la négociation, la personne responsable des achats pourrait s'y référer inconsciemment tout au long de la discussion et mesurer ses gains en fonction de cette annonce.

Une tactique différente consiste à englober des éléments qui ne sont pas compris dans l'offre de la concurrence. Si, pour reprendre un exemple mentionné précédemment, vous incluez une année de maintenance dans votre prix alors que votre concurrence la facture séparément, la comparaison devient moins évidente et plus favorable pour vous.

L'utilisation de décimales constitue une autre méthode qui peut s'avérer efficace. Au lieu de diminuer le prix de 4%, vous pourriez le baisser de 3,89%. Cela envoie deux messages à votre vis-à-vis: d'une part, que chaque concession est calculée avec rigueur et d'autre part, que vous avez sérieusement réfléchi à votre réduction de prix avant de la soumettre.

La résistance au prix survient souvent lorsque la clientèle ne discerne pas encore pleinement la valeur du produit ou du service que vous vendez par rapport à son coût. Souvenez-vous que la valeur perçue par la clientèle est la différence entre les bénéfices et les sacrifices, comme la somme qu'elle devra verser (Töytäri et Rajala, 2015). Une manière d'opérationnaliser cette perception de la valeur est de la quantifier, en démontrant un retour sur l'investissement. Cette approche rejoint le «N» de la méthode SPIN, soit la nécessité de rendement. Le prix constitue ce que la clientèle paie dans l'immédiat, mais le coût représente les dépenses à long terme. Si votre produit ou service génère des économies sur plusieurs mois ou années, utilisez cet argument pour rehausser la valeur perçue. Rappelez-vous aussi que la relation elle-même est un élément de valeur. Si vous croyez être personnellement plus fiable et plus crédible que votre concurrence, mentionnez cette qualité de la façon la plus humble possible. Cela peut particulièrement fonctionner si vous avez déjà bâti un excellent rapport avec l'individu en face de vous. L'encadré 5.1 présente quatre réflexions essentielles sur la gestion des prix, afin de vous convaincre de ne jamais les réduire à moins d'y être contraint.

ENCADRÉ 5.1 / **Quatre réflexions sur les prix**

1. Une firme concurrente qui propose un prix moins cher existe presque toujours. Même si vous réduisez le vôtre, vous perdrez malgré tout des commandes. La véritable question est donc la suivante : jusqu'où pouvez-vous ajuster vos prix sans heurter votre image de marque ni compromettre la rentabilité de votre organisation ?
2. Vous réaliserez des ventes malgré votre prix plus élevé. Certaines personnes comprendront la valeur de votre offre et choisiront votre solution, en dépit d'un prix supérieur. Demandez-vous si, comme consommateur, vous avez déjà choisi un véhicule plus cher qu'un autre ou préféré le spectacle d'un artiste, même si le prix des billets vous semblait exorbitant. Si la réponse est oui, alors votre clientèle peut elle aussi faire ce type de choix si la valeur perçue est bien démontrée.
3. Si vous devez réduire trop souvent vos prix, vous n'êtes peut-être pas suffisamment préparé pour le défendre. Rappelez-vous que votre vis-à-vis a probablement suivi des formations sur la négociation, particulièrement sur celle des prix. Et vous, à quand remonte votre dernière formation en négociation tarifaire ?
4. Enfin, n'oubliez jamais que la rentabilité est essentielle. Vous devez générer des profits pour récompenser votre prise de risques, assurer une rémunération équitable à votre personnel, et réinvestir dans vos équipements et vos processus. Chaque point de pourcentage concédé lors de la négociation se traduira par de l'argent qui vous manquera pour le développement de votre organisation.

À un certain moment, il se peut que vous deviez appliquer une hausse de prix, encouragés par une dynamique de marché favorable comme une réduction de l'offre, une flambée du coût des matières premières ou une augmentation des prix de la concurrence (Smith, 2018). Cet exercice doit comprendre trois actions pour soutenir l'équipe de développement des affaires. D'abord, la hausse doit être explicitement annoncée sur l'ensemble des canaux de communications disponibles. Elle doit être accompagnée de justifications claires, expliquant les facteurs concrets qui la motivent. Ces explications devront être répétées et renforcées par l'équipe de vente lors de ses échanges avec la clientèle, afin d'assurer une compréhension et une acceptation maximales. La deuxième intervention consiste à rester ferme sur cet accroissement de prix. L'entreprise doit être disposée à accepter une éventuelle perte de volume, un message qui doit aussi être relayé à la force de vente. Enfin, Smith (2018) recommande de maintenir une approche de vente basée sur la valeur, en veillant à ne concéder une réduction de la hausse de prix qu'en échange d'une contrepartie tangible. Cela pourrait inclure une diminution des délais d'entreposage ou de paiement, une prolongation de la durée du contrat, ou encore une augmentation des volumes d'achats.

Un mécanisme particulier en *Business-to-Business* (B2B) peut influencer négativement les prix et la profitabilité, à savoir les enchères inversées sur des plateformes numériques. Dans ce type d'enchères, les fournisseurs se font concurrence en abaissant leur prix jusqu'à ce qu'une offre soit acceptée. Toutes les firmes qui se font concurrence voient le prix le plus bas soumis, mais ignorent quelle entreprise l'a proposé, ce qui les empêche d'ajuster leur offre en fonction de la qualité ou des caractéristiques distinctives de leurs produits ou services. De plus, la firme qui organise l'enchère ne s'engage pas à acheter auprès du soumissionnaire le moins cher. Une fois son choix effectué, rien ne l'empêche de continuer à négocier pour tirer encore plus de bénéfices. Les enchères inversées sont particulièrement avantageuses pour la clientèle qui obtient des réductions de prix substantielles. Cependant, pour le fournisseur en place, ce procédé constitue une menace majeure pour sa rentabilité et remet en question les investissements réalisés tout au long de la relation d'affaires. Pour les autres participants et participantes, en revanche, l'enchère peut offrir une opportunité d'entrer en contact avec une nouvelle entreprise cliente et de se positionner sur le marché.

4/ La négociation à distance

La négociation par courriel occupe une place de plus en plus importante dans les relations entre les achats et les ventes, notamment en raison de l'essor des outils de communication numériques et des habitudes de travail instaurées pendant la pandémie, puis consolidées par la suite. Le courriel offre plusieurs avantages. D'abord, il permet une négociation asynchrone, c'est-à-dire que chaque partie peut répondre au moment qui lui convient. Il offre aussi l'avantage de laisser une trace écrite, ce qui facilite le suivi des échanges et le partage de l'information avec d'autres membres de l'équipe. En revanche, cette constante disponibilité et cette flexibilité peuvent également entrainer un allongement des négociations, qui risquent de s'éterniser. Le courriel présente aussi un désavantage majeur, car il réduit le canal de communication à la seule composante verbale, c'est-à-dire que seul du texte est utilisé sans pouvoir ajouter le ton de la voix et les expressions du visage pour transmettre les émotions (Singh et coll., 2020).

L'emploi de la visioconférence s'est beaucoup propagé depuis la pandémie, mais les études au sujet de la négociation qui utilisent ce mode demeurent peu répandues. On notera tout de même que la visioconférence se rapproche de la négociation en face-à-face. En revanche, le contact visuel peut être plus difficile, selon la position de la caméra par rapport à celle de l'écran. Les signaux non verbaux, distincts de ceux du visage, sont habituellement invisibles en raison du cadrage choisi (Prat, 2023). On rapporte aussi que la visioconférence entraine une fatigue plus rapide. D'autres technologies devront éventuellement être étudiées en lien avec la négociation à distance comme la réalité virtuelle qui emploie des avatars pour représenter les parties en présence.

La négociation de prix à distance autorise l'intervention de l'intelligence artificielle afin de renforcer l'aide à la décision pour la personne en développement des affaires. Une expérience menée par Karlinsky-Shichor et Netzer en 2024 a vu la profitabilité des commandes augmenter de 11% quand un algorithme établissait les prix de façon indépendante. En revanche, lorsqu'il s'agit d'établir des prix complexes ou de formuler des soumissions atypiques, un modèle hybride combinant l'expertise humaine de la personne en vente et la puissance de l'algorithme permettait un meilleur résultat. Il ne fait aucun doute que l'intelligence artificielle sera de plus en plus utilisée dans les négociations entre les services des achats et les ventes.

5 / Obtenir la commande

Si vous avez suivi attentivement les étapes des méthodes SPIN ou Challenger, il se peut que l'autre partie vous propose spontanément de conclure une entente ou de passer une commande. La plupart du temps, vous devrez l'encourager à franchir ce dernier pas. Si vous êtes rendu à cette étape, c'est que la personne croit que vous ajoutez une réelle valeur, que vous démontrez une grande maitrise de votre offre et de votre entreprise, et, surtout, que vous êtes une personne digne de confiance.

Développer une habileté à identifier les signaux verbaux et non-verbaux qui dévoilent l'intérêt de la personne pour votre proposition est essentiel pour obtenir la commande. Cette phase de la vente peut engendrer du stress, parce que le risque de se faire dire «non» demeure. Mais même un refus apparent peut représenter une occasion de creuser davantage, de découvrir de nouvelles préoccupations ou de répondre à des objections non formulées jusqu'ici. Si l'autre partie rejette d'abord la conclusion d'une entente, vous pouvez poursuivre la conversation, en guettant les signaux d'intérêts, présentés à l'encadré 5.2. Si vous faites preuve de persuasion, le moment venu, vous pourrez encourager une nouvelle fois la personne à se compromettre.

ENCADRÉ 5.2 / **Les signaux d'achat à surveiller**

Verbaux

- La personne décrit des situations dans lesquelles elle utilise le bien ou le service: «Nos employés pourraient certainement faire plus d'entretien préventif si le robot nettoyeur effectuait d'autres tâches à leur place.»
- La personne s'intéresse à des détails de votre offre: «Dans combien de temps pourriez-vous installer le logiciel et former notre équipe de 12?»

Non-verbaux

- Les muscles du visage et la position du corps de l'autre vous indiquent que la personne est détendue
- La personne manipule l'échantillon ou la documentation

Malgré votre succès au cours de la négociation, vous pouvez ressentir que la personne responsable des achats hésite à conclure une entente. Plusieurs raisons peuvent expliquer une telle réticence. Prendre une décision implique toujours de renoncer à d'autres options, ce qui peut faire craindre de commettre une erreur. Ne pas agir peut dans certains cas conforter la personne. Elle pourrait peut-être aussi avoir à justifier son choix auprès de ses collègues. Parfois, la décision est retardée par manque de temps pour finaliser tous les documents ou en raison de la complexité bureaucratique pour recueillir des approbations à l'intérieur ou à l'extérieur de la centrale d'achats. Vous devrez alors exercer une pression adéquate tout en tenant compte de ses intérêts et en démontrant une confiance envers vous-mêmes et votre offre.

Une personne en développement des affaires peut employer différentes techniques pour conduire habilement vers une entente. Avant de recourir à ces méthodes, elle peut poser des questions pour sonder la position de son interlocuteur, telles que «Sommes-nous bien d'accord sur ce point?», ou «Aimeriez-vous obtenir plus d'informations sur différents aspects de ma proposition?»

Un premier procédé nommé la *demande directe* peut être choisi si vous croyez que la personne responsable des achats est prête à ratifier un contrat. Par exemple, vous pourriez dire: «Pourriez-vous signer au bas de cette page, pour que nous puissions réserver vos dates à notre centre de conférence?» Vous pourriez subtilement la confronter à un choix qui confirmera implicitement son engagement envers votre offre de cette façon: «Souhaitez-vous une livraison le 15 ou le 30 octobre?» Vous pourriez aussi lui octroyer un petit bonus sans grand impact sur vos coûts comme: «Je pourrais ajouter une caisse de croustilles gratuites à votre commande de trois palettes de boissons gazeuses.»

Si vous sentez que l'autre personne ne veut pas encore se commettre, vous devriez récapituler de façon extensive les bénéfices. Cela lui fera prendre conscience qu'elle doit accepter votre offre dans les plus brefs délais pour en profiter pleinement. Une autre tactique à utiliser lorsque le responsable des achats hésite, est celle des pour et contre, une approche réservée aux personnes en vente expérimentée. Cette tactique consiste à laisser l'autre créer sa propre liste des pour et des contre. Évidemment, la personne en vente aura déjà prévu un inventaire d'arguments favorables nettement plus étoffé, afin de bonifier la réflexion de l'acheteur ou l'acheteuse au besoin.

Si, après de multiples séances de négociation, vous croyez être sur le point de conclure un important contrat, vous pourrez vous faire accompagner par votre direction au prochain rendez-vous. En avisant en avance de la venue de votre collègue, vous encouragerez aussi l'autre partie à inviter sa direction. Cette ultime rencontre pourrait en définitive sceller favorablement l'issue de la négociation.

Rien ne vous empêche de combiner deux tactiques ou plus pour maximiser vos chances de finaliser les discussions. N'oubliez pas l'effet puissant du silence : il donne du temps à votre vis-à-vis pour réfléchir, tout en générant une certaine tension propice à la prise de décision.

Lorsque vous obtenez finalement le contrat ou la commande, remerciez la personne et félicitez-la pour son choix et continuez de la rassurer sur sa décision. Elle devra peut-être se défendre auprès de collègues, alors vous lui fournirez les arguments qu'elle pourra utiliser au besoin. Vous pourriez aller plus loin en lui envoyant une lettre ou un courriel de remerciement quelques semaines plus tard ou encore, demandez à votre patron de l'appeler pour connaître sa satisfaction à l'égard du processus d'acquisition du produit. Dans certaines industries, comme l'assurance vie, on sollicite souvent la clientèle pour recueillir deux ou trois références de gens qui pourraient profiter d'une couverture d'assurance. Cette approche est délicate et ne fonctionnera peut-être pas avec les grandes sociétés, sauf si vous sollicitez le contact de collègues œuvrant dans d'autres divisions de la même entreprise.

L'euphorie qui suit la signature d'un accord d'une valeur de millions de dollars de vente peut vous rendre aveugle et vous faire omettre de lire attentivement les clauses des contrats qui vous sont proposés. Pourtant vous devriez attribuer beaucoup d'importance à cette lecture comme l'explique l'encadré 5.3.

ENCADRÉ 5.3 / **Les clauses de contrat à surveiller**

Au cours de ma carrière, j'ai souvent eu l'occasion de travailler de près avec des juristes lorsqu'une compagnie importante nous offrait de signer un contrat d'approvisionnement. Les juristes m'incitaient à faire une première lecture de l'entente avant de leur en envoyer une copie. Ils m'ont ainsi appris à repérer deux clauses fréquentes particulièrement dommageables en cas de litige. La première concerne la propriété intellectuelle. On pourrait vous demander de la céder ou d'en partager les retombées financières simplement parce que vous avez suggéré une innovation qui améliore un procédé ou un produit du client. Même si cette exigence vous semble anodine aujourd'hui, qui peut prédire la portée future de l'une de vos créations? L'autre clause qui faisait immédiatement réagir le service juridique est celle sur la responsabilité illimitée en cas de défaut de fabrication ou d'exécution. Par exemple, si vous fabriquez des emballages pour des produits pharmaceutiques, une erreur de votre part dans la posologie inscrite sur l'emballage pourrait vous obliger, non seulement à remplacer les boites, disons au coût de 5000 dollars, mais aussi à indemniser votre client pour les éventuels préjudices de millions de dollars qu'il pourrait avoir à verser aux victimes de cette erreur.

Si la négociation a échoué, vous pouvez remercier la personne responsable des achats de vous avoir donné l'occasion de participer à son processus d'achat. Vous devriez aussi tenter de vous faire expliquer les raisons de votre échec. Si vous conservez ces informations dans un dossier auquel les autres membres de votre organisation ont accès, vous pourrez vous ajuster pour de futures négociations avec cette entreprise ou encore pour des pourparlers avec d'autres, surtout si ce sont les mêmes obstacles à la conclusion qui reviennent. Une réunion de votre équipe est de mise pour réfléchir aux points qui pourraient être améliorés dans votre offre, votre présentation ou votre négociation. Procédez également à un examen de votre performance personnelle. Avez-vous été à la hauteur? Avez-vous fait preuve d'écoute? Ce rejet constitue une formidable occasion d'apprendre.

6 / La pérennisation de l'engagement par le service à la clientèle et le partenariat

Dans cette phase de l'engagement, la partie du service à la clientèle qui relève du développement des affaires occupe une place déterminante. Si vous croyez réaliser une autre vente importante auprès d'une compagnie d'ici quelques semaines, mois ou années, vous devriez entretenir le lien, surtout si vous vendez des produits qui commandent un prix plus élevé.

Longtemps, le service à la clientèle a été perçu comme une simple étape postérieure à la vente, où les entreprises se contentaient d'assurer l'entretien ou la réparation des produits vendus. Aujourd'hui, cette vision est dépassée. Le service à la clientèle doit soutenir les membres de la centrale d'achats tout au long de leur parcours, en les aidant à résoudre les difficultés ou les problèmes rencontrés et en leur procurant une excellente expérience (Sheth et coll., 2023).

Les entreprises clientes ont des attentes très élevées, et les fournisseurs se livrent une concurrence féroce pour gagner ou conserver des parts de marché. Le service à la clientèle peut donc être défensif pour se prémunir contre une attrition de la clientèle. Il relève aussi d'une stratégie offensive, car il fournit une occasion de plus de rencontrer les membres de la centrale d'achat, de leur communiquer des informations additionnelles, de comprendre l'évolution de leurs besoins et ultimement, de réaliser de nouvelles ventes. Pour l'équipe de développement des affaires, le service à la clientèle peut prendre plusieurs aspects, comme l'organisation d'une séance de formation, un sondage de satisfaction ou la visite de courtoisie d'un membre du personnel plus élevé dans la hiérarchie.

6.1 / Le traitement des plaintes

La façon de répondre aux plaintes constitue un autre élément important du service à la clientèle. La plainte représente une opportunité parce qu'elle fournit une rétroaction sur le produit ou le service.

Une gestion formelle des plaintes doit être mise sur pied afin d'opérer un processus d'amélioration continue. Pour l'équipe de développement des affaires, la plainte permet de démontrer la qualité du service à la clientèle comme l'illustrent les deux situations vécues présentées à l'encadré 5.4.

ENCADRÉ 5.4 / **Les bénéfices d'un traitement efficace de plaintes**

Deux anecdotes marquantes survenues durant ma carrière en gestion des ventes me reviennent en tête au sujet du traitement des plaintes.

D'abord, je me souviens d'une entreprise nouvellement cliente pour laquelle nous ne cessions de commettre des bévues, une commande après l'autre. À un certain moment, cette comédie d'erreurs m'a incité à demander à l'acheteur pourquoi il continuait de nous donner des contrats. Il m'a répondu que la façon dont nous adressions ses plaintes en disait beaucoup sur notre volonté à bien faire. Il avait confiance que nous appliquerions des processus pour éviter que de telles erreurs se reproduisent.

Dans un autre cas, il s'agissait d'une réclamation suffisamment grave et urgente pour mettre en péril un accord annuel de plusieurs millions de dollars. Nous avons nolisé un petit avion pour déplacer cinq personnes pour une rencontre de quelques heures à l'usine de la firme cliente, entre autres pour démontrer notre sérieux. Cette décision s'est avérée critique pour la suite des choses et nous avons pu maintenir cette importante relation. Et pour compléter cette histoire, notre appareil a été le dernier de la journée à décoller de l'aéroport local, qui devait fermer en raison d'une tempête de neige. Je salue les nerfs d'acier de nos pilotes.

D'un point de vue tactique, le traitement des plaintes repose sur une méthode éprouvée. La première étape consiste à écouter activement les récriminations de la personne, sans l'interrompre, pour lui permettre d'exprimer à la fois les faits et les émotions qu'elle a pu ressentir. Remerciez-la ensuite de vous en avoir informé et exprimez vos regrets. Les excuses ne doivent pas à coup sûr comporter une reconnaissance d'obligations de votre part. Vous pourriez simplement vous excuser pour la situation vécue par la personne. Avant d'admettre une quelconque responsabilité, prenez soin de recueillir la version de votre équipe. Peut-être serez-vous choqué d'apprendre que le client n'a pas toujours raison? C'est un fait. Une réclamation peut être infondée et vous voulez éviter d'accepter votre responsabilité trop tôt. Assurez-vous donc d'avoir bien compris ce qui s'est passé. Selon l'ampleur de la plainte ou la valeur de la transaction, certaines situations pourront être résolues immédiatement, sans enquête exhaustive. D'autres, en revanche, nécessitent un délai raisonnable pour analyser les faits de votre côté. Dans certains cas, après votre enquête, et une fois que vous aurez obtenu toutes les autorisations nécessaires pour trancher, vous pourriez demander ce que serait un règlement satisfaisant. Des personnes proposent parfois des solutions moins onéreuses que celles que vous aviez envisagées. Assurez-vous de la mise en place de cette solution une fois que vous serez parvenu à un arrangement. Enfin, ne laissez pas à la personne la responsabilité de faire un suivi si le problème persiste. Prenez plutôt le temps de la recontacter vous-même pour vous assurer que tout est rentré dans l'ordre.

6.2 / La clientèle clé et le partenariat

Si vous avez connu un succès manifeste en développement des affaires, vous devrez peut-être accorder une attention particulière à certaines compagnies, en particulier celles dont les achats représentent une part significative de vos ventes. On les désigne souvent comme clientèle clé. Cela dit, une compagnie qui contribue largement à votre chiffre d'affaires ne mérite pas automatiquement un traitement privilégié. Parmi les éléments qui distinguent véritablement une clientèle clé, on note qu'elle exige des services personnalisés, qu'elle centralise ses fonctions logistiques dans une équipe spécialisée maitrisant votre secteur et localisée dans un même lieu, qu'elle aspire à une relation à long terme, qu'elle offre un potentiel de croissance intéressant et qu'elle jouit d'une influence notable dans son industrie.

Les gestionnaires de clientèle clé consacrent généralement moins de temps aux activités traditionnelles de vente que les personnes en vente elles-mêmes. Leur rôle s'inscrit dans une approche plus stratégique. Ils se distinguent par leurs habiletés à piloter des interactions avec différentes fonctions, à faire preuve d'inventivité et à comprendre et utiliser les éléments de création de valeur. Ils sont également en mesure de coordonner les interventions de plusieurs membres du personnel de leur propre firme et, ultimement, d'établir une culture de vente en équipe.

Comme la clientèle clé privilégie les relations durables, cette collaboration peut aboutir à la formation d'un partenariat où certains services de votre organisation pourraient être implantés directement dans les installations de la firme cliente. Il s'agit là d'un niveau d'engagement avancé, puisque votre personnel aura accès à des informations qui ne seront pas à la portée de votre concurrence.

En résumé

Pour que l'engagement se concrétise, les personnes en développement des affaires doivent planifier en détail chaque étape de la négociation : bien cerner les forces en présence, définir les objets de la négociation, identifier leur meilleure option à une entente négociée (BATNA) et délimiter la zone d'entente possible. Durant la négociation, la personne en vente doit savoir se prémunir contre les tactiques de l'autre partie en suivant un scénario précis afin d'éviter les concessions hâtives ou excessives. Le prix demeure un élément incontournable de presque toutes les négociations et certaines techniques peuvent éviter de faire perdre des profits au fournisseur. La négociation à distance, que ce soit par courriel ou visioconférence, prend de l'ampleur. Dans ce contexte, l'intelligence artificielle pourra rendre de précieux services aux négociateurs et négociatrices. L'engagement se scelle officiellement à la signature du contrat, dont chaque clause doit être examinée avec minutie. Pour qu'il perdure, le fournisseur devra miser sur un service à la clientèle exemplaire, une gestion proactive des plaintes, et envisager des partenariats solides avec la clientèle stratégique.

Engel, J. F., Blackwell, R. D. et Miniard, P. W. (1995). *Consumer behavior* (8e édition). Dryden Press.

Fino et Bang Marketing (2024). Étude sur les tendances marketing B2B au Québec. Comprendre les tendances pour bâtir une stratégie gagnante. https://mailchi.mp/bang-marketing/etude-marketing-b2b-quebec

Fischer, H., Seidenstricker, S. et Poeppelbuss, J. (2022). The triggers and consequences of digital sales : A systematic literature review. *Journal of Personal Selling & Sales Management, 43*(1), 5-23. https://doi.org/10.1080/08853134.2022.2102029

Fishbein, M. et Ajzen, I. (1975). *Belief, attitude, intention, and behavior : An introduction to theory and research.* Addison-Wesley Pub. Co.

Fisher, R., Ury, W. et Patton, B. (2011). *Getting to yes : Negotiating agreement without giving in* (3e *édition*). Penguin Books.

Fleming, D. E. et Hawes, J. M. (2017). The negotiation scorecard : A planning tool in business and industrial marketing. *Journal of Business & Industrial Marketing, 32*(4), 519-524. https://doi.org/10.1108/jbim-06-2015-0120

Friestad, M. et Wright, P. (1994). The persuasion knowledge model : How people cope with persuasion attempts. *Journal of Consumer Research, 21*(1), 1-31.

Gabler, C. B., Landers, V. M. et Itani, O. S. (2023). Sustainability and professional sales : A review and future research agenda. *Journal of Personal Selling & Sales Management, 43*(4), 336-353. https://doi.org/10.1080/08853134.2023.2244675

Gartner (2023). *Gartner expects 60% of seller work to be executed by generative AI technologies within five years.* https://www.gartner.com/en/newsroom/press-releases/2023-09-21-gartner-expects-sixty-percent-of-seller-work-to-be-executed-by-generative-ai-technologies-within-five-years#:%E2%88%BC:text=One%20of%20the%20Most%20Transformative,%2C%20according%20to%20Gartner%2C%20Inc

Gilliam, D. A. et Flaherty, K. E. (2015). Storytelling by the sales force and its effect on buyer-seller exchange. *Industrial Marketing Management, 46*, 132-142. https://doi.org/10.1016/j.indmarman.2015.01.013

Good, V., Hughes, D. E., Kirca, A. H. et McGrath, S. (2022). A self-determination theory-based meta-analysis on the differential effects of intrinsic and extrinsic motivation on salesperson performance. *Journal of the Academy of Marketing Science, 50*(3), 586-614. https://doi.org/10.1007/s11747-021-00827-6

Habel, J., Alavi, S. et Heinitz, N. (2023). A theory of predictive sales analytics adoption. *AMS Review, 13*(1-2), 34-54. https://doi.org/10.1007/s13162-022-00252-0

Habel, J., Alavi, S. et Linsenmayer, K. (2021). Variable compensation and salesperson health. *Journal of Marketing, 85*(3), 130-149. https://doi.org/10.1177/0022242921993195

Halb, F. et Seebacher, U. G. (2021). User experience and touchpoint management : A touchpoint performance management toolkit for the buyer journey. Dans U. G. Seebacher (dir.), *B2B marketing. Management for professionals* (p. 285-315). Springer, Cham.

Hartmann, N. N., Wieland, H., Gustafson, B. et Habel, J. (2023). Research on sales and ethics : Mapping the past and charting the future. *Journal of the Academy of Marketing Science, 52*(3), 653-671. https://doi.org/10.1007/s11747-023-00961-3

Hinterhuber, A., Snelgrove, T. C. et Stensson, B.-I. (2021). Value first, then price : The new paradigm of B2B buying and selling. *Journal of Revenue and Pricing Management, 20*(4), 403-409. https://doi.org/10.1057/s41272-021-00304-3

Ingram, T. N., Avila, R. A., Schwepker, C. H., Williams, M. R. et Shannahan, K. (2015). *Sell* (2e édition canadienne). Nelson Education.

Interbrand (2025). *Best Global Brands 2024*. https://interbrand.com/best-global-brands/

Jolicoeur, M. (2024). Immobilier : sanction record contre la courtière vedette Christine Girouard et son conjoint. *Le Journal de Montréal*. https://www.journaldemontreal.com/2024/09/10/immobilier-sanction-record-contre-la-courtiere-vedette-christine-girouard-et-son-conjoint#:~:text=La%20courti%C3%A8re%20en%20immobilier%20Christine,d'achat%20bidon%20pour%20cr%C3%A9er3

Jones, D. G. B. et Tadajewski, M. (2016). *The Routledge companion to marketing history*. Routledge Taylor & Francis Group.

Kane, N. et Zappulla, J. (2016). *Critical selling : How top performers accelerate the sales process and close more deals*. John Wiley & Sons.

Karlinsky-Shichor, Y. et Netzer, O. (2024). Automating the B2B salesperson pricing decisions : A human-machine hybrid approach. *Marketing Science, 43*(1), 138-157. https://doi.org/10.1287/mksc.2023.1449

Kotler, P. et Pfoertsch, W. (2006). *B2B brand management*. Springer.

Lanzrath, A. I., Homburg, C. et Ruhnau, R.-C. M. (2023). Women's underrepresentation in business-to-business sales : Reasons, contingencies, and solutions. *Journal of the Academy of Marketing Science*. https://doi.org/10.1007/s11747-023-00988-6

La Presse (2018). A220 : Airbus force ses fournisseurs à baisser leurs prix de 20%. https://www.lapresse.ca/affaires/economie/transports/201807/28/01-5191147-a220-airbus-force-ses-fournisseurs-a-baisser-leurs-prix-de-20-.php

Lasrado, F., Thaichon, P. et Nyadzayo, M. W. (2022). Exploring the role of relationship management and relationship quality in B2B : Empirical insights and future research directions. *Journal of Business & Industrial Marketing, 38*(5), 1055-1086. https://doi.org/10.1108/jbim-05-2021-0267

Lemon, K. N. et Verhoef, P. C. (2016). Understanding customer experience throughout the customer journey. *Journal of Marketing, 80*(6), 69-96. https://doi.org/10.1509/jm.15.0420

Leroux, É. (2015). Le métier d'imprimeur au Québec : 200 ans d'évolution (1764-1960). *Documentation et bibliothèques, 51*(2), 107-116. https://doi.org/10.7202/1030092ar

Manning, G. L., Ahearne, M., Reece, B. L. et MacKenzie, H. F. (2016). *Selling today : Partnering to create value* (7e édition canadienne). Pearson Canada.

McClure, C. E., Epler, R. T., Schmitt, L. et Rangarajan, D. (2024). AI in sales : Laying the foundations for future research. *Journal of Personal Selling & Sales Management, 44*(2), 108-127. https://doi.org/10.1080/08853134.2024.2329905

McGowan, P. (2020). Sales failure : A review and future research directions. *International Journal of Logistics Research and Applications, 24*(1), 23-50. https://doi.org/10.1080/13675567.2020.1726306

Mehrabian, A. et Ferris, S. R. (1967). Inference of attitudes from nonverbal communication in two channels. *Journal of Consulting Psychology, 31*(3), 248-252. https://doi.org/10.1037/h0024648

Merrill, D. W. et Reid, R. H. (1981). *Personal styles and effective performance : Make your style work for you*. Chilton Book.

Minerbo, C., Kleinaltenkamp, M. et Brito, L. A. L. (2021). Unpacking value creation and capture in B2B relationships. *Industrial Marketing Management, 92*, 163-177. https://doi.org/10.1016/j.indmarman.2020.11.011

Moncrief, W. C. et Marshall, G. W. (2005). The evolution of the seven steps of selling. *Industrial Marketing Management, 34*(1), 13-22. https://doi.org/10.1016/j.indmarman.2004.06.001

Morgan, R. M. et Hunt, S. D. (1994). The commitment-trust theory of relationship marketing. *Journal of Marketing, 58*(3), 20-38. https://doi.org/10.2307/1252308

Mullins, R. et Agnihotri, R. (2022). Digital selling : Organizational and managerial influences for frontline readiness and effectiveness. *Journal of the Academy of Marketing Science, 50*(4), 800-821. https://doi.org/10.1007/s11747-021-00836-5

Nguyen, M., Sharma, P., Kingshott, R. et Malik, A. (2024). Culturally diverse teams and inter-organizational knowledge sharing behavior : The role of perceived morality and relationship orientation. *Industrial Marketing Management, 116*, 120-129. https://doi.org/10.1016/j.indmarman.2023.12.003

O'Donnell, E. et Marsh, L. (2021). The impact of compensation structure on salesperson perceptions and behaviors : Insights from the sales literature. *Compensation & Benefits Review, 54*(1), 3-11. https://doi.org/10.1177/08863687211043441

Paschen, J., Kietzmann, J. et Kietzmann, T. C. (2019). Artificial intelligence (AI) and its implications for market knowledge in B2B marketing. *Journal of Business & Industrial Marketing, 34*(7), 1410-1419. https://doi.org/10.1108/jbim-10-2018-0295

Paschen, J., Wilson, M. et Ferreira, J. J. (2020). Collaborative intelligence : How human and artificial intelligence create value along the B2B sales funnel. *Business Horizons, 63*(3), 403-414. https://doi.org/10.1016/j.bushor.2020.01.003

Peasley, M. C., Hochstein, B., Britton, B. P., Srivastava, R. V. et Stewart, G. T. (2020). Can't leave it at home? The effects of personal stress on burnout and salesperson performance. *Journal of Business Research, 117*, 58-70. https://doi.org/10.1016/j.jbusres.2020.05.014

Peesker, K. M., Kerr, P. D., Bolander, W., Ryals, L. J., Lister, J. A. et Dover, H. F. (2022). Hiring for sales success : The emerging importance of salesperson analytical skills. *Journal of Business Research, 144*, 17-30. https://doi.org/10.1016/j.jbusres.2022.01.070

Perez-Vega, R., Hopkinson, P., Singhal, A. et Mariani, M. M. (2022). From CRM to social CRM : A bibliometric review and research agenda for consumer research. *Journal of Business Research, 151*, 1-16. https://doi.org/10.1016/j.jbusres.2022.06.028

Prat, M. (2023). *Videoconference buyer-seller negotiations : A systematic literature review*. https://www.researchsquare.com/article/rs-3394981/v1

Prior, D. D. (2021). Mapping purchase situations. Dans *Organisational buying : A multidisciplinary perspective* (p. 21-39). Palgrave Macmillan, Cham.

Purmonen, A., Jaakkola, E. et Terho, H. (2023). B2B customer journeys : Conceptualization and an integrative framework. *Industrial Marketing Management, 113*, 74-87. https://doi.org/10.1016/j.indmarman.2023.05.020

Rackham, N. (1988). *SPIN selling*. McGraw-Hill.

Robinson, P. J., Faris, C. W., Wind, Y. et Marketing Science Institute (1967). *Industrial buying and creative marketing*. Allyn & Bacon.

Schmitt, L., Casenave, E. et Pallud, J. (2021). Salespeople's work toward the institutionalization of social selling practices. *Industrial Marketing Management, 96*, 183-196. https://doi.org/10.1016/j.indmarman.2021.05.008

Selling Power (2010). *John Henry Patterson : The father of professional selling*. https://www.sellingpower.com/blog/john-henry-patterson-the-father-of-professional-selling-part-2

Sheth, J. N., Jain, V. et Ambika, A. (2023). The growing importance of customer-centric support services for improving customer experience. *Journal of Business Research, 164*. https://doi.org/10.1016/j.jbusres.2023.113943

Singh, S. K., Marinova, D. et Singh, J. (2020). Business-to-business e-negotiations and influence tactics. *Journal of Marketing, 84*(2), 47-68. https://doi.org/10.1177/0022242919899381

Siskind, B. (1997). *The power of exhibit marketing* (4e édition). Self-Counsel business series.

Smith, J. M. (2018). *The price negotiation playbook - A practical B2B guide for winning your best price with confidence*. Bradley Publishing.

Spiro, R. L. et Weitz, B. A. (1990). Adaptive selling : Conceptualization, measurement, and nomological validity. *Journal of Marketing Research, 27*(1), 61-69. https://doi.org/10.2307/3172551

Sujan, H., Weitz, B. A. et Sujan, M. (1988). Increasing sales productivity by getting salespeople to work smarter. *The Journal of Personal Selling and Sales Management, 9*(2), 9-19.

Terho, H., Giovannetti, M. et Cardinali, S. (2022). Measuring B2B social selling : Key activities, antecedents and performance outcomes. *Industrial Marketing Management, 101*, 208-222. https://doi.org/10.1016/j.indmarman.2021.12.016

Terho, H., Haas, A., Eggert, A. et Ulaga, W. (2012). « It's almost like taking the sales out of selling » - Towards a conceptualization of value-based selling in business markets. *Industrial Marketing Management, 41*(1), 174-185. https://doi.org/10.1016/j.indmarman.2011.11.011

Thaichon, P., Surachartkumtonkun, J., Quach, S., Weaven, S. et Palmatier, R. W. (2018). Hybrid sales structures in the age of e-commerce. *Journal of Personal Selling & Sales Management, 38*(3), 277-302. https://doi.org/10.1080/08853134.2018.1441718

Töytäri, P. et Rajala, R. (2015). Value-based selling : An organizational capability perspective. *Industrial Marketing Management, 45*, 101-112. https://doi.org/10.1016/j.indmarman.2015.02.009

Vitali, V., Bazzani, C., Gimigliano, A., Cristani, M., Begalli, D. et Menegaz, G. (2022). Trade show visitors and key technological trends : From a literature review to a conceptual framework. *Journal of Business & Industrial Marketing, 37*(13), 142-166. https://doi.org/10.1108/jbim-10-2021-0461

Webster, F. E. et Wind, Y. (1972). A general model for understanding organizational buying behavior. *Journal of Marketing, 36*(2), 12-19. https://doi.org/10.2307/1250972

Weitz, B. A., Sujan, H. et Sujan, M. (1986). Knowledge, motivation, and adaptive behavior : A framework for improving selling effectiveness. *Journal of Marketing, 50*(4), 174-191. https://doi.org/https://doi.org/10.2307/1251294

Wieland, D. A. C., Ivens, B. S., Kutschma, E. et Rauschnabel, P. A. (2024). Augmented and virtual reality in managing B2B customer experiences. *Industrial Marketing Management, 119*, 193-205. https://doi.org/10.1016/j.indmarman.2024.04.007

Witell, L., Kowalkowski, C., Perks, H., Raddats, C., Schwabe, M., Benedettini, O. et Burton, J. (2020). Characterizing customer experience management in business markets. *Journal of Business Research*, *116*, 420-430. https://doi.org/10.1016/j.jbusres.2019.08.050

Woodruff, R. B. (1997). Customer value : The next source for competitive advantage. *Journal of the Academy of Marketing Science*, *25*, 139-153. https://doi.org/10.1007/BF02894350

MARQUIS
Québec, Canada

MIXTE
Papier | Pour une gestion forestière responsable
FSC® C103567
FSC www.fsc.org

PCF

BIO GAZ ÉNERGIE

PERMANENT

Ce livre est imprimé sur des matériaux issus de forêts
bien gérées certifiées FSC® et de matériaux recyclés.